Dr. Amit Rawat is a seasoned academician and researcher in Computer Science & Engineering with over 14 years of teaching and research experience. He is currently serving as an Assistant Professor at the Faculty of Engineering & Technology, Agra College, Agra, since June 2013. Dr. Rawat holds a Ph.D. from the National Institute of Technology, Patna, along with an M.Tech from Uttarakhand Technical University and a B.Tech from R.B.S. College, Agra.

His research contributions include publications in high-impact SCI/SCOPUS-indexed journals such as Image and Vision Computing, IETE Journal of Research, and Applied Soft Computing. His work focuses on applications of neural networks and image processing for medical diagnosis and agricultural automation. He also holds a patent for a design titled "Single-Handed Showing and Fertilizing Machine." Dr. Rawat has guided multiple M.Tech and B.Tech projects and has delivered expert lectures on topics like NEP-2020, research methodology, and Indian culture. He has actively organized and participated in national conferences and workshops along with several academic and administrative roles. His academic instruction spans foundational and advanced subjects in computer science including Digital Logic, Automata, Cryptography, and Software Engineering. Dr. Rawat's profile reflects a strong blend of teaching, research, institutional service, and administrative leadership.

Dr. Amit Dutta, a distinguished academic and administrator with over 26 years of experience in the field of technical education, currently serves as Director at the All India Council for Technical Education (AICTE), New Delhi, under the Ministry of Education, Government of India. He currently heads the Policy and Planning Bureau at AICTE and has played a pivotal role in shaping and implementing several national educational initiatives. With a doctoral specialization in Neural Networks, complemented by an MCA, M.Sc., and a CMI Level 5 Certificate in Leadership and Management from the AICTE-UKIERI Technical Leadership Programme (Dudley College, London), Dr. Dutta has contributed significantly to the fields of AI, data mining, and academic governance.

He has served in several key positions, including Deputy Director (ATAL/NEP/Awards) Cell, Regional Officer, AICTE Mumbai, and Coordinator for major initiatives such as the Stanford University Survey in India, RUSA, and AISHE. His leadership has been instrumental in the successful execution of landmark AICTE programs, including the Smart India Hackathon, ATAL Faculty Development Programs, AICTE Idea Lab, and the AICTE Productisation Fellowship. Dr. Dutta has served as Member Secretary in national curriculum development committees on emerging technologies like Quantum Computing, Defense, Space, VLSI, and

Nuclear Technologies. He holds governance roles in premier institutions including IIM Bodhgaya, IIM Sambalpur, and BOAT Mumbai, and is actively associated with various statutory and academic boards.

An accomplished researcher and author, Dr. Dutta has published 4 books, 26 international journal papers, and participated in numerous national and international conferences. He also serves on the editorial boards of reputable journals and holds memberships in professional bodies, including the IEEE, the Computer Society of India, and the MIR Lab in the USA. Dr. Dutta's multi-faceted contributions reflect his deep commitment to innovation-driven, inclusive, and future-ready technical education in India.

Information Technology and Panch Parivartan

Framework for Sustainable Bharat 2047

Mahesh Manchanda M.Tech (CSE), M.Phil (CS) Ph.D (CSE)
Computer Science and Engineering Department
Graphic Era Hill University
Dehradun-248002, Uttarakhand, India

Gagan Matta M.Sc., Ph.D
Department of Zoology and Environmental Science
Gurukula Kangri (Deemed to be University)
Haridwar - 249404, Uttarakhand, India

Amit Rawat M.Tech, Ph.D
Department of Computer Science & Engineering
Faculty of Engineering & Technology
Agra College, Agra-282001
Uttar Pradesh, India

Amit Dutta MCA.,M.Sc, Ph.D
All India Council for Technical Education (AICTE)
New Delhi

NIPA® GENX ELECTRONIC RESOURCES & SOLUTIONS P. LTD.
New Delhi-110 034

NIPA® GENX ELECTRONIC
RESOURCES & SOLUTIONS P. LTD.
101,103, Vikas Surya Plaza, CU Block
L.S.C. Market, Pitam Pura, New Delhi-110 034
Ph : +91-11-43860225, Mob.: +91 9717133558, 9540816132
E-mail: newindiapublishingagency@gmail.com
Website: www.nipaersources.com

Print ISBN: 978-93-58872-12-5

ebook ISBN: 978-93-58878-87-5

Composed and Designed by NIPA®.

Information Technology and Panch Parivartan

NIPA® GENX ELECTRONIC RESOURCES & SOLUTIONS P. LTD.
New Delhi-110 034

About the Authors

Dr. Mahesh Manchanda is a distinguished academician and administrator with over 22 years of experience in the field of Computer Science and Information Technology. He currently serves as Professor in the Department of Computer Science & Engineering at Graphic Era Hill University, Dehradun, and holds several key administrative positions, including University IQAC Coordinator and Nodal Officer for national regulatory bodies such as UGC, AISHE, NIRF, and SAMARTH.

Dr. Manchanda earned his Ph.D. in Computer Science and Engineering from Gurukul Kangri University, Haridwar. He has authored more than 59 research papers in reputed international journals and conferences and holds 13 patents in emerging areas such as Web Mining, Machine Learning, and Software Sustainability. He has also authored over 16 academic textbooks for various Indian universities and open learning institutions.

A committed academic leader, Dr. Manchanda has successfully led initiatives in institutional quality assurance and accreditation and has organized IEEE-sponsored international conferences including ICICDA 2023 and ICMSLE 2024. He is the recipient of a government-funded research grant for his project on "Digital Literacy and Cybersecurity Awareness for Rural Youth." A Life Member of CSI and a Senior Member of IEEE, he is frequently invited as a session chair and expert reviewer in national and international academic forums.

Dr. Gagan Matta is a renowned environmental scientist and academician with over 18 years of research experience and 14 years of teaching. He holds a Ph.D. in Environmental Science from Gurukul Kangri University, Haridwar, India, where he is an Associate Professor in the Department of Zoology and Environmental Science. Specializing in water quality indexing, limnology, aquatic ecology, toxicology, and environmental pollution, Dr. Matta's research spans hydrology, green audits, and environmental communication. He has authored six books, edited 18, published over 90 journal articles, and contributed 18 book chapters. His work has earned 14 research grants and seven copyrights. Dr. Matta has been recognized with prestigious awards, including the "Scientist of the Year Award – 2023." He has mentored 8 Ph.D. candidates and 31 M.Sc. dissertations while contributing to developing academic programs and environmental policies.

As an advocate for science communication, Dr. Matta works on initiatives to raise awareness of water hygiene and environmental sustainability. His global presence, invited talks in the UK, Italy, and China, and his roles in various academic bodies solidify his status as a leading environmental scientist committed to research, teaching, and policy development.

Foreword

Nature is formed of the five elements: water, air, ether, fire, and earth. Human beings, too, are made of these very same elements, which is why we are inseparable from nature. Animals, plants, and even the inanimate world are integral parts of this great whole. Bhartiya philosophy teaches that the same essence pervades all things, living or non-living. Matter and consciousness, therefore, are interdependent, bound together in a web of connection. To preserve and nurture the natural world, this holistic perspective is indispensable.

यही भारतीय दृष्टि है। **This is the Indian vision.**

The *Gita*, the *Upanishads*, and other ancient scriptures describe how the Creator fashioned nature and then entered into it Himself. Nature, therefore, is sacred, worthy of reverence, devotion, and protection.

यही भारतीय दृष्टि है। **This is the Indian vision.**

For human beings to become truly sensitive towards nature, the reasoning outlined above is more than sufficient. Sensitivity is the first and most essential quality of being human. Without it, neither the protection nor the flourishing of nature is possible. In the view of Bhart, reverence and reality are inseparable: essence itself is the foundation of reverence, while reality must become its basis. This vision is not meant for Bhart alone but for the entire world, to free it from corruption and rescue it from destruction. Yet today we see a distortion of reverence: the worship of the land on one hand and its exploitation on the other, and devotion to the Ganga on one side, alongside the pollution of rivers on the other. The pollution of water, air, fire, earth, and sky has now spiralled beyond control. Environmental degradation threatens our very survival. Where reverence has been lost, selfishness has taken root. And with selfishness has come the appetite to consume without restraint, fuelling unchecked exploitation of nature. The Bhartiya vision, however, speaks not of abuse but of responsible use. True reverence inspires care and moderation; once reverence is corrupted, exploitation inevitably begins.

Nature is the creation of the Creator, and creation cannot be separated from its source. Man, too, is part of this creation. The living being, the universe, and the Divine are bound in a profound and indivisible relationship. This unity is the foundation of Bhartiya philosophy. Nature, therefore, is not merely our companion; she is our Mother.

यही भारतीय दृष्टि है। **This is the Indian vision.**

It is the absence of such a vision that has brought humanity face-to-face with today's destructive challenges. Global warming, climate change, and pollution pose significant threats to health, biodiversity, and sustainable development worldwide. Yet the Bhartiya vision offers pathways to solutions. If we adapt the lifestyle and work ethic that have sustained us for centuries, refined through modern research, to meet present needs, answers will emerge. But rapid industrialisation and urbanisation are already causing grave damage. In most cities, air pollution has triggered a rise in respiratory diseases. Water pollution has raised concerns about the availability of clean drinking water. Agriculture suffers from water scarcity, while soil erosion and deforestation threaten both harvests and wildlife. Global warming, meanwhile, invites natural disasters and undermines environmental stability. This distorted sense of reverence towards nature has forced us into crisis. If unchecked, it could even lead to humanity's extinction.

The time has therefore come for the world to unite, with one mind and one resolve, under the urgent motto *"now or never."* Protection and restoration of the environment must begin at every level: individual, family, institution, community, nation, and the global stage. Alongside Bhart, every country must act swiftly. By embracing the principle of 'एक पृथ्वी – एक परिवार' *"One Earth, One Family,"* humanity can rise above competition to build relationships rooted in harmony and coexistence. The Vedic mantra "एक पृथ्वी – एक परिवार" *"Mata Bhumi, Putroham Prithivya" "The Earth is our Mother, and we are her children"* must become the lived reality of our age.

By 2030, Bhart has set ambitious goals: to increase its non-fossil fuel capacity to 500 gigawatts, to meet half of its energy needs from renewable sources, to reduce carbon emissions by one billion tonnes, and to reduce the carbon intensity of its economy by 45%. To succeed, the government must also ensure affordable and high-quality education, healthcare, and transportation. Only with a sensitive government and a responsible society can such aspirations be fulfilled. At the personal level, each of us has duties that range from abandoning single-use plastics to planting trees to absorb the carbon dioxide we release and replenish the oxygen we breathe. We must pledge that not a single drop of water, a single grain of food, or a single unit of electricity will be wasted. Every household, institution, and community should adopt solar power to meet their energy needs.

As Tulsidas wrote: 'क्षिति जल पावक गगन समीरा, पंच रहित अति अधम सरीरा' *"The body without the five elements of earth, water, fire, ether, and air is utterly degraded."* These words hold eternal truth and must be universally acknowledged. Through such commitments, Bhart can present itself as a living example of an environmentally conscious nation, offering the world not only a warning but also a vision to overcome the crises of our time. As Bharat stands at a pivotal crossroads in its developmental journey, caught between the weight of its

Preface

The pursuit of sustainability in the contemporary Indian context demands not only innovation and policy reform but also a radical rethinking of development paradigms in light of accelerating technological change. As the world negotiates an era marked by climate instability, socioeconomic inequality, resource stress, and digital disruption, India must simultaneously confront and capitalise on these transformative currents. This book, *Information Technology and Panch Parivartan: A Framework for Sustainable Bharat 2047,* addresses this dual challenge by conceptualising a multidimensional framework for national transformation, where digital technologies are not only tools of economic progress but also foundational instruments for systemic sustainability. The central thesis rests on the proposition that achieving a "Viksit Bharat" by 2047, India's centenary of Independence, necessitates a deliberate integration of information technology (IT) with principles of sustainability across all sectors of society and governance.

The book is anchored in the core idea of "Panch Parivartan," a five-fold framework that captures the transformative potential of IT when applied holistically across Policy, People, Processes, Products, and Platforms. These five dimensions are not discrete compartments but deeply intertwined nodes in a systemic transition. The Panch Parivartan concept is rooted in the recognition that digital technologies are enablers of profound structural change—reshaping how decisions are made, how services are delivered, how resources are managed, how social inclusion is realised, and how innovation ecosystems evolve. By operationalising these five dimensions, this framework provides a model to integrate digital systems into the core of India's sustainability journey, aligning technological advancements with inclusive, equitable, and regenerative development. India's development trajectory presents unique challenges and opportunities. With a demographic dividend, a rapidly growing digital economy, and a complex institutional structure, India stands at the cusp of what could be a transformative era. However, the persistence of developmental deficits in access to quality education, healthcare, water, energy, waste management, and social security continues to constrain progress. Moreover, ecological degradation, rapid urbanisation, and climate vulnerability compound the difficulty of achieving sustainable outcomes. In this context, IT emerges as a unifying force that can bridge institutional silos, democratise knowledge, enhance public participation, and optimise resources through real-time data, predictive analytics, and intelligent automation. The Panch Parivartan framework, as developed in this book, responds to these challenges by advocating for systemic digital integration at all levels of governance, planning, and implementation.

The rationale for such a framework is not based on technology for its own sake but on the strategic convergence of IT and sustainability as mutually reinforcing domains. Digital transformation, when aligned with ecological and social objectives, can substantially improve the adaptive capacity of institutions and communities. It enables transparent and accountable governance, facilitates targeted service delivery, and enhances the efficiency of infrastructure systems. Moreover, it supports real-time monitoring and evidence-based policymaking, both of which are crucial for complex systems such as energy grids, agricultural networks, urban mobility, and climate adaptation mechanisms. This book illustrates how IT can move beyond its conventional role as a support function and become a catalytic force in reengineering development itself.

The Panch Parivartan model distinguishes itself with a systemic orientation, treating IT interventions as a framework that positions digital technology as an embedded layer within larger sociotechnical systems. For example, policy transformation is not limited to e-governance tools but involves reimagining decision-making processes through open data ecosystems, participatory platforms, and decentralised digital institutions. Similarly, the empowerment of people through IT is viewed through the lens of universal digital access, skills democratisation, and inclusive platform economies that transcend urban-rural and gender divides. Process transformation involves not only the digitisation of workflows but also the restructuring of institutional logic toward flexibility, transparency, and resilience. Sustainable product innovation and eco-design, in this view, are not standalone outcomes but part of a digitally managed lifecycle that links production, consumption, and reuse. Lastly, digital platforms are presented not merely as service layers but as intelligent ecosystems capable of enabling multi-stakeholder collaboration and co-governance.

The integrated nature of this model holds significant implications for both theory and practice. On the theoretical front, it contributes to the emerging discourse on digital sustainability, providing a context-specific articulation relevant to the Global South. It recognises that sustainable development in India cannot simply replicate Western models but must evolve through Indigenous innovation, institutional hybridity, and cultural inclusiveness. On the practical front, the book provides policy designers, technocrats, urban planners, and civil society actors with a roadmap to design interventions that are both scalable and context-sensitive. The emphasis on interoperability, user-centric design, open standards, and real-time feedback loops ensures that digital systems remain adaptive to evolving developmental needs.

The book also situates its arguments within the wider national and global policy architecture. It aligns closely with India's aspirations under the *Digital India, Make in India*, and *Aatmanirbhar Bharat* missions while also reflecting the commitments under the UN Sustainable Development Goals, the Paris Climate Accord, and global dialogues on digital public infrastructure. By emphasising locally grounded, globally resonant strategies, the authors underline the potential of India not only as a beneficiary of global digital trends but also as a contributor to shaping ethical and sustainable digital futures. One of the most compelling aspects of this work is its future-oriented vision. It anticipates technological trajectories, such as the rise of decentralized governance systems, quantum computing, bio-digital convergence, and AI for environmental prediction, and calls for anticipatory governance to prepare for the opportunities and disruptions these might bring. This foresight is essential in a world where technology cycles are accelerating and where developmental inertia can have long-term consequences. The framework thus balances immediate implementation pathways with long-range planning horizons, advocating for a national innovation ecosystem that is resilient, inclusive, and ethically grounded.

The scientific rigor of this book lies in its evidence-based methodology, cross-sectoral analysis, and scenario-based thinking. While rooted in theoretical frameworks, the arguments are supported with empirical insights and practical illustrations that demonstrate feasibility and relevance. Importantly, the book resists the temptation to offer silver-bullet solutions. Instead, it advances a pluralistic, layered approach—acknowledging the diversity of India's geographies, cultures, institutional capacities, and resource constraints. This recognition of complexity strengthens the credibility of its proposals and enhances their adaptability to real-world conditions.

Information Technology and Panch Parivartan: Framework for Sustainable Bharat 2047 is a landmark contribution to the discourse on sustainable transformation in the digital age. It presents a robust synthesis of vision and pragmatism, combining technological imagination with developmental ethics. By proposing an integrated model for national advancement, the book challenges policymakers, scholars, and citizens alike to rethink the future not as a linear extension of the past but as a space of deliberate, systemic innovation. As India prepares to mark its centennial milestone, this work presents both a framework and a moral imperative urging us to co-create a nation that is technologically empowered, ecologically secure, socially inclusive, and truly sustainable.

Authors

Contents

Introduction

India's aspiration to emerge as a developed nation by 2047, encapsulated in the vision of "Viksit Bharat," demands a strategic alignment of technological advancements and transformative governance reforms. Central to this journey are the twin pillars of Information Technology (IT) and the "Panch Parivartan" (Five Transformations), which collectively aim to revolutionize India's socioeconomic fabric while upholding cultural integrity and national unity.

Role of Information Technology in Shaping Viksit Bharat: Information Technology is the backbone of modern development, offering scalable solutions to address India's multifaceted challenges. India's IT sector has long been a driver of economic growth, contributing significantly to GDP and employment. However, its role in achieving Viksit Bharat extends beyond economic metrics to societal transformation. Major initiatives like the Digital India Mission, Aadhaar-enabled services, and the Unified Payments Interface (UPI) have democratized access to financial and digital resources, empowering even the most marginalized sections of society. Emerging technologies such as Artificial Intelligence (AI), Quantum Computing, Blockchain, and the Internet of Things (IoT) are being deployed to solve critical problems in healthcare, agriculture, education, and urban development. For instance:

- **Healthcare**: AI-powered diagnostic tools and telemedicine platforms have improved access to quality healthcare in rural areas. AI algorithms can predict disease outbreaks, optimize treatment plans, and enhance patient outcomes.
- **Agriculture**: IoT-enabled smart farming systems and blockchain-based supply chain management ensure efficiency and transparency, addressing food security issues and farmer profitability.
- **Education**: E-learning platforms and adaptive learning technologies are bridging the digital divide, ensuring equitable access to quality education for all.

The **National Quantum Mission** and the **Cyber-Physical Systems Initiative** underscore the government's commitment to technological innovation. These programs aim to establish India as a global leader in advanced sciences, reduce import dependencies, and foster a robust research and development ecosystem.

Framework of Panch Parivartan: Panch Parivartan, or Five Transformations, is a visionary framework designed to address India's systemic inefficiencies and catalyze sustainable growth. While the specifics of this framework vary in interpretation, it resonates with the "Panch Pran" pledges articulated by the leadership:

- **Developed Nation Goal**: Setting an unwavering commitment to economic progress, infrastructural excellence, and social equity.
- **Eradication of Colonial Legacy**: Reforming governance models, judiciary systems, and societal norms that reflect colonial hangovers, fostering a distinctly Indian development path.
- **Cultural Renaissance**: Celebrating and integrating India's rich heritage and cultural diversity into the global narrative.
- **Unity and Inclusion**: Promoting social cohesion by addressing disparities in caste, gender, and regional development.
- **Responsible Citizenship**: Instilling a sense of duty and collective responsibility among citizens, driving grassroots participation in national progress.

Synergizing IT and Panch Parivartan: Integrating IT with Panch Parivartan accelerates the transformation of India's socioeconomic landscape. Key areas of impact include:

- **Governance and Transparency**: IT-enabled e-governance platforms like the Public Financial Management System (PFMS) and the Government e-Marketplace (GeM) have revolutionized public service delivery. These platforms enhance transparency, reduce corruption, and enable real-time monitoring of government programs, aligning with eradicating inefficiencies.
- **Social Inclusion**: Technology has proven to be a great equalizer. Initiatives like **Jan Dhan-Aadhaar-Mobile (JAM) Trinity** have facilitated financial inclusion for millions. Rural internet connectivity programs such as BharatNet also bridge the urban-rural digital divide.
- **Education and Skill Development**: IT tools address India's educational challenges. Students across socioeconomic strata access world-class

resources through digital classrooms, Massive Open Online Courses (MOOCs), and mobile learning apps. Skill development initiatives like **Skill India** leverage AI to match market demands with workforce capabilities.

- **Environment and Sustainability**: IT contributes to sustainable development through smart urban planning, renewable energy management, and environmental monitoring. For instance, AI-driven analytics in climate modelling aids disaster preparedness, while IoT-based systems optimize energy consumption in smart cities.
- **Cultural Preservation and Global Positioning**: Digital archives and AI-powered linguistic tools preserve India's cultural heritage. Platforms showcasing Indian art, music, and literature promote cultural pride and enhance India's global soft power.

Achievements and Milestones: India's progress in integrating IT and governance reforms reflects its commitment to Viksit Bharat. As of 2024:

- **Economic and Innovation Leadership**: India ranks 39th in the Global Innovation Index, demonstrating its growing emphasis on research and technology-driven solutions.
- **Gender Inclusivity in R&D**: Female participation in STEM fields has risen, with women comprising 18.6% of the research workforce.
- **Startup Ecosystem**: India is home to over 140,000 startups, making it the third-largest startup ecosystem globally. These enterprises are at the forefront of innovation, creating solutions for both domestic and global challenges.

Vision for 2047

Despite advancements, challenges persist. Bridging the digital divide, ensuring cybersecurity, fostering R&D funding, and addressing ethical concerns around emerging technologies is critical. Establishing the Anusandhan National Research Foundation (ANRF) is a promising step, aiming to institutionalize research funding and collaboration. The ethical deployment of technologies, especially in AI and data privacy, is another pressing concern. Striking a balance between innovation and regulation will be crucial for maintaining public trust and ensuring equitable benefits.

By 2047, India envisions a future where every citizen benefits from the confluence of IT and Panch Parivartan. Smart cities powered by renewable energy, inclusive financial systems, universal healthcare access, and a thriving innovation ecosystem are key components of this vision. Additionally, India's leadership in global technology governance will reflect its commitment to ethical and sustainable development. The journey to Viksit Bharat 2047 is both an opportunity and a responsibility. By aligning technological prowess with transformative governance reforms, India can pioneer a development model that is inclusive, sustainable and rooted in its rich cultural ethos. This dual approach of leveraging IT and embracing the Panch Parivartan ensures that India stands as a beacon of progress and equity on the global stage.

Framing the Vision for Viksit Bharat 2047

India's vision for "Viksit Bharat 2047" marks a transformative chapter in its journey towards becoming a developed nation by the centenary of its independence. This ambitious goal is deeply intertwined with the ideals of Atmanirbhar Bharat (Self-Reliant India), a framework that emphasizes resilience, innovation, and self-sufficiency across economic, social, and environmental dimensions. Central to this vision is the imperative of sustainable development, ensuring that India's growth trajectory meets present demands and preserves future generations' well-being. This chapter delves into the synergistic relationship between self-reliance and sustainability as foundational pillars of Viksit Bharat 2047, elucidating how these aspirations can shape India's developmental narrative.

Aspirations for Atmanirbhar Bharat: Atmanirbhar Bharat represents a clarion call for self-sufficiency, resilience, and global competitiveness. Announced in 2020 as a response to the challenges posed by the COVID-19 pandemic, this initiative aims to reduce India's dependence on imports, foster domestic production, and create an ecosystem that empowers local industries to thrive globally. At its core, Atmanirbhar Bharat is built on five pillars:

- **Economy**: Transitioning from incremental growth to quantum leaps in economic output, India envisions a $5 trillion economy driven by robust domestic industries and exports.
- **Infrastructure**: Building next-generation infrastructure to support urbanization, connectivity, and industrial growth.
- **Technology-Driven Systems**: Leveraging digital and technological advancements, such as artificial intelligence (AI), blockchain, and robotics, to modernize governance and industrial processes.
- **Demography**: Harnessing the potential of India's youthful population through education, skill development, and entrepreneurship.
- **Demand**: Creating a robust domestic market ecosystem that supports local industries and sustains economic momentum.

The **Production-Linked Incentive (PLI)** schemes have been a cornerstone of Atmanirbhar Bharat, incentivizing domestic manufacturing in electronics, automotive, pharmaceuticals, and renewable energy sectors. For instance, India has emerged as a global leader in smartphone production, with companies like Apple and Samsung expanding their manufacturing footprints within the country.

Sustainable Development as the Backbone of Viksit Bharat: Sustainable development is not merely a complementary goal for Viksit Bharat; it is an essential prerequisite for achieving long-term prosperity. As defined by the United Nations, sustainable development focuses on meeting the needs of the present without compromising the ability of future generations to meet their own needs. For India, this principle encompasses three key dimensions:

- **Economic Sustainability**: Building a robust economy resilient to global shocks, inclusive of diverse stakeholders, and innovative in addressing developmental challenges.
- **Social Sustainability**: Promoting equitable access to resources, education, and healthcare and fostering social cohesion in a nation as diverse as India.
- **Environmental Sustainability**: Protecting natural resources, reducing carbon emissions, and addressing climate change to ensure ecological balance.

India's commitment to sustainability is reflected in its international obligations, including the Paris Agreement and the Sustainable Development Goals (SDGs). For instance, India has pledged to achieve net-zero carbon emissions by 2070 and has made significant progress in renewable energy deployment, ranking fourth globally in installed wind power capacity.

Synergy Between Atmanirbhar Bharat and Sustainable Development: The intersection of self-reliance and sustainability presents a unique opportunity for India to reimagine its growth paradigm. The following areas highlight the synergy between these two aspirations:

- **Renewable Energy and Green Economy**
  - India's renewable energy capacity has surpassed 175 GW as of 2024, with ambitious targets to reach 500 GW by 2030. Solar and wind energy form the cornerstone of this transition, with programs like the National Solar Mission and the KUSUM scheme supporting decentralized and grid-connected power generation.
  - Initiatives promoting electric vehicles (EVs), green hydrogen, and biofuels align with Atmanirbhar Bharat's industrial goals and sustainability targets, reducing reliance on fossil fuels and lowering greenhouse gas emissions.

- **Circular Economy and Waste Management**
    - A circular economy approach focuses on minimizing waste and maximizing resource efficiency. The Swachh Bharat Mission 2.0 emphasizes solid waste management through recycling, composting, and waste-to-energy technologies, creating a self-reliant resource utilisation model.
    - Innovations in sustainable packaging, such as biodegradable alternatives and e-waste recycling, have gained traction under Atmanirbhar Bharat's manufacturing and entrepreneurship initiatives.
- **Sustainable Agriculture**
    - India's agricultural sector, a significant contributor to GDP and employment, is transitioning towards sustainability through organic farming, precision agriculture, and water-efficient irrigation systems.
    - The Pradhan Mantri Krishi Sinchayee Yojana and the Soil Health Card Scheme are policies that integrate self-reliance with ecological stewardship, enhancing productivity while conserving natural resources.
- **Green Urbanization and Infrastructure**
    - Urbanization is inevitable for India's growth, but sustainable planning ensures that cities become engines of innovation and resilience. Smart city initiatives integrate energy efficiency, green mobility, and smart grids, aligning with the twin goals of Atmanirbhar Bharat and sustainability.
    - Infrastructure projects like the Delhi-Mumbai Industrial Corridor (DMIC) incorporate green technologies and promote eco-industrial parks that reduce environmental footprints.

Barriers, Policy and Institutional Support

- **Resource Dependency**: Achieving self-reliance in critical sectors like electronics and semiconductors requires importing raw materials like lithium and cobalt, creating external dependencies.
- **Environmental Degradation**: Rapid industrialization, a cornerstone of Atmanirbhar Bharat, must navigate the delicate balance of economic growth and ecological preservation.
- **Technology Transfer and Innovation**: Building indigenous capabilities in advanced technologies, such as quantum computing and biotechnology, requires significant investment in R&D and collaboration with global partners.

- **Social Equity**: Ensuring that self-reliance initiatives benefit all sections of society, particularly marginalized communities, remains a critical challenge in achieving inclusive development.
- **Anusandhan National Research Foundation (ANRF)**: Established under the ANRF Act 2023, this foundation seeks to institutionalize research funding and foster collaborations between academia, industry, and government.
- **National Green Hydrogen Mission**: This mission supports India's leadership in green hydrogen production, reducing dependency on fossil fuels and positioning the country as a global hub for sustainable energy solutions.
- **PLI Schemes for Green Technologies**: Targeted incentives for renewable energy manufacturing, EVs, and energy storage systems encourage sustainable industrial growth.

Road to 2047

Framing the vision for Viksit Bharat 2047 for Atmanirbhar Bharat and sustainable development highlights the transformative potential of aligning self-reliance with ecological responsibility. As India navigates its growth journey, the integration of these principles will not only secure its position as a global leader but ensure its citizens' prosperity and well-being. India is poised to script an inclusive, resilient, and sustainable development story through visionary policies, technological innovation, and grassroots participation. This synthesis of aspirations and actions will serve as a blueprint for other developing nations, establishing India as a model for 21st-century progress. The following milestones are envisaged:

- **Carbon Neutrality**: Accelerating renewable energy adoption and green industrial practices to achieve net-zero emissions ahead of the 2070 target.
- **Global Leadership in Innovation**: Establishing India as a hub for technological innovation, focusing on sustainable solutions for global challenges.
- **Social Equity and Inclusion**: Bridging the urban-rural divide and ensuring that growth benefits every segment of society.
- **Resilient and Green Infrastructure**: Building smart, sustainable cities and industrial corridors that reflect a commitment to both economic progress and environmental preservation.

Panch Parivartan Framework

The **Panch Parivartan Framework**: Policy, People, Processes, Products, and Platforms represents a comprehensive strategy for driving transformational change across socioeconomic and environmental domains. Rooted in the principles of inclusivity, innovation, and sustainability, this framework seeks to align India's growth trajectory with the global paradigms of technological advancement and ecological responsibility. By fostering synergies between Information Technology (IT) and sustainability, the Panch Parivartan Framework offers a roadmap for creating a resilient and equitable future. This chapter delves into the five pillars of the framework and explores how their interplay shapes transformative outcomes for a sustainable and self-reliant India.

Policy – Catalyst for Structural Transformation: Policies form the bedrock of systemic change, providing direction and coherence to developmental initiatives. In the Panch Parivartan Framework, policy interventions are designed to be forward-looking, inclusive, and adaptive.

- **Framework for Sustainable Policies**
    - Policies like the National Action Plan on Climate Change (NAPCC) and the National Energy Policy reflect India's commitment to integrating sustainability into governance.
    - Digital policies such as the Digital India Mission and the National Cybersecurity Policy provide the regulatory backbone for fostering IT innovation while ensuring data security and privacy.
- **Incentivizing Green Practices**
    - Policy mechanisms like the Perform, Achieve, Trade (PAT) scheme under the Energy Conservation Act promote energy efficiency across industries.
    - Production-linked incentives (PLIs) for green technologies, such as renewable energy equipment and electric vehicles, align sustainability goals with industrial growth.
- **Global Policy Alignment**
    - India's participation in international agreements like the Paris Agreement and its leadership in initiatives like the International Solar Alliance (ISA) showcases its policy-driven commitment to sustainability.

People – Core of Transformation: People are at the heart of the Panch Parivartan Framework, both agents and beneficiaries of change. Empowering individuals and communities through education, skill development, and participatory governance is pivotal for achieving sustainable outcomes.

- **Building Human Capital**
  - Initiatives like Skill India and Atal Innovation Mission focus on enhancing the employability of India's workforce in emerging technologies such as AI, machine learning, and robotics.
  - Public awareness campaigns around climate change and sustainable practices foster a culture of environmental stewardship.
- **Inclusivity and Equity**
  - Programs like the Pradhan Mantri Jan Dhan Yojana (PMJDY) and the Jeevan Jyoti Bima Yojana promote financial and social inclusion, ensuring growth benefits the marginalized.
  - Women's participation in STEM fields and green industries is encouraged through targeted scholarships and policy incentives, fostering gender equity.
- **Grassroots Participation**
  - Decentralized governance models, such as those enabled by the Panchayati Raj System, integrate local perspectives into decision-making processes, ensuring relevance and efficacy.

Processes – Streamlining Systems for Efficiency: Processes refer to the methodologies and systems that operationalize policies and empower people. In the Panch Parivartan Framework, reengineering processes with IT-enabled solutions is essential for achieving efficiency, transparency, and sustainability.

- **E-Governance**
  - Platforms like the Aadhaar-enabled Public Distribution System (PDS) and the Government e-Marketplace (GeM) have streamlined public service delivery, reducing inefficiencies and leakages.
  - Real-time data monitoring through systems like the Public Financial Management System (PFMS) ensures accountability in resource allocation.
- **Sustainable Practices in Industries**
  - Adopting green manufacturing processes, supported by IoT and AI, minimizes resource consumption and waste generation.
  - The National Water Mission emphasizes efficient water usage in agriculture and industry through drip irrigation and wastewater recycling.

- **Resilience in Supply Chains**
  - Blockchain technology enhances transparency and traceability in supply chains, particularly in agriculture and pharmaceuticals.
  - Powered by AI, predictive analytics optimizes logistics, reducing energy consumption and operational costs.

Products – Innovation for Sustainability: Products, as tangible outputs of industrial and creative processes, represent the transformative potential of innovation. In the Panch Parivartan Framework, the emphasis is on developing sustainable, high-quality, and competitive products that meet the needs of a globalized economy.

- **Green Technologies**
  - India's advancements in solar photovoltaic panels, wind turbines, and battery storage systems reflect its leadership in renewable energy products.
  - The development of electric vehicles (EVs), supported by indigenous battery technology, exemplifies the synergy between IT and sustainability.
- **Smart Products for Smart Cities**
  - IoT-enabled devices, such as smart meters and intelligent traffic management systems, contribute to energy conservation and urban efficiency.
  - Modular housing solutions using recyclable materials address urban housing demands while reducing environmental impact.
- **Healthcare and Biotech Innovations**
  - Integrating AI in diagnostics, telemedicine, and personalized medicine enhances healthcare access and affordability.
  - Biodegradable medical devices and sustainable pharmaceutical packaging reflect a commitment to reducing environmental footprints.

Platforms: Enablers of Integration and Scalability: Platforms are the digital and physical ecosystems facilitating collaboration, innovation, and scalability. In the Panch Parivartan Framework, platforms act as enablers, connecting policy, people, processes, and products.

- **Digital Platforms**
  - The **IndiaStack** ecosystem, including Aadhaar, UPI, and DigiLocker, exemplifies how integrated digital platforms simplify citizen access to services.

- Platforms like the National Digital Health Mission (NDHM) centralize health records, improving healthcare delivery and policymaking.

- **Research and Development (R&D)**
  - The Anusandhan National Research Foundation (ANRF) fosters collaboration between academia, industry, and government, accelerating innovation.
  - Technology clusters, such as those supported by the Atal Tinkering Labs, provide incubation spaces for startups and entrepreneurs.

- **Public-Private Partnerships (PPPs)**
  - PPP models have been instrumental in building infrastructure, from highways to renewable energy plants, ensuring efficient resource utilization and risk-sharing.
  - Collaborative platforms like the International Solar Alliance (ISA) position India as a leader in global sustainability efforts.

Synergies Between IT and Sustainability: The intersection of IT and sustainability is a cornerstone of the Panch Parivartan Framework. IT drives efficiency and innovation and facilitates monitoring, measurement, and optimization of sustainability initiatives.

- **Data-Driven Decision-Making**
  - Big data analytics enables precise forecasting of energy demand, resource allocation, and environmental risks.
  - Satellite imagery and AI-powered tools monitor deforestation, water levels, and agricultural productivity, guiding sustainable interventions.

- **Green IT Solutions**
  - Cloud computing and virtualization reduce the energy footprint of data centres, supporting green IT practices.
  - Smart grids, powered by IoT, optimize energy distribution and integrate renewable sources seamlessly into the grid.

- **Carbon Footprint Reduction**
  - IT-enabled systems for telecommuting and virtual collaboration reduce the need for physical travel, lowering carbon emissions.

- Digital platforms for e-governance and e-commerce minimize the environmental impact of traditional processes.

- **Sustainability Reporting**:
  - Blockchain technology ensures sustainability reporting transparency, tracking carbon credits, and adherence to environmental standards.

Role of IT in Shaping a Developed Bharat

The Information Technology (IT) revolution has been one of the most transformative forces shaping India's trajectory toward becoming a developed nation. From the advent of computerization in the late 20th century to the contemporary wave of advanced technologies like artificial intelligence (AI), blockchain, and quantum computing, IT has consistently acted as a catalyst for systemic change across domains. The journey from Digital India, launched in 2015, to the seamless adoption of advanced technology underscores the centrality of IT in India's socioeconomic evolution. This chapter explores the historical evolution of IT in India, its role as a growth enabler, and its significance in accelerating India's transition to Viksit Bharat 2047.

Historical Evolution of IT in India: The roots of India's IT revolution trace back to the 1980s, a period marked by the liberalization of the computer industry and the establishment of institutions that laid the groundwork for a tech-driven economy.

- **Early Beginnings**
  - The establishment of the Department of Electronics (DoE) in 1970 and the founding of the National Informatics Centre (NIC) in 1976 signalled India's early foray into IT.
  - Computerization of government departments began in the 1980s, enabling data management and administrative efficiency.

- **Economic Liberalization and IT Boom**
  - The economic reforms 1991 catalyzed the IT sector, encouraging foreign investment and entrepreneurship. The establishment of software parks, like the Software Technology Parks of India (STPI), provided the infrastructure necessary for IT exports.
  - Companies like TCS, Infosys, and Wipro emerged as global players, positioning India as a hub for IT services.

- **Digital Revolution**
  - The turn of the millennium saw the rise of e-governance initiatives, including the Aadhaar project in 2009, which revolutionized identity management for over a billion citizens.
  - The launch of **Digital India** in 2015 marked a paradigm shift, focusing on universal digital access, digital literacy, and IT-enabled governance.

Digital Revolution – From Digital India to Advanced Technology Adoption: The Digital India Mission was foundational in democratizing technology and creating a digitally empowered society. Its success has paved the way for adopting advanced technologies that are now reshaping India's economic and social landscape.

- **Infrastructure Development**
  - Initiatives like BharatNet aim to connect 250,000 gram panchayats with high-speed broadband, bridging the digital divide between urban and rural India.
  - Mobile connectivity, with over 1.2 billion subscribers, and the rollout of 5G networks underscore India's robust IT infrastructure.
- **Digital Economy**
  - The Unified Payments Interface (UPI) has transformed India into one of the fastest-growing digital payment markets globally, processing over 10 billion monthly transactions as of 2024.
  - The Open Network for Digital Commerce (ONDC) promotes a decentralized e-commerce ecosystem, empowering small businesses and local artisans.
- **Advanced Technologies**
  - AI and machine learning are being deployed in agriculture (precision farming), healthcare (AI diagnostics), and governance (smart cities).
  - Blockchain technology is revolutionizing supply chain management, land record digitization, and financial services.

IT as a Catalyst for Systemic Change Across Domains: IT's transformative power lies in its ability to act as a cross-cutting enabler, driving systemic change across multiple sectors

- **Governance**
  - E-governance platforms like the Pradhan Mantri Jan Dhan Yojana (PMJDY), DigiLocker, and UMANG app ensure transparency, reduce corruption, and enhance service delivery.

- Integrating real-time monitoring tools like the Public Financial Management System (PFMS) enables efficient public expenditure management.

- **Healthcare**
  - The National Digital Health Mission (NDHM) creates a unified digital health infrastructure, improving access to healthcare services and enabling telemedicine in remote areas.
  - AI-driven tools assist in predictive analytics for disease outbreaks, resource allocation, and personalized medicine.
- **Education**
  - Digital learning platforms like DIKSHA and SWAYAM expand access to quality education, particularly in underserved regions.
  - AI-powered adaptive learning tools personalize education to cater to individual student needs, bridging learning gaps.
- **Agriculture**
  - IT-enabled platforms like eNAM (National Agriculture Market) connect farmers with markets, enhancing price discovery and reducing intermediaries.
  - Satellite imaging and IoT devices facilitate precision agriculture, optimizing resource use and boosting yields.
- **Urban Development**
  - Smart city initiatives integrate IT solutions for traffic management, waste disposal, and energy efficiency, improving urban living standards.
  - IoT-enabled utilities, such as smart meters, reduce energy wastage and promote sustainable urbanization.
- **Sustainability and Environment**
  - IT supports environmental monitoring through tools like Geographic Information Systems (GIS) to track deforestation, air quality, and water resources.
  - Blockchain-based carbon credit systems incentivize businesses to adopt sustainable practices.

Digital Ecosystem as a Growth Enabler: India's digital ecosystem is a dynamic enabler of economic growth, social equity, and global competitiveness. The following elements highlight its role in shaping a developed Bharat:

- **Start-up Ecosystem**
    - India boasts over 90,000 start-ups, with over 100 unicorns as of 2024, making it the third-largest start-up ecosystem globally. Platforms like Startup India and Atal Innovation Mission provide funding and mentorship for entrepreneurs.
    - IT-driven innovations in fintech, edtech, health tech, and agritech create employment opportunities and address socioeconomic challenges.
- **Financial Inclusion**
    - The combination of Aadhaar, UPI, and mobile banking has brought millions into the formal financial system, enabling direct benefit transfers (DBTs) and reducing leakages.
    - Digital microfinance platforms empower small businesses and self-help groups, particularly in rural India.
- **Global IT Leadership**
    - India's IT services exports, valued at over $200 billion annually, showcase its global leadership in software development, cybersecurity, and cloud solutions.
    - Collaboration in international initiatives, such as the Quantum Alliance and AI for Good, positions India as a key player in emerging technologies.
- **Digital Literacy**
    - Programs like the Pradhan Mantri Gramin Digital Saksharta Abhiyan (PMGDISHA) aim to digitally literate 60 million rural households, ensuring inclusivity in the digital revolution.
    - Public-private partnerships foster IT skill development, preparing India's workforce for Industry 4.0.

The Way Forward

Opportunities lie in fostering innovation ecosystems, strengthening public-private partnerships, and leveraging IT for global competitiveness. Emerging technologies, such as quantum computing, metaverse applications, and edge computing, offer avenues for India to lead in the next wave of digital transformation. While IT has proven to be a powerful enabler of growth, several challenges must be addressed to maximize its potential:

- **Bridging the Digital Divide**
  - Despite progress, disparities in digital access persist, particularly in rural and underserved regions.
  - Addressing the affordability and availability of devices and connectivity is crucial for equitable growth.
- **Data Security and Privacy**:
  - The increasing digitization of sensitive data requires robust cybersecurity frameworks and adherence to data privacy norms, such as the **Digital Personal Data Protection Act 2023.**
- **Skill Development**
  - Preparing India's workforce for AI, robotics, and blockchain jobs requires continuous investment in education and reskilling.

The role of IT in shaping a developed Bharat cannot be overstated. From its historical evolution to its current status as a driver of systemic change, IT has redefined the contours of governance, economy, and social equity. As India progresses toward its vision of Viksit Bharat 2047, the digital revolution must continue to prioritize inclusivity, sustainability, and global competitiveness. By addressing challenges and capitalizing on opportunities, IT can propel India into a future of innovation, resilience, and prosperity. The seamless integration of IT across domains will transform India into a developed nation and establish it as a model for the world in leveraging technology for human and environmental advancement.

Understanding "Viksit Bharat" for Global Sustainability Goals

India's vision of "Viksit Bharat" (Developed India) by 2047 reflects an ambitious, multi-dimensional framework to achieve socioeconomic prosperity, ecological sustainability, and global leadership. The vision aligns with the international mandate of the United Nations Sustainable Development Goals (SDGs), a comprehensive agenda to end poverty, protect the planet, and ensure prosperity for all by 2030. India's aspirations for self-reliance, embodied in the Atmanirbhar Bharat initiative, further underscore its commitment to sustainable development, inclusive growth, and leadership in addressing global challenges. This chapter explores how the Viksit Bharat vision aligns with the SDGs, India's progress and challenges in achieving sustainability goals, and the country's growing role on the global stage.

Alignment with the UN Sustainable Development Goals: The SDGs encompass 17 interconnected goals, addressing critical areas such as poverty eradication, education, health, gender equality, clean energy, climate action, and economic growth. India's development strategies, particularly under the Viksit Bharat framework, align intrinsically with these objectives.

1. Poverty and Hunger Eradication (SDG 1 and SDG 2)

- India has made significant strides in poverty alleviation, with programs like the Pradhan Mantri Garib Kalyan Anna Yojana (PMGKAY) providing free food grains to over 800 million people during the COVID-19 pandemic.
- Sustainable agricultural practices, supported by the Soil Health Card Scheme and initiatives like e-NAM (National Agriculture Market), aim to enhance productivity and ensure food security.

2. Quality Education and Health (SDG 3 and SDG 4)

- The National Education Policy (NEP) 2020 emphasizes equitable access to education, integration of digital learning, and skill development, particularly for rural and underserved populations.
- India's health initiatives, such as Ayushman Bharat and the National Health Mission, aim to provide universal healthcare, address maternal and child mortality, and combat non-communicable diseases.

3. Gender Equality and Social Inclusion (SDG 5 and SDG 10)

- Programs like Beti Bachao, Beti Padhao and Pradhan Mantri Ujjwala Yojana empower women by improving access to education, healthcare, and clean energy.
- India's efforts to uplift marginalized communities through targeted policies ensure inclusivity and reduce inequalities.

4. Clean Energy and Climate Action (SDG 7 and SDG 13)

- India's leadership in renewable energy is evident in its ambitious targets under the National Solar Mission and the International Solar Alliance (ISA).
- The transition to electric mobility and promotion of energy-efficient appliances through initiatives like UJALA and FAME India align with the global commitment to climate action.

5. Sustainable Economic Growth and Innovation (SDG 8 and SDG 9)

- Infrastructure development under the **Gati Shakti National Master Plan** and support for start-ups through **Startup India** drive economic growth and innovation.
- Investments in green technologies, smart cities, and digital infrastructure highlight India's focus on sustainable industrialization and urbanization.

6. Life on Land and Below Water (SDG 14 and SDG 15)

- Conservation programs such as Project Tiger, Namami Gange, and afforestation initiatives promote biodiversity and ecological balance.
- Marine ecosystem preservation is addressed through measures to combat overfishing and pollution.

India's Aspirations for Self-Reliance (Atmanirbhar Bharat): India's vision of self-reliance, encapsulated in the Atmanirbhar Bharat initiative, complements the goals of Viksit Bharat and the SDGs by focusing on building resilience, reducing dependencies, and fostering sustainable development.

1. Economic Self-Sufficiency

- India's efforts to boost manufacturing under the Make in India program align with SDG 8 (Decent Work and Economic Growth) by generating employment and fostering innovation.
- The Production-Linked Incentive (PLI) schemes incentivize domestic production in key electronics, pharmaceuticals, and renewable energy sectors.

2. Energy Independence

- Investments in renewable energy, particularly solar and wind power, reduce reliance on imported fossil fuels while addressing SDG 7 (Affordable and Clean Energy).
- Initiatives like National Hydrogen Mission promote green hydrogen as a clean fuel alternative, positioning India as a global leader in energy innovation.

3. Agricultural Resilience

- Modernizing agriculture through precision farming, crop diversification, and irrigation technologies ensures food security and rural livelihoods.
- Promoting organic farming and agroforestry aligns with SDG 12 (Responsible Consumption and Production).

4. Technology and Digital Transformation

- The Digital India initiative empowers citizens through digital inclusion, e-governance, and innovation, contributing to SDG 9 (Industry, Innovation, and Infrastructure).
- Digital platforms like the Unified Payments Interface (UPI) and DigiLocker enhance transparency and efficiency in governance and economic transactions.

India's Role in Global Sustainability Leadership: India's commitment to global sustainability extends beyond its borders, as demonstrated by its proactive role in international forums and collaborative efforts.

1. International Climate Leadership

- India's ambitious targets under the Nationally Determined Contributions (NDCs), such as achieving net-zero carbon emissions by 2070, reflect its commitment to SDG 13 (Climate Action).
- As a founding member of the International Solar Alliance (ISA), India promotes solar energy adoption in developing countries.

2. Technology and Knowledge Sharing

- India's IT prowess positions it as a global leader in offering cost-effective technology solutions for education, healthcare, and governance.
- India provides technical assistance and capacity building to other developing nations through the South-South Cooperation.

3. Humanitarian Assistance

- Under the Vaccine Maitri initiative, India's vaccine diplomacy during the COVID-19 pandemic highlights its leadership in global health.
- Disaster relief efforts, such as those coordinated through the Coalition for Disaster Resilient Infrastructure (CDRI), underscore India's commitment to resilience and recovery.

The vision of "Viksit Bharat" is not merely a national aspiration but a global commitment to fostering a sustainable, equitable, and resilient future. By aligning its developmental goals with the UN Sustainable Development Goals, India demonstrates its dedication to addressing global challenges while pursuing self-reliance and inclusive growth. As India marches toward 2047, its ability to harmonize economic prosperity, social equity, and environmental sustainability will define its legacy as a leader in the global sustainability agenda. Through innovation, collaboration, and a steadfast commitment to its people and the planet, India is poised to emerge as a model for sustainable development and global leadership in the 21st century.

1

Foundation of Sustainability in Digital Age for Viksit Bharat

Introduction

Sustainability in the digital age represents a transformative approach to addressing the critical challenges of economic growth, environmental stewardship, and social equity. For India's vision of "Viksit Bharat" (Developed India) by 2047, integrating sustainable practices within the digital transformation framework is pivotal. This foundation has three primary pillars: technological innovation, inclusive development, and environmental responsibility.

Role of Technological Innovation: Digital technologies have emerged as essential drivers of change, reshaping traditional methods of governance, education, healthcare, and industrial processes. Initiatives such as Digital India have laid the groundwork for building a robust digital infrastructure, empowering millions of Indians, especially in rural areas, with access to essential services. Artificial intelligence (AI), the Internet of Things (IoT), and blockchain technology are transforming industries by enabling smarter decision-making, improving resource efficiency, and promoting transparency. For instance, AI-powered systems are being used to optimize energy grids, reduce waste in supply chains, and monitor environmental conditions. These advancements directly contribute to sustainability by minimizing resource consumption and mitigating environmental degradation. IoT-enabled smart cities enhance urban planning, improve waste management systems, and facilitate efficient energy use. India's Smart Cities Mission exemplifies the integration of such technologies, targeting sustainable urbanization to accommodate the growing population without overburdening the environment.

Inclusive Digital Development: A sustainable future for India hinges on ensuring equitable access to digital resources. Bridging the digital divide is crucial for empowering marginalized communities, including women, farmers, and economically disadvantaged groups. The deployment of high-speed internet in rural areas and the proliferation of affordable smartphones have enabled millions to participate in the digital economy.

Agriculture, a backbone of the Indian economy, benefits significantly from digital tools that provide real-time weather updates, soil analysis, and market trends. Platforms like eNAM (National Agriculture Market) allow farmers to access a unified online trading portal, ensuring fair prices and reducing intermediaries. This improves livelihoods and fosters sustainable agricultural practices by promoting precision farming. Similarly, integrating digital technologies in healthcare ensures access to affordable medical services for remote and underserved regions. Telemedicine platforms and AI-driven diagnostic tools reduce dependency on physical infrastructure, addressing health inequities while minimizing environmental impacts associated with traditional healthcare delivery models.

Environmental Responsibility in the Digital Era: While digital advancements offer opportunities for sustainability, they also pose challenges, particularly in energy consumption and electronic waste. The proliferation of data centres, fueled by increasing digital activity, contributes significantly to global carbon emissions. To mitigate this impact, India must prioritize energy-efficient technologies, renewable energy integration, and green data centres.

The circular economy offers a sustainable pathway to address e-waste. Initiatives that promote the recycling and reuse of electronic components are critical to minimizing the environmental footprint of the digital sector. Companies like Recykal are leveraging technology to streamline waste management and recycling processes, demonstrating how digital platforms can foster sustainable practices. India's ambitious targets for renewable energy adoption, such as achieving 50% of its energy requirements from non-fossil fuel sources by 2030, align with its commitment to sustainable development. The deployment of smart grids, powered by digital technologies, facilitates the efficient distribution and storage of renewable energy, reducing reliance on fossil fuels.

Policy Framework for Sustainable Digital Growth: Realizing the vision of Viksit Bharat necessitates a robust policy framework that integrates sustainability into every aspect of digital transformation. Policies promoting electric vehicles (EVs), green buildings, and adopting renewable energy are essential for creating an ecosystem supporting sustainable growth. Incentives for startups and enterprises to adopt environmental, Social, and Governance (ESG) principles further reinforce aligning economic activities with sustainability goals. The government's emphasis on fostering innovation through programs like Startup India and Atal Innovation Mission ensures that technological advancements are geared toward addressing pressing sustainability challenges. Public-private partnerships can be pivotal in scaling these solutions, ensuring widespread adoption and impact.

Current Scenario of Sustainability

Sustainability has emerged as one of the most critical global imperatives as societies, economies, and environments face unprecedented pressure from rapid industrialization, climate change, and resource depletion. As the world continues to evolve, the need to define and structure sustainability becomes more pressing. Sustainability is commonly understood as meeting the needs of the present without compromising the ability of future generations to meet their own needs. However, this seemingly simple definition encapsulates complex interdependencies between three primary pillars: environmental, social, and economic sustainability. Each of these pillars is essential to understanding the broader sustainability agenda and its applications of rapid technological advancements.

Defining Sustainability and Pillars: At its core, sustainability seeks to harmonize environmental health, social equity, and economic vitality to create systems that are not only resilient in the short term but also capable of flourishing over the long term. The three pillars of sustainabilityenvironmental, social, and economicserve as the foundation for sustainable development. These pillars interact with one another in complex ways, and a holistic approach to sustainability requires addressing all of them simultaneously.

- **Environmental Sustainability**: This pillar emphasizes protecting and preserving natural resources, ecosystems, and biodiversity. Environmental sustainability is concerned with reducing human impact on the planet, mitigating climate change, conserving biodiversity, and managing resources to ensure their availability for future generations. Central to environmental sustainability is reducing the ecological footprint, ensuring that human activity remains within the planet's biophysical limits. Key actions in this area include adopting renewable energy sources, promoting sustainable agriculture and land use practices, reducing waste, and curbing emissions of greenhouse gases.
- **Social Sustainability**: Social sustainability focuses on ensuring the well-being of individuals and communities, with a particular emphasis on social equity, justice, and access to opportunities. It seeks to promote inclusive societies where all individuals can lead healthy, fulfilling lives. This includes ensuring equitable access to resources like education, healthcare, and economic opportunities and fostering social cohesion and justice. Social sustainability also involves protecting human rights, reducing inequality, and creating social systems that adapt to changing demographics and societal needs. Rapid technological advancements for social sustainability entail addressing issues such as digital inclusion, privacy, and the social impacts of automation and AI.
- **Economic Sustainability**: Economic sustainability ensures long-term economic growth and stability while respecting ecological limits and social equity. This pillar advocates for creating resilient economies that can endure

fluctuations in markets, technology, and global crises without jeopardizing the well-being of future generations. Economic sustainability calls for innovation and diversification, moving away from a dependence on finite resources like fossil fuels toward circular economies, green technologies, and sustainable industries. It also emphasizes the importance of fair distribution of economic benefits, including fair wages, opportunities for all, and the reduction of poverty and inequality.

Each of these pillars must be carefully balanced, as overemphasising one can undermine the others. For example, focusing on economic growth without regard to environmental protection can lead to resource depletion, while prioritizing environmental sustainability without addressing economic inequalities can exacerbate poverty. The "triple bottom line" conceptwhich seeks to measure success through the lens of people, planet, and profitreflects the need for an integrated approach to sustainability.

Challenges of Achieving Sustainability in Rapid Technological Advancement: While technological advancement offers a range of solutions to help achieve sustainability goals, it also presents numerous challenges, many of which complicate balancing the three pillars of sustainability. As technology evolves exponentially, several key challenges arise in achieving the sustainability agenda, particularly in environmental conservation, social equity, and economic stability.

- **Environmental Challenges in the Technological Age**: Rapid technological advancements, while offering new tools for sustainability, also introduce significant environmental challenges. The expansion of digital technologies, including data centers, cloud computing, and the proliferation of electronic devices, has increased energy consumption. For instance, the global energy demand from data centers is projected to double within the next decade, potentially exacerbating issues of carbon emissions if the energy is sourced from non-renewable sources. Additionally, the rise of the Internet of Things (IoT) and the growth of digital infrastructures contribute to mining rare-earth metals, leading to environmental degradation from extraction practices. The environmental impact of technologies like artificial intelligence (AI) and blockchain, which require vast computational power, remains a significant concern regarding energy use and e-waste.

  The problem of e-waste is another growing environmental concern. With the rapid pace of technological innovation, devices such as smartphones, computers, and other electronics are quickly rendered obsolete, contributing to increasing volumes of waste. In 2019, the world generated 53.6 million metric tons of e-waste, a number expected to rise in the coming years. The environmental impact of this waste, including harmful chemicals leaching into soil and water, poses significant challenges to environmental sustainability.

- **Social Challenges: Digital Divide, Job Displacement, and Privacy**: On the social front, rapid technological advancement raises concerns about the equitable distribution of benefits. One of the most pressing issues is the digital dividethe gap between those without access to modern technologies. This divide is particularly evident in developing countries, rural areas, and marginalized communities, where the internet, computers, and smartphones are limited. The digital divide exacerbates existing social inequalities, as those without access to digital tools are excluded from economic opportunities, education, and healthcare services. Bridging this gap is essential for achieving social sustainability but requires significant investment in infrastructure, education, and digital literacy programs.

  Another social challenge is the potential for job displacement due to automation and AI. While these technologies promise to increase efficiency and productivity, they also threaten to displace millions of jobs, particularly in manufacturing, logistics, and customer service. Without corresponding efforts in retraining and reskilling, the displacement of workers can deepen social inequality and increase unemployment rates, creating social instability. The emergence of gig economies and precarious work further compounds these challenges, as workers in non-traditional employment are often deprived of social safety nets such as healthcare and retirement benefits. The rise of digital technologies introduces significant concerns around privacy and data security. Tech companies' collection, storage, and use of personal data raise issues of surveillance, data breaches, and misuse of information. In a world where personal data has become a valuable commodity, maintaining privacy and ensuring robust data protection are critical components of social sustainability in the digital age.

- **Economic Challenges – Unintended Consequences of Technological Progress**: From a financial perspective, while technology has the potential to drive growth and innovation, it also introduces challenges related to inequality and market disruption. The rapid pace of technological advancement can disproportionately benefit wealthy individuals and corporations, exacerbating wealth inequality. The concentration of technological power in a few large corporations such as Amazon, Google, and Facebook has raised concerns about monopolies, market competition, and the concentration of economic power.

- Technological advancements can disrupt traditional industries, leading to economic instability. For example, the rise of e-commerce has significantly impacted brick-and-mortar retail businesses, and the automation of manufacturing processes has led to the decline of specific industries. This disruption often results in the loss of jobs, especially in low-skilled sectors, while creating new job opportunities in high-tech industries. Ensuring a fair transition for workers affected by these changes through policies such as universal basic income (UBI), retraining programs, and job creation in emerging sectors is crucial for maintaining economic stability and achieving long-term sustainability. The costs associated with implementing green technologies and transitioning to a low-carbon economy can also be substantial. Although renewable energy technologies such as solar and wind power offer long-term economic benefits, the initial investment required for infrastructure development can be a barrier for many countries and businesses. Economic sustainability, therefore, requires a balance between investing in green technologies and ensuring that the costs of these transitions do not hinder economic growth.

Pathways to Sustainability in the Technological Age: Despite these challenges, technological advancements can also provide solutions to many sustainability challenges. The key lies in harnessing the potential of technology while mitigating its negative impacts. One promising pathway is the circular economy concept, which emphasizes the recycling, repurposing, and reusing of materials to minimize waste and reduce reliance on finite resources. Digital technologies, such as blockchain and AI, can facilitate the creation of circular supply chains by improving tracking, transparency, and efficiency. In the realm of social sustainability, technological innovations in education, healthcare, and financial inclusion have the potential to reduce inequalities significantly. Mobile platforms, for example, have revolutionized access to banking services in underserved communities, while telemedicine and e-health solutions can improve healthcare delivery in remote regions. Bridging the digital divide through infrastructure development, training, and policy interventions is essential to distribute these benefits widely. Promoting green technologies, such as renewable energy, electric vehicles, and sustainable agriculture practices, for economic sustainability offers growth opportunities that align with environmental goals. Governments, businesses, and individuals must work collaboratively to foster innovation, create jobs, and develop sustainable industries.

The rapid pace of technological advancement presents opportunities and challenges for achieving sustainability in the modern world. While digital technologies have the potential to drive innovation, improve efficiency, and reduce environmental impacts, they also raise significant concerns related to environmental degradation, social inequality, and economic disruption. Achieving sustainability A comprehensive and integrated approach is required to address sustainability's environmental, social, and economic pillars. By leveraging technology to promote green solutions, inclusivity, and equitable economic growth, societies can build a sustainable future that benefits all stakeholders and future generations.

Sustainability Pillars for India: Economic Growth, Environmental Conservation, and Social Equity

As one of the world's fastest-growing economies, India faces the critical challenge of aligning rapid economic development with environmental conservation and social equity. The pursuit of sustainability in India is grounded in three essential pillars: economic growth, environmental conservation, and social equity. While individually significant, these pillars must be integrated to achieve a balanced, resilient, and inclusive future. This integration is especially critical for a country like India, where developmental pressures are high, and the demand for resources continues to escalate, even as the government strives to address deep-seated social inequities and environmental degradation.

Economic Growth – Engine of Development: India's economic growth is central to its development strategy. As the world's fifth-largest economy by nominal GDP, India's growth trajectory is underpinned by its burgeoning population, young labour force, and significant global presence in information technology (IT), pharmaceuticals, and services. Economic growth in these industries is pivotal in alleviating poverty, providing jobs, and improving living standards. However, for development to be sustainable, it needs to be both inclusive and environmentally responsible.

Economic growth in India has historically been associated with industrialization, infrastructure development, and an increasing reliance on fossil fuels. While these factors have contributed to economic expansion, they have also led to environmental challenges such as air pollution, water scarcity, and land degradation. The need to decouple economic growth from environmental degradation is urgent. To this end, India strives towards a green economy by emphasizing the importance of renewable energy, clean technologies, and sustainable industrial practices. Policies like the National Action Plan on Climate Change (NAPCC) and the Make in India initiative aim to create a framework for sustainable industrialization that is resource-efficient and low-carbon.

One of the primary challenges in balancing economic growth with sustainability is the overexploitation of natural resources. India's rapid industrial development has often come at the expense of forests, water bodies, and mineral resources. The mining sector, for instance, has led to severe environmental degradation in various regions, displacing local communities and impacting biodiversity. To overcome this, India is focusing on sustainable development strategies such as the circular economy, which emphasizes recycling, reusing, and reducing waste, alongside increasing energy efficiency. Another aspect of economic growth is the push towards urbanization. As more people move into cities, the pressure on infrastructure, housing, and natural resources grows. Urban sustainability, which includes building smart cities, improving public transportation, and reducing the urban heat island effect, is critical to ensuring that economic growth does not compromise the environment or the quality of life.

Environmental Conservation – Growing Necessity: Environmental conservation is an essential pillar of sustainability, especially for a country like India, which is rich in biodiversity but faces enormous environmental challenges. India's environmental challenges are profound and multifaceted, ranging from air and water pollution to deforestation, biodiversity loss, and climate change. With its large population and rapidly expanding industrial base, the country is at a crossroads, needing to balance economic growth with the imperative to protect the environment.

India's environmental policies emphasize resource conservation, sustainable agriculture, forest protection, and renewable energy. The government has set ambitious renewable energy targets, aiming to achieve 500 gigawatts (GW) of renewable energy capacity by 2030, and committed to reducing its economy's carbon intensity. This push towards clean energy is driven by policy incentives, technological advancements, and international agreements, such as the Paris Agreement, which India is committed to upholding. Solar and wind energy and hydropower offer significant potential to reduce reliance on fossil fuels, major contributors to greenhouse gas emissions. Achieving true environmental sustainability requires more than just switching to renewable energy. It necessitates addressing waste management, sustainable water use, soil degradation, and biodiversity protection. The Ganga Action Plan, the National Clean Energy Fund, and the Swachh Bharat Mission are examples of national initiatives aimed at addressing these challenges. The promotion of sustainable agricultural practices is another crucial strategy for environmental conservation. With agriculture being a primary source of income for millions, farming practices must move away from conventional, resource-intensive methods towards more sustainable approaches, including organic farming, water-efficient techniques, and agroforestry.

India's biodiversity is under threat due to habitat destruction, poaching, and pollution. The country has established numerous protected areas, wildlife sanctuaries, and national parks to counter these threats. However, the encroachment of human settlements and industrial activity threatens these natural habitats. Integrating biodiversity conservation into broader development policies is essential for ensuring that economic growth does not come at the expense of the country's natural heritage.

Climate change poses another significant environmental challenge. India is already experiencing the effects of climate change, including extreme weather events such as floods, droughts, and heatwaves. These events threaten agricultural productivity, water availability, and human health. India has implemented various adaptation and mitigation strategies to address climate change, including promoting climate-resilient crops, improving water conservation techniques, and enhancing disaster management capabilities. India's Nationally Determined Contributions (NDCs) under the Paris Agreement aim to reduce its carbon emissions intensity and promote sustainable practices in key sectors like energy, transportation, and waste management.

Social Equity – Cornerstone of Sustainable Development: Social equity is the third pillar of sustainability, ensuring that development benefits are distributed fairly and no group is left behind. India is a country of immense diversity, with significant disparities in income, education, healthcare access, and opportunities. Social equity must be at the heart of development policies for sustainability to be achieved. India's focus on social equity has resulted in numerous initiatives to improve the welfare of its most marginalized populations. Programs such as the Mahatma Gandhi National Rural Employment Guarantee Act (MGNREGA), Pradhan Mantri Awas Yojana (PMAY), and Pradhan Mantri Jan Dhan Yojana (PMJDY) are designed to address poverty, provide employment, ensure housing, and promote financial inclusion. These programs are integral to ensuring that the benefits of economic growth are shared across all segments of society.

Education is another vital aspect of social equity. The government has implemented several schemes to promote access to quality education for all children, particularly in rural and underserved areas. The Right to Education Act, launched in 2009, guarantees free and compulsory education for children between 6 and 14. This focus on education is crucial for empowering individuals, reducing inequality, and ensuring that all citizens have the skills and opportunities necessary to participate in the digital economy. Significant challenges remain in ensuring social equity. Despite substantial progress, India still grapples with deep-rooted issues such as caste-based discrimination, gender inequality, and regional disparities in wealth and development. Women, for instance, continue to face barriers to economic participation and access to education and healthcare. Addressing these inequalities is essential for ensuring that development benefits are inclusive and sustainable.

India's social security system also faces challenges, particularly in the informal sector, where millions of workers lack job security, healthcare, and pension benefits. The rapid pace of technological advancement, particularly automation and artificial intelligence (AI), poses additional challenges to social equity, as it risks displacing low-skilled workers and exacerbating unemployment. Policies that support workforce transition, reskilling, and inclusive growth are crucial in ensuring that technological progress benefits all sections of society.

Balancing Rapid Development with Sustainability: India's pursuit of rapid economic growth faces several challenges in balancing it with sustainability. The tension between development and sustainability is particularly acute in the manufacturing, energy, and infrastructure sectors. Industrialization, essential for economic growth, has led to environmental degradation, especially in air pollution, deforestation, and water contamination. Rapid urbanization, while contributing to economic dynamism, places significant pressure on cities, leading to overcrowding, inadequate infrastructure, and environmental pollution.

One of the key challenges is the energy sector, which is dominated by fossil fuels like coal and natural gas. While India has made significant progress in increasing the share of renewables in its energy mix, the transition is still in its nascent stages. The reliance on coal for electricity generation remains high, and India continues to face the challenge of ensuring energy security while reducing its carbon footprint. Agriculture, another cornerstone of India's economy, also faces sustainability challenges. Overusing chemical fertilizers, pesticides, and water-intensive crops has led to soil degradation, water scarcity, and health issues. Transforming India's agricultural sector to adopt sustainable farming practices is critical to environmental conservation and social equity. However, the transition to sustainable agriculture faces resistance from entrenched interests and the economic pressures of feeding a growing population.

Aligning National Policies with Global SDGs: India's sustainability efforts must be aligned with global sustainability frameworks, particularly the United Nations' Sustainable Development Goals (SDGs). The SDGs provide a comprehensive blueprint for addressing the world's most pressing challenges, from poverty and hunger to climate change and inequality. India has made significant strides in aligning its policies with the SDGs, but challenges remain in ensuring that the targets are achieved equitably across all regions and sectors. India's national policies, such as the National Action Plan on Climate Change (NAPCC), the National Mission for Clean Ganga, and the National Rural Employment Guarantee Scheme (MGNREGA), are directly aligned with the SDGs. However, the challenge lies in implementing these policies at the grassroots level, where regional disparities, bureaucratic inefficiencies, and resource constraints often undermine their effectiveness.

India's commitment to achieving the SDGs requires further strengthening its institutional frameworks, fostering public-private partnerships, and ensuring that sustainable development becomes a key priority across all sectors of the economy. Integrating sustainability principles into economic planning, industrial development, and social policies is essential for long-term sustainability.

Role of Technology in Sustainability

Information technology (IT) has become an indispensable tool for driving sustainable development in the modern era. The rapid advancement of IT has brought about transformative changes across various sectors, from agriculture and healthcare to energy and transportation. As nations and organizations strive to meet global sustainability goals, the potential of IT to foster environmental, social, and economic sustainability is becoming increasingly evident. Technology, particularly IT, plays a crucial role in enabling resource optimization, promoting equity, and facilitating innovationall of which are vital components of a sustainable future. By strategically leveraging IT, it is possible to achieve holistic development that ensures the planet's well-being and its inhabitants' prosperity.

IT as a Tool for Achieving Holistic Development: Holistic development addresses the interconnected needs of society, the economy, and the environment. Achieving sustainable development requires a comprehensive approach integrating environmental stewardship, economic growth, and social equity. When applied correctly, information technology is a powerful tool to incorporate these three pillars, thus fostering holistic development.

IT enables the collection, processing, and analysis of vast amounts of data that can inform policy decisions, drive innovation, and enhance the efficiency of public and private sector operations. For example, smart city technologies leverage IT to optimize urban resource management, from water distribution and waste management to energy consumption and transportation systems. This optimization reduces environmental impact, enhances quality of life, and increases the efficiency of urban infrastructure, contributing to sustainable urban development. Similarly, IT solutions like Geographic Information Systems (GIS) have been invaluable in managing natural resources, from tracking deforestation to mapping biodiversity hotspots.

IT also supports holistic development by improving access to essential services. Technology has effectively bridged education, healthcare, and financial inclusion gaps in remote or underserved areas. The widespread adoption of mobile phones and internet access has brought telemedicine, online education, and mobile banking to millions in rural regions, thus fostering social equity and reducing inequalities. IT-based solutions for governance, such as e-Governance platforms, enable more transparent, efficient, and participatory systems that help ensure that all citizens, regardless of location, benefit from development initiatives. Digital platforms enable a more inclusive economy by providing small businesses and startups access to markets, funding, and tools once limited to large corporations. Through e-commerce, online marketing, and digital payment systems, even the smallest enterprises can scale and thrive in a globalized economy. The use of IT to support inclusive economic growth ensures that sustainable development does not leave bchind any segment of society.

IT as a Tool for Resource Optimization and Efficiency: One of the most significant ways IT supports sustainability is by enabling resource optimization. The efficient use of resources is essential to minimizing waste, conserving energy, and reducing environmental degradation. IT systems, such as the Internet of Things (IoT), big data analytics, and artificial intelligence (AI), have revolutionized how resources are managed across sectors, from energy consumption and water usage to waste disposal and agricultural practices. In the energy sector, IT has enabled the development of smart grids that optimize electricity distribution by balancing supply and demand, reducing energy losses, and integrating renewable energy sources. Smart meters, powered by IoT, allow consumers to track their energy usage in real-time, enabling them to make more informed decisions and reduce consumption. AI-driven algorithms are used to predict energy demand and optimize generation, which helps minimize waste and improves the reliability of energy systems.

In agriculture, IT is crucial in optimizing resource use through precision farming. Satellite imaging, drones, and sensors allow farmers to accurately monitor soil conditions, weather patterns, and crop health. This enables them to apply only fertilizers, water, and pesticides where and when needed, thus reducing waste and minimizing environmental impact. Precision farming increases crop yields and reduces input costs, making agriculture more economically and environmentally sustainable. IT also aids in waste management by facilitating the tracking, sorting, and recycling of materials. Municipalities can optimize waste collection routes, monitor waste disposal practices, and encourage recycling efforts through digital platforms. Big data analytics can identify trends in waste generation, helping policymakers design more effective recycling programs and waste diversion strategies. Additionally, blockchain technology has the potential to enhance transparency and accountability in waste management by ensuring that recyclable materials are properly processed and not diverted into landfills.

In the industrial sector, IT drives resource optimization through digital twins and virtual models of physical systems that can simulate real-world processes. By analyzing the performance of machinery, production lines, and supply chains in real-time, digital twins enable businesses to identify inefficiencies and optimize operations. This leads to reduced resource consumption, lower emissions, and enhanced sustainability. AI and machine learning algorithms also optimize supply chain management, reducing waste and improving logistics efficiency. IT plays a crucial role in optimizing water usage. Advanced sensor networks and data analytics are being used to monitor water consumption in real-time, detecting leaks and inefficiencies in irrigation systems, municipal water supplies, and industrial processes. These systems enable more efficient water use, which is particularly critical in regions facing water scarcity. IT also supports the development of water purification and desalination technologies, helping to address global water shortages.

IT as a Tool for Promoting Equity and Inclusion: One of the most potent aspects of IT in sustainability is its ability to promote equity and inclusion. As a pillar of sustainable development, social sustainability ensures that all individuals have equal access to opportunities and resources regardless of their background or socioeconomic status. Information technology plays a pivotal role by creating new pathways for social inclusion and reducing inequalities in access to essential services like education, healthcare, and financial services.

IT has democratized access to learning opportunities through online platforms, e-learning tools, and digital libraries in education. These technologies have made education more accessible to people in remote areas, allowing them to learn new skills and improve their employability. Initiatives such as massive open online courses (MOOCs), e-learning platforms like Coursera, Khan Academy, and edX, and digital literacy programs have expanded access to quality education, helping to bridge the digital divide and promote lifelong learning. Telemedicine and mobile health applications are bringing medical services to underserved populations. With mobile phones, individuals in rural or remote areas can consult doctors, receive health advice, and access diagnostic services without needing long-distance travel. IT-driven solutions like electronic health records and health management systems are improving the efficiency of healthcare delivery, enabling better patient care and reducing health disparities. Additionally, AI is used to diagnose diseases, track epidemics, and predict healthcare trends, ensuring that resources are allocated efficiently and equitably.

Financial inclusion, another critical aspect of social equity, has been dramatically enhanced by IT. Mobile banking, digital payment systems, and blockchain-based financial services have provided millions of people in developing countries access to banking services. These technologies have enabled individuals to store money securely, make transactions, and access credit without the need for traditional banking infrastructure. This has empowered individuals and stimulated economic growth in previously underserved regions. IT facilitates participatory governance and citizen engagement. E-Government platforms, where citizens can access services, submit grievances, and participate in decision-making, have enhanced transparency and accountability in the public sector. This democratization of information and decision-making processes ensures that all segments of society, including marginalized and vulnerable groups, have a voice in shaping policies that affect their lives.

Enabling Sustainable Growth through Innovation: Innovation, driven by IT, is a cornerstone of sustainable growth. Technological advancements allow the development of new products, services, and business models that prioritize sustainability while fostering economic growth. IT enables industries to embrace sustainable practices by providing tools for monitoring and reducing environmental impacts, improving efficiency, and developing green technologies.

One of the most significant innovations in recent years is the rise of the circular economy, which is supported by digital technologies. The circular economy model aims to reduce waste, extend the lifespan of products, and promote the reuse and recycling of materials. IT facilitates this model by enabling businesses to track the lifecycle of products, optimize supply chains, and identify opportunities for product repair, refurbishment, and recycling. For instance, blockchain technology ensures the traceability of materials in supply chains, ensuring that products are sourced sustainably and disposed of responsibly.

IT also supports sustainable entrepreneurship by providing startups and small businesses with the tools to innovate and scale sustainably. Digital platforms offer companies access to global markets, funding opportunities, and tools for collaboration, allowing them to develop new products and services that meet the demands of a sustainability-conscious consumer base. The role of IT in driving sustainable innovation is particularly evident in developing clean technologies, such as solar and wind energy, electric vehicles (EVs), and energy storage solutions. These innovations are essential to reducing reliance on fossil fuels and curbing greenhouse gas emissions. IT is instrumental in advancing the efficiency of these technologies through data analytics, AI, and automation. For example, AI algorithms can optimize the operation of solar and wind farms, increasing energy production and reducing costs.

IT as a Strategic Enabler of Sustainable Development: As a strategic enabler of sustainable development, IT helps integrate sustainability into business models, government policies, and social systems. IT helps create more efficient, equitable, and environmentally responsible systems by providing the tools to monitor, analyse, and optimise resources. IT supports sustainability goals and drives the transition to a sustainable economy by fostering innovation, enabling collaboration, and providing access to critical resources and information. Governments, businesses, and civil society organizations must collaborate to ensure that IT is used effectively to achieve sustainability objectives. This involves investing in digital infrastructure, ensuring equitable access to technology, and promoting digital literacy. Policies that encourage innovation in green technologies, support sustainable business practices, and integrate sustainability into decision-making processes are essential for leveraging IT as a driver of sustainable development.

Scenario for Viksit Bharat 2047 – Rise of AI and IoT in Enabling Smart, Sustainable Systems

As India envisions becoming a *Viksit Bharat* (Developed India) by 2047, the country stands at the cusp of transformative changes in technology, particularly with the rise of artificial intelligence (AI) and the Internet of Things (IoT). These technologies are set to play a critical role in shaping India's path towards sustainable development, enabling the creation of smart, efficient, and resilient systems. Integrating AI and IoT into various sectorsenergy, transportation, healthcare, agriculture, and urban developmentcan revolutionise how India addresses its development challenges. These technologies promise to accelerate growth that aligns with global sustainability frameworks, such as the United Nations' Sustainable Development Goals (SDGs). By 2047, Viksit Bharat is poised to leverage the synergies between AI, IoT, and sustainability, setting a benchmark for other nations striving for green and equitable development.

Role of AI and IoT in Enabling Smart, Sustainable Systems: AI and IoT, when combined, offer a powerful toolkit for creating smart, sustainable systems that optimize resources, reduce waste, and enhance overall efficiency. The Internet of Things, which connects physical devices to the internet, allows for real-time data collection and communication between machines, infrastructure, and humans. Artificial intelligence, on the other hand, processes and analyzes this vast amount of data to make intelligent decisions that improve operational efficiency and enable predictive actions. Together, AI and IoT form the backbone of smart cities, smart agriculture, energy management systems, and environmental monitoring tools.

Smart Cities – Sustainable Urban Development: The concept of smart cities has emerged as one of the most promising applications of AI and IoT in urban development. IoT sensors are embedded in the urban infrastructure of a smart city, including transportation systems, buildings, energy grids, and public services. These sensors continuously collect data on air quality, traffic congestion, energy usage, waste management, and water consumption. AI processes this data in real-time to optimize city operations, reduce energy consumption, and minimize pollution. For instance, AI-powered traffic management systems can dynamically adjust traffic signals based on real-time traffic flow, reducing congestion and lowering carbon emissions.

Additionally, AI and IoT contribute to more sustainable and efficient building management. Smart buildings with IoT sensors can monitor temperature, lighting, and ventilation systems, adjusting them automatically to optimize energy use. Such buildings are essential for reducing the carbon footprint of urban environments. The integration of AI further enhances energy efficiency by predicting demand patterns and optimizing energy distribution in real-time. By 2047, as India becomes a Viksit Bharat, most cities are expected to adopt smart city technologies, leading to more efficient urban management and a significant reduction in resource consumption.

Sustainable Agriculture – Precision Farming: India's agricultural sector, which employs a large portion of its population, faces significant sustainability challenges, including soil degradation, water scarcity, and the overuse of chemical fertilizers. AI and IoT are poised to transform agriculture into a highly efficient, resource-conscious industry, allowing India to address these challenges while improving food security and farmers' livelihoods.

IoT devices, such as soil moisture sensors, weather stations, and crop health monitoring systems, provide real-time data to help farmers make informed decisions about irrigation, fertilizer use, and pest control. AI algorithms can analyze this data to offer actionable insights, optimizing resource usage and enhancing crop yields. For example, AI-powered systems can predict the best time for planting and harvesting crops based on weather patterns and soil conditions, reducing waste and maximizing productivity. Enabled by IoT and AI, precision farming allows more efficient water usage, reducing the strain on India's already scarce water resources. By 2047, India's agricultural sector is expected to fully embrace these technologies, leading to more sustainable farming practices and improved food production without further environmental degradation.

Energy Management – Transitioning to Renewable Energy: India's energy sector is one of the most critical areas for achieving sustainability. With an increasing population and rapid industrialization, the energy demand is expected to rise substantially in the coming decades. However, India's reliance on coal and other fossil fuels for energy generation has resulted in high greenhouse gas emissions, contributing to climate change and air pollution. AI and IoT are transforming the energy sector by facilitating the transition to renewable energy sources such as solar, wind, and hydropower. IoT devices, such as smart meters and energy sensors, are already being deployed to monitor and manage energy consumption in real-time. These devices enable utilities to track energy usage patterns, detect inefficiencies, and reduce transmission losses. AI, when applied to this data, can predict energy demand fluctuations, optimize energy production, and manage the distribution of electricity to maximise the use of renewable energy sources.

The integration of AI and IoT in smart grids, which are designed to manage and distribute energy more efficiently, is pivotal in reducing the carbon footprint of India's energy sector. By 2047, India is expected to have a grid that integrates a high percentage of renewable energy, supported by AI systems that balance supply and demand, predict energy consumption patterns, and ensure the efficient distribution of electricity. The deployment of AI-powered energy storage systems will play a crucial role in enhancing the reliability of renewable energy sources. Energy storage solutions, such as batteries, allow surplus energy generated by solar or wind power to be stored and used during periods of high demand or low renewable energy production. AI algorithms can optimize these batteries' charge and discharge cycles, ensuring the most efficient use of energy storage and improving the grid's stability.

Environmental Monitoring and Conservation: AI and IoT technologies also contribute to environmental conservation efforts in India. With the country facing deforestation, water pollution, and biodiversity loss, these technologies offer powerful tools for monitoring and mitigating environmental damage. IoT sensors can be deployed in forests, rivers, and wildlife habitats to collect data on environmental conditions, such as air and water quality, soil health, and biodiversity levels.

AI can then analyze this data to detect patterns and predict environmental changes, such as pollution spikes, habitat destruction, or illegal activities like poaching. For example, AI algorithms can analyze satellite images to detect deforestation in real-time, allowing authorities to take swift action. Similarly, IoT sensors can monitor water quality in rivers and lakes, alerting authorities to potential contamination and enabling proactive measures to protect water resources. AI is also being used to model the impacts of climate change on ecosystems, helping policymakers develop strategies for mitigating these effects. For instance, AI-powered climate models can predict the potential impacts of rising temperatures on agriculture, water resources, and biodiversity, allowing India to plan for climate adaptation strategies.

Integration of IT into Global Sustainability Frameworks: As India strives to become a Viksit Bharat by 2047, its commitment to global sustainability frameworks, such as the United Nations Sustainable Development Goals (SDGs), will be critical in guiding its development trajectory. Integrating AI, IoT, and other emerging technologies into these global frameworks will be essential for aligning national efforts with international sustainability objectives. The SDGs provide a comprehensive blueprint for addressing global challenges, including poverty, hunger, inequality, climate change, and environmental degradation. India's adoption of AI and IoT technologies can support achieving many of these goals. For example, AI can enhance efforts to achieve SDG 2 (Zero Hunger) by improving food security through precision agriculture. IoT-enabled monitoring systems can support SDG 6 (Clean Water and Sanitation) by ensuring efficient water use in agriculture and urban environments. AI-driven smart grids can help achieve SDG 7 (Affordable and Clean Energy) by facilitating the transition to renewable energy sources.

Additionally, India's efforts to reduce carbon emissions, enhance energy efficiency, and protect its ecosystems align with SDG 13 (Climate Action). AI and IoT to monitor and reduce emissions in industrial sectors, optimize transportation networks, and manage energy use will be crucial in mitigating the effects of climate change. AI and IoT can play a vital role in supporting SDG 9 (Industry, Innovation, and Infrastructure) by fostering the development of sustainable, innovative industries. By leveraging these technologies, India can transition to a green economy that promotes economic growth and prioritizes environmental protection and social equity. Integrating IT into global sustainability frameworks requires collaboration between governments, businesses, and civil society. By aligning national policies with the SDGs and adopting emerging technologies, India can lead the global transition towards a more sustainable, equitable, and resilient world. The country's focus on smart, sustainable systems driven by AI and IoT will enhance its development and contribute to the broader global effort to achieve the United Nations 2030 Agenda for Sustainable Development.

2

Panch Parivartan Framework for IT-Led Change for Transformation

Introduction

Panch Parivartan, which translates to "Five Transformations," is an innovative framework that uses information technology (IT) to catalyze a profound and holistic transformation across multiple sectors of society, governance, and the economy. This framework aims to create a sustainable and inclusive model for development by leveraging IT to drive change in key areas, including governance, economic growth, social equity, education, and environmental sustainability. In India's development, particularly in the drive towards a *Viksit Bharat* (Developed India) by 2047, the Panch Parivartan framework outlines a roadmap where IT is the backbone for achieving transformative change. Through the strategic deployment of IT solutions, the framework seeks to address complex challenges, unlock new opportunities, and pave the way for a more equitable, efficient, and sustainable society.

1. Parivartan in Governance – eGovernance and Transparency: The first pillar of Panch Parivartan focuses on transforming governance through adopting e-Governance. E-Governance refers to using digital platforms and IT tools to enhance the delivery of government services, improve transparency, and enable better communication between government entities and citizens. In India, the governance model has long been characterized by inefficiency, bureaucratic red tape, and limited access to essential services, particularly in rural areas. The advent of IT solutions has the potential to radically transform this landscape by streamlining administrative processes, making government services more accessible, and ensuring that citizens are at the center of governance.

The implementation of e-Governance initiatives can enhance transparency by automating bureaucratic procedures and creating systems that allow citizens to track the progress of their requests and applications. For example, initiatives like the Digital India program, which aims to digitize government services, have brought online portals for various public services such as passport applications, income tax filings, and welfare programs. IT-driven platforms can provide real-time updates on the status of applications, reducing corruption and ensuring timely service delivery. Additionally, technologies like blockchain can provide secure and transparent record-keeping systems, further promoting accountability and reducing the likelihood of fraud. By integrating IT into governance, the Panch Parivartan framework aims to foster a participatory democracy where citizens have easy access to government services, better avenues to voice concerns and greater confidence in public institutions. It is anticipated that by 2047, India will have a fully digitized government system that enables seamless interactions between citizens and the state, ensuring that the benefits of governance are equitably distributed across all segments of society.

2. Parivartan in Economy – Digital Empowerment and Financial Inclusion: The second pillar of the Panch Parivartan framework is transforming the economy through digital empowerment and financial inclusion. Information technology, especially digital banking, mobile payments, and financial technology (fintech), has the potential to reshape India's economic landscape radically. With a significant portion of India's population remaining unbanked or underbanked, financial inclusion is a critical area where IT can drive substantial change.

Digital financial services, including mobile wallets, online banking, and microfinance platforms, have already begun to break down barriers to financial access. Mobile payment systems like *UPI* (Unified Payments Interface) and digital lending platforms have allowed millions of previously underserved individuals and small businesses to access banking services, saving time and money previously spent travelling to physical banks. IT-driven solutions enable users to make instant payments, access loans, and build credit histories, which were once out of reach for many in rural and semi-urban areas. The push for financial inclusion is aligned with India's broader vision of creating a *cashless economy* and reducing the informal economy that often operates outside of regulatory frameworks. Digital literacy programs can empower citizens, especially women and marginalized communities, to engage with the financial system, thereby improving their economic participation and boosting overall economic growth.

The transformation of the economy through IT under Panch Parivartan extends beyond banking to include sectors like e-commerce, digital manufacturing, and skill development. By integrating IT into business processes, small and medium enterprises (SMEs) can increase their market reach, optimize their operations, and improve competitiveness in both national and international markets. India's focus on *Make in India* and *Startup India* initiatives is further bolstered by IT, as digital tools foster innovation, entrepreneurship, and job creation, which are critical for sustainable economic growth by 2047.

3. Parivartan in Society – Digital Literacy and Social Equity: The third pillar of Panch Parivartan emphasizes the transformation of society through digital literacy and social equity. Social transformation, as envisioned by the framework, requires inclusive access to technology, equitable opportunities for participation, and the empowerment of marginalized groups, including women, rural populations, and low-income communities. IT can catalyze the narrowing the socioeconomic divide and create opportunities for social mobility. Digital literacy is foundational to this transformation. In a rapidly digitalizing world, accessing and using technology is crucial for participation in modern society. Digital literacy programs that equip individuals with the skills to use computers, mobile phones, and the internet are vital for creating an inclusive culture where everyone can thrive. Government initiatives such as *Pradhan Mantri Gramin Digital Saksharta Abhiyan* (PMGDISHA) are designed to improve digital literacy in rural India, enabling individuals to participate more fully in the digital economy.

Social equity is also supported through IT-driven solutions that address critical areas such as healthcare, education, and public welfare. Telemedicine platforms, for example, have provided millions of people in rural and remote areas with access to healthcare services that were previously unavailable. E-learning platforms empower students across India to access quality education, particularly in underserved areas. Through mobile applications, social welfare programs can be more effectively delivered, ensuring that benefits reach intended recipients without leakage or inefficiency. The transformation of society through IT under Panch Parivartan aims to ensure that all individuals, regardless of their socioeconomic status or geographical location, have equal access to the benefits of technological progress. By promoting digital literacy and equitable access, the framework seeks to create a society where IT is not a barrier to advancement but a tool for empowerment.

4. Parivartan in Education – Bridging the Knowledge Gap: The fourth pillar of the Panch Parivartan framework is the transformation of education through the integration of IT. India's education system faces numerous challenges, including poor access to quality education in rural areas, teacher shortages, and inadequate infrastructure. The use of IT in education, through e-learning platforms, digital classrooms, and virtual tutoring, can significantly improve the quality of education while addressing these gaps.

Digital education has the potential to make learning more accessible, flexible, and engaging. Platforms like *SWAYAM*, *DIKSHA*, and *eVidya* provide students access to online courses, video lectures, and learning materials. AI and machine learning algorithms can personalize learning experiences, tailoring educational content to students' needs, thereby improving learning outcomes. Additionally, virtual classrooms and video conferencing tools enable students in remote areas to attend lectures and engage with teachers, bridging the rural-urban divide in education. The framework also envisions integrating IT into vocational training and skill development programs, ensuring that students and young professionals have the skills required for a modern, digital economy. By 2047, it is expected that India will have a well-integrated, IT-enabled education system that offers high-quality education to every citizen, contributing to creating a highly skilled workforce capable of driving the nation's growth.

5. Parivartan in Environment – Sustainable Development through Technology: The fifth pillar of Panch Parivartan emphasizes environmental sustainability, an increasingly urgent global issue. IT has a critical role in enabling the sustainable use of resources, mitigating climate change, and conserving biodiversity. Technologies like remote sensing, IoT, big data analytics, and AI can provide real-time insights into environmental conditions, enabling more informed decision-making and better management of natural resources. For example, IoT sensors and satellite data can monitor air and water quality, track deforestation, and predict weather patterns, helping policymakers take timely actions to protect the environment. Smart grids and AI-powered energy management systems can optimize energy use, reducing reliance on fossil fuels and promoting renewable energy sources such as solar and wind. IT-based solutions for waste management, such as smart waste collection systems, can reduce pollution and promote recycling. Sustainability through IT is also critical for agriculture. Powered by IoT and AI, precision farming technologies allow farmers to optimize water usage, reduce chemical inputs, and increase crop yields. These technologies can help achieve food security while reducing the environmental impact of farming practices. By 2047, it is envisioned that India will have integrated environmental monitoring systems powered by IT, enabling the country to manage its natural resources efficiently and transition to a sustainable, low-carbon economy.

Theoretical Framework and Relevance to India's Context

Panch Parivartan, which translates to "Five Transformations," is a novel and comprehensive framework conceptualized to leverage the potential of information technology (IT) as a transformative tool for sustainable socioeconomic development. This framework aims to catalyze large-scale change by addressing key sectors of governance, economy, society, education, and the environment. By leveraging the disruptive power of IT, Panch Parivartan envisions a future where technology acts not just as an enabler of development but as a central agent of transformation. The core idea behind this framework is to focus on five interconnected areas that can drive India toward a holistic and inclusive transformation: governance, economic growth, social equity, education, and environmental sustainability. The use of IT is not confined to improving efficiency or accessibility; rather, it aims to radically change the way society operates, empowering individuals and communities and fostering a participatory, transparent, and accountable state.

The theoretical underpinnings of Panch Parivartan are rooted in the principles of digital inclusion, sustainability, and equity. The framework draws inspiration from several developmental theories, including the Technology Acceptance Model (TAM), Diffusion of Innovations (DOI) theory, and Sustainable Development Goals (SDGs), which all highlight the potential of technological adoption to drive societal and economic change. Additionally, Panch Parivartan aligns closely with the Digital India initiative and the concept of a Smart Nation, where the seamless integration of IT solutions fosters inclusive development while promoting social, economic, and environmental sustainability. In the Indian context, the framework is particularly relevant, given the nation's ongoing efforts to leverage technology to address its vast socioeconomic challenges and transition towards a more sustainable, inclusive, and prosperous society. This introduction aims to explore the theoretical foundations of Panch Parivartan, contextualize its relevance to India, and outline its significance for achieving comprehensive national development.

Theoretical Foundations of Panch Parivartan: At its core, Panch Parivartan is built upon several interrelated theories that form the foundation of its approach to IT-driven transformation. These include frameworks that focus on technological adoption, the role of innovation in social change, and sustainable development.

- **Technology Acceptance Model (TAM)**: The Technology Acceptance Model is central to understanding how individuals and organizations accept and use new technologies. TAM posits that perceived ease of use and usefulness are the primary determinants of technology adoption. Panch Parivartan incorporates these principles by promoting IT solutions designed to be accessible and beneficial for all layers of society. By ensuring that the technologies used for governance, economic development, education, and environmental sustainability are user-friendly and directly address real-world needs, the framework facilitates the widespread acceptance and application of these tools across various sectors.
- **Diffusion of Innovations (DOI)**: Diffusion of Innovations theory, developed by Everett Rogers, emphasizes how new ideas, practices, or technologies spread across societies. This model identifies factors such as the perceived advantage of innovation, compatibility with existing values, and simplicity of use as key elements in the adoption process. Panch Parivartan leverages these insights by introducing IT solutions that are not only technologically advanced but also culturally and socially compatible with India's diverse contexts. For instance, mobile-based platforms for governance services or financial inclusion are designed to cater to India's large rural population, many of whom are more familiar with mobile phones than computers. The ease of accessing and using digital technologies and their clear benefits support their adoption and integration into various sectors.

- **Sustainable Development Goals (SDGs)**: The United Nations' SDGs are pivotal in shaping the global development agenda. These goals emphasize the need for economic growth that is socially inclusive and environmentally sustainable. Panch Parivartan is closely aligned with SDGs, particularly those related to quality education (SDG 4), decent work and economic growth (SDG 8), reduced inequalities (SDG 10), and climate action (SDG 13). By promoting digital tools for education, financial inclusion, governance, and environmental management, the Panch Parivartan framework helps India fulfil its commitments to these goals. It enables India to transition toward a more sustainable, equitable, and inclusive development model that addresses its unique challenges.
- **Digital Transformation and Smart Nations**: Digital India's concept is the most direct framework that informs Panch Parivartan. This national initiative aims to transform India into a digitally empowered society and knowledge economy. Digital India focuses on building digital infrastructure, delivering government services digitally, and enhancing digital literacy. Panch Parivartan extends these goals by promoting a comprehensive approach to digital governance and economic growth, ensuring that digital tools are used for holistic national development. The framework envisions India as a Smart Nation where IT solutions enable smarter governance, efficient resource use, and better citizen participation. Panch Parivartan addresses the challenge of the digital divide and champions inclusivity by focusing on areas that promote accessibility for all sectors of society.

Relevance of Panch Parivartan in the Indian Context: India's socioeconomic landscape is marked by stark disparities in income, access to resources, and opportunities for social mobility. The country's large rural population and diverse cultural, linguistic, and regional dynamics present unique challenges and opportunities for sustainable development. The relevance of Panch Parivartan lies in its recognition of these challenges and its comprehensive approach to addressing them through IT-led transformation.

- **Addressing Governance Challenges**: India's governance system has long been characterized by bureaucratic inefficiencies, corruption, and a lack of transparency, which have hindered the effective delivery of public services. Panch Parivartan envisions using IT tools to address these issues by promoting e-Governance, digital transparency, and accountability. Initiatives like the Public Financial Management System (PFMS), Digital India, and Aadhaar have already significantly addressed governance challenges. By expanding and refining these initiatives, Panch Parivartan aims to ensure that citizens can access public services efficiently and transparently. Additionally, the framework envisions enhancing the citizen-government interface through digital platforms that foster participatory governance, allowing people to engage directly in decision-making processes.

- **Economic Empowerment Through Digital Inclusion**: Economic growth and development in India have been constrained by limited access to financial services, education, and employment opportunities, especially for marginalized groups. Panch Parivartan addresses these barriers by promoting financial inclusion through IT tools like mobile banking, e-wallets, and digital education platforms that provide learning opportunities to underserved communities. The rise of digital payments and fintech has already enabled millions of Indians to gain access to banking services, create credit histories, and participate in the formal economy. By further expanding the reach of these services and integrating them with India's broader development strategies, Panch Parivartan aims to create an inclusive economy where all citizens can benefit from growth.
- **Social Equity and Empowerment**: India's large population of women, youth, and marginalized communities faces barriers to full social participation. Panch Parivartan seeks to address these issues by promoting digital literacy and ensuring these groups have the same opportunities as the rest of the population. Digital platforms can provide employment, skill development, and entrepreneurship opportunities, empowering individuals and improving their social and economic status. Initiatives to increase women's participation in the digital economy, such as training programs and digital entrepreneurship, align with the framework's goal of achieving **gender equality** and **social equity**.
- **Education as a Catalyst for Change**: The transformation of India's education system is central to its overall development. The framework emphasizes the use of IT to enhance access to education, particularly for children in rural and underserved areas. E-learning platforms, MOOCs (Massive Open Online Courses), and digital classrooms are revolutionizing education, providing millions of students with the tools they need to succeed. By expanding the reach of these digital tools and integrating them with India's educational policies, Panch Parivartan aims to create a knowledge-driven society, which is essential for innovation and sustainable development.
- **Environmental Sustainability**: As India continues to industrialize and urbanize, environmental sustainability has become a pressing concern. Panch Parivartan highlights the role of IT in enabling smarter environmental management. The framework envisions a more sustainable approach to

managing water, energy, and waste through IoT sensors, AI-powered resource management, and big data analytics. Digital tools can help optimize energy consumption, reduce carbon footprints, and monitor environmental degradation, thus contributing to India's climate action and environmental preservation goals.

Five Dimensions of Transformation: Policy, People, Processes, Products, and Platforms

The rapid advancement of Information Technology (IT) is reshaping societies and economies globally, offering new possibilities for governance, innovation, and development. Within India's **ambition to become a *Viksit Bharat* (Developed India) by 2047, the framework of transformation through IT is structured around five key dimensions: Policy, People, Processes, Products, and Platforms. These dimensions are integral to the holistic application of technology** in creating sustainable and inclusive growth while addressing the inherent challenges of India's vast demographic and developmental diversity. Each dimension encapsulates an essential aspect of digital transformation, contributing to the effectiveness and equity of the ongoing national development agenda.

1. Policy – IT in Governance: The foundation of any transformation driven by IT lies in formulating effective and adaptive policies. In India, where governance challenges range from bureaucratic inefficiencies to corruption and poor service delivery, policies integrating IT into governance are critical. The role of digital governance in modernizing public administration is pivotal, enabling governments to serve citizens more efficiently, transparently, and inclusively. India's Digital India program, launched in 2015, marks a significant shift in how governance and public services are delivered. Through e-Governance initiatives, the government aims to streamline processes, improve transparency, and ensure that services are accessible to all citizens, regardless of geographic location. For instance, digital initiatives such as e-Districts, Aadhaar, and PMGDISHA (Pradhan Mantri Gramin Digital Saksharta Abhiyan) have played a transformative role in enabling digital access for rural and marginalized populations, reducing corruption, and making governance more accountable.

Effective policy formulation also requires an understanding of how IT can enhance decision-making. Big Data Analytics and AI-driven insights allow governments to make more informed and timely decisions in areas such as public health, urban planning, and disaster management. These tools help understand trends, forecast needs, and optimise resource allocation. However, for these policies to be impactful, they must be inclusive, ensuring that all citizens, especially those from vulnerable or underserved communities, share the benefits of digital transformation. Policy frameworks need to be adaptive to rapidly evolving technologies. This includes formulating robust data protection laws, ensuring cybersecurity, and addressing privacy and digital rights concerns. India's Personal Data Protection Bill (PDPB), for instance, is an effort to ensure that personal data is handled responsibly and ethically, providing citizens with the confidence to engage with digital platforms. By developing comprehensive policies that not only promote digitalization but also protect citizens' interests, India can create a governance ecosystem that is both innovative and accountable.

2. People – Bridging the Digital Divide: One of the most pressing challenges facing India in its digital transformation is bridging the digital dividethe gap between those who have access to and can effectively use digital technologies and those who do not. India's vast and diverse population has stark disparities in access to digital infrastructure, education, and literacy. Addressing the digital divide is central to ensuring that the benefits of IT-led transformation are equitably distributed across society. The first step toward bridging this divide is promoting digital literacy. India has made significant strides in increasing internet penetration, but digital skills remain a key barrier. For instance, the Pradhan Mantri Gramin Digital Saksharta Abhiyan aims to train millions of rural Indians in basic digital literacy, enabling them to use mobile phones, computers, and the internet for personal and professional purposes. Such initiatives are essential in ensuring that all individuals, regardless of socioeconomic status or geographic location, can participate in the digital economy.

Access to affordable internet is another critical factor. India's ambitious BharatNet project is working to connect rural areas with high-speed internet, providing the infrastructure needed to make digital services accessible in remote locations. In parallel, mobile technology plays a crucial role, as mobile phones are often the most accessible point of access to the internet in rural India. By addressing these infrastructural challenges, India can give its citizens the tools to participate in digital governance, access healthcare, education, and financial services, and engage in the digital economy. To truly bridge the digital divide, however, it is not enough to focus on access alone. Digital inclusion also requires addressing the social, cultural, and economic factors that limit certain groups, particularly women, lower-income communities, and marginalized groups, from accessing and benefiting from IT. Initiatives such as women's digital empowerment programs, skill development for underserved communities, and targeted programs to reduce the gender gap in technology adoption are essential to creating an equitable digital ecosystem.

3. Processes – Operational Efficiency in Systems: One of the most transformative impacts of IT in governance and development is its ability to optimize processes and enhance operational efficiency across public administration, private enterprises, and civil society. IT is revolutionizing how systems and institutions operate, enabling automation, reducing redundancy, and increasing transparency. In India, where inefficiencies in bureaucratic systems often hinder development,

IT presents a powerful tool for improving service delivery. For example, the introduction of e-Governance systems has streamlined various administrative functions. Digital platforms such as Direct Benefit Transfers (DBT) and Public Distribution Systems (PDS) have reduced the time and cost of delivering government benefits, eliminating corruption, and ensuring that benefits reach the intended recipients. Unified Payments Interface (UPI) has further revolutionized the payment ecosystem, providing a fast, secure, and efficient system for financial transactions.

The digitization of records has greatly reduced paperwork and manual errors, ensuring that government services are more efficient and reliable. For instance, digitising land records under initiatives like SVAMITVA (Survey of Villages and Mapping with Improvised Technology in Village Areas) has improved property ownership clarity and reduced land disputes in rural areas. In the private sector, IT-driven systems have enabled Supply Chain Optimization, Automated Customer Relationship Management (CRM) systems, and AI-driven predictive analytics that enhance decision-making and operational efficiency. IoT (Internet of Things) and Artificial Intelligence (AI) have enabled more efficient resource management, inventory control, and demand forecasting in manufacturing, agriculture, and retail sectors, leading to lower costs and improved service delivery. Thus, transforming processes through IT is not merely about technological implementation but also fundamentally reshaping how organizations and institutions operate to provide greater value to society.

4. Products – Sustainable Innovation: The fourth dimension of transformation is the development of sustainable products that promote environmental sustainability while driving economic growth. IT can foster innovation across various sectors by providing tools for efficiency and sustainability. The move toward green technologies and sustainable innovation is crucial for addressing the challenges of climate change and resource depletion.

For example, smart grids, solar power technologies, and battery storage solutions are revolutionizing how energy is produced, distributed, and consumed in the energy sector. These innovations make energy more efficient and reduce reliance on fossil fuels, contributing to a cleaner, more sustainable energy landscape. Similarly, IT enables precision farming techniques that optimize water use, fertilizers, and pesticides, thus improving agricultural yields while minimizing environmental impact. IT-driven platforms can promote circular economy models, where products and resources are reused, refurbished, and recycled to minimize waste. For example, digital platforms for product lifecycle management (PLM) and waste tracking can help companies design products with longer life spans and reduced environmental footprints. AI, machine learning, and big data analytics foster sustainable innovation. These technologies can be employed to predict trends, identify inefficiencies, and design products and solutions that are more sustainable, thus contributing to both economic growth and environmental preservation.

5. Platforms – Enabling Collaboration: The final dimension of transformation is the role of platforms in enabling collaboration, innovation, and the sharing of resources across various stakeholders. IT platforms create ecosystems where governments, businesses, and civil society can collaborate more effectively, breaking down silos and fostering greater coordination and efficiency. In the governance context, platforms like MyGov and IndiaStack have created opportunities for citizens to participate in decision-making, give feedback, and engage directly with policymakers. By enabling collaborative governance, these platforms enhance transparency, accountability, and public participation, empowering citizens to impact policies and programs directly affecting their lives.

In the private sector, open innovation platforms and crowdsourcing have enabled businesses to collaborate with startups, researchers, and consumers to co-create solutions that meet evolving market demands. Cloud computing and shared platforms have allowed organizations to reduce costs, increase collaboration, and accelerate innovation by providing scalable and flexible infrastructures. In education and healthcare, digital platforms have also enabled widespread collaboration across institutions, experts, and service providers, making sharing knowledge, research, and best practices easier. Telemedicine platforms, for example, have facilitated the exchange of medical expertise and services across regions, improving healthcare delivery in rural and underserved areas. By creating these collaborative platforms, IT can help foster a more connected and efficient ecosystem where resources, ideas, and opportunities are shared for mutual benefit, driving sustainable development across sectors.

Vision for 2047 – Holistic IT-Enabled Transformation Strategy for India

As India approaches its centennial of independence in 2047, the country envisions a future where technology, particularly Information Technology (IT), plays a central role in driving national development. The vision for India's transformation into a sustainable, equitable, and prosperous nation is rooted in the potential of a holistic IT-enabled transformation strategy. This strategy must be integrated with India's larger national goals, leveraging the synergies between the five dimensions of development: economic growth, social equity, environmental sustainability, technological innovation, and governance. A key element of this transformation strategy is the integration of IT with India's Panch Parivartan modelrepresenting the five key pillars of change: Panchayati Raj Institutions (PRI), agriculture, education, health, and infrastructure. These pillars collectively form the foundation for achieving the nation's long-term goals of inclusive growth, job creation, poverty alleviation, and sustainable development.

Holistic IT-Enabled Transformation Strategy: The vision for 2047 requires a comprehensive and unified approach to technology adoption. IT is not simply an enabler but an integrated tool that must permeate every aspect of societal development. The strategy for achieving this transformation involves not just deploying advanced technologies but embedding them into the fabric of Indian society, ensuring that every citizen benefits from the advancements in IT. This transformation will be achieved through government policies, technological innovations, strategic partnerships, and local, regional, and national capacity building. The heart of this transformation lies in the convergence of key technological components like Artificial Intelligence (AI), Internet of Things (IoT), blockchain, 5G, and big data. These technologies will form the backbone of India's digital infrastructure, enabling real-time decision-making, creating smarter cities, and facilitating delivering essential services to the masses. India's Vision for 2047 envisions using these technologies to promote economic efficiency, improve public sector service delivery, enhance citizen engagement, and ensure equitable distribution of resources.

A fundamental part of this strategy is digital inclusivity, where IT becomes a tool for bridging the digital divide between urban and rural populations, different socioeconomic groups, and regions with varying infrastructure levels. Ensuring that marginalized communities, such as women, the elderly, and low-income groups, are brought into the fold of digital empowerment is paramount for achieving broad-based development. This requires a two-pronged approach: widespread internet penetration and a robust digital literacy program to ensure citizens can effectively engage with digital technologies.

Synergies Between the Five Dimensions: The vision for 2047 anticipates synergies between the five core dimensions of developmenteconomic growth, social equity, environmental sustainability, technological innovation, and governance. These synergies will enable India to harness the full potential of IT, integrating it into its strategic development plans while achieving national goals.

- **Economic Growth and IT Integration**: The first dimension, economic growth, forms the foundation of India's vision for 2047. A strong, technology-driven economy will be built on the pillars of IT, which will fuel innovation, improve productivity, and enhance competitiveness. India will be a global leader in emerging technologies, with industries such as FinTech, EdTech, AgriTech, and HealthTech driving economic dynamism. The national economy will undergo a digital transformation underpinned by digital financial services, online commerce, and AI-based services that enable new economic activity and employment avenues. IT will also streamline supply chains, improve manufacturing, and optimize agricultural outputs, making India a global hub for technology-driven production and services.

The country will embrace the Internet of Things (IoT) for real-time data analysis, improving operational efficiency across industries. Through the digitalization of traditional sectors like agriculture and manufacturing, India will witness the creation of millions of new jobs in emerging technologies, e-commerce, and digital services, transforming the national workforce into a skilled, tech-savvy labour force.

- **Social Equity and IT Integration**: The second pillar, social equity, is fundamental to India's vision for 2047. The future of India will be defined by the inclusivity of its digital economy, where IT enables marginalized and underserved populations to access education, healthcare, and economic opportunities. IT solutions will be deployed to bridge the gaps in service delivery, ensuring equitable access to essential services such as healthcare, education, and housing. Technology-enabled education will reduce disparities in learning outcomes, ensuring that quality education is accessible to all, irrespective of geographical location or socioeconomic status. AI and machine learning will enable personalized learning pathways for students, especially in rural areas, overcoming infrastructure challenges by delivering education through mobile platforms and satellite systems. Health technology, driven by telemedicine and AI-based diagnostics, will address the shortage of medical professionals and enable affordable and accessible healthcare for every Indian, especially in rural and remote areas. Similarly, digital platforms for skill development will provide marginalized populations with tools to enhance their employability and entrepreneurial capabilities, aligning with India's national goal of increasing job opportunities and reducing poverty.

- **Environmental Sustainability and IT Integration**: The third dimension, environmental sustainability, presents one of the most pressing challenges and opportunities for IT in India's vision for 2047. As climate change becomes a growing concern, IT solutions will play a crucial role in achieving sustainability goals. Smart grids, IoT-based energy systems, and AI-driven waste management solutions will optimize the use of natural resources and reduce India's carbon footprint. Technologies like AI and big data analytics will monitor and manage environmental variables, track deforestation, and monitor water usage. Renewable energy solutions, such as solar and wind energy, will be integrated into smart energy networks, ensuring the efficient use of resources and reducing dependence on fossil fuels. IT-driven solutions will also enable India to meet international environmental targets, such as those outlined in the Paris Agreement. In agriculture, precision farming technologies powered by AI, drones, and IoT will help optimize water, fertilizers, and pesticides, ensuring that India's agricultural practices are efficient and sustainable. Digital platforms will encourage waste reduction, recycling, and circular economy practices, reducing waste generation and improving resource management across industries.

- **Technological Innovation and IT Integration**: Technological innovation is a cornerstone of India's vision for 2047. IT will fuel technological advances that enhance productivity and solve some of India's most pressing challenges, such as urbanization, healthcare, and climate change. AI, blockchain, and quantum computing will be at the heart of the digital economy, driving innovation across sectors. The next wave of technological breakthroughs will include autonomous vehicles, robotics, 5G-powered systems, and advanced data analytics, reshaping industries and societal processes. India will also foster an innovation ecosystem where start-ups, research institutions, and enterprises collaborate to develop scalable, sustainable, and context-specific solutions. These innovations will cater to the domestic market and serve as a model for developing nations worldwide, positioning India as a leader in technological innovation.
- **Governance and IT Integration:** Effective governance is a critical pillar for realizing India's vision for 2047. The digital transformation of governance will increase transparency, improve efficiency, and make public service delivery more citizen-centric. Blockchain, AI, and data analytics will ensure government systems are more accountable, equitable, and responsive. E-governance platforms, powered by AI and machine learning, will help the government provide real-time services to citizens, from welfare disbursement to grievance redressal. These systems will enhance the efficiency of bureaucratic processes, reducing corruption and improving service delivery. IT-enabled data analytics will allow the government to make informed decisions on national planning, development, and resource allocation, ensuring that policies are data-driven and impactful.

Panch Parivartan and IT Integration: The integration of IT with Panch Parivartan will be pivotal to achieving India's 2047 vision. Panch Parivartan, as a model of localized transformation, emphasizes the importance of grassroots-level development. It revolves around five key sectors: Panchayati Raj Institutions (PRIs), agriculture, education, healthcare, and infrastructure. IT will be the thread that connects these five pillars, enabling them to work synergistically to achieve national goals.

- **Panchayati Raj Institutions (PRIs)** will be empowered by IT to improve governance at the local level, enabling real-time data access and decision-making. Digital platforms will strengthen PRI participation in local governance, ensuring local communities have a say in decision-making and that development is aligned with grassroots needs.

- **Agriculture**, IT-driven precision farming techniques, mobile applications, and digital marketplaces will enhance productivity, optimize resource use, and provide farmers with direct access to global markets, making them more competitive and sustainable.
- **With digital platforms providing quality education across rural and urban areas, IT will revolutionise education**, ensuring every child can access learning opportunities and skill-building resources. E-learning systems, AI-powered tutors, and digital content will enable personalized and inclusive education at scale.
- **Healthcare** will benefit from telemedicine, digital health records, and AI-driven diagnostics, ensuring that healthcare services are more accessible, affordable, and effective, especially in underserved regions.
- **Infrastructure development**, especially in urban planning, transport, and housing, will be transformed by smart cities, where IT will optimize resource use, reduce environmental impacts, and improve the quality of life for citizens.

3

IT in Policy Transformation

Introduction

Information Technology (IT) is indispensable in transforming national policies, governance frameworks, and administrative structures, making them more efficient, transparent, and responsive. In policy transformation, IT is a powerful enabler that can streamline bureaucratic processes, improve access to public services, facilitate data-driven decision-making, and enhance citizen engagement. As countries strive to modernize their governance systems, integrating IT into policy frameworks is vital in ensuring inclusive growth, social equity, and sustainable development. This integration is especially critical for a country like India, which, with its diverse population, vast geographical expanse, and complex socioeconomic challenges, requires innovative policy tools that are both scalable and adaptable. The potential of IT in policy transformation is multifaceted and can significantly improve the efficiency, accountability, and inclusivity of public governance.

At the core of IT's role in policy transformation lies its ability to digitize governance processes. Traditional systems often involve cumbersome paperwork, manual records, and face-to-face interactions, which lead to inefficiencies, delays, and corruption. In contrast, IT solutions such as e-Governance, cloud computing, big data analytics, and artificial intelligence (AI) have revolutionized how public services are delivered, enhancing the quality and speed of service delivery. By implementing IT systems, governments can automate routine tasks, digitally store and manage information, and provide real-time access to data. This not only improves the effectiveness of public administration but also minimizes human error, reduces corruption, and increases transparency in government operations.

In India, the adoption of IT in policy transformation has been spearheaded by various initiatives aimed at improving administrative efficiency and citizen-centric governance. Programs such as Digital India, Make in India, and Startup India exemplify the transformative role of IT in reshaping national policies. For instance, Digital India seeks to connect rural areas with the digital infrastructure needed to improve governance and provide public services, ranging from education and healthcare to financial inclusion and e-Governance. IT enables these policies to be implemented at scale, ensuring that even remote and underserved regions benefit from government services, which in the past may have been inaccessible due to logistical constraints. Using biometric-based identification through Aadhaar, direct benefit transfers (DBT), and online grievance redressal mechanisms are all examples of how IT is embedded in policy transformation to enhance accessibility, efficiency, and equity.

One of the key areas where IT has significantly impacted policy transformation is the management of public resources. Through big data and data analytics, governments can make informed decisions about resource allocation, monitor policy implementation in real-time, and assess the effectiveness of various initiatives. For example, public financial management systems, which integrate IT for budget tracking, expenditure monitoring, and financial transparency, help ensure that public funds are allocated efficiently and are used for their intended purposes. IT tools also allow governments to collect and analyze vast amounts of data, enabling them to identify patterns, predict future needs, and tailor policies to meet the evolving demands of society. In this way, IT becomes a crucial tool for evidence-based policymaking, allowing governments to make decisions grounded in empirical data rather than political considerations or outdated models.

IT plays a vital role in enhancing public participation in the policymaking process. Traditionally, policy development has been an elite-driven process, often excluding large population segments from contributing to or understanding the decision-making process. However, IT has opened new avenues for democratizing policy and enabling greater citizen engagement. Digital platforms such as MyGov facilitate direct interaction between the government and citizens, allow the public to provide feedback on government policies, participate in online consultations, and suggest reforms. This inclusive approach helps build public trust in government decisions and leads to more informed and responsive policies. By using social media platforms, online petitions, and crowdsourced data, governments can tap into the collective wisdom of citizens, ensuring that policies are not only top-down but also bottom-up, reflecting the needs and concerns of diverse communities.

IT also plays a transformative role in policy transparency. Governments can use digital tools to publish and disseminate policy documents, budgets, and regulatory frameworks in formats that are easily accessible and understandable to the public. Open data platforms and digital repositories allow citizens, businesses, and civil society organizations to track the progress of policy initiatives and hold policymakers accountable. For example, Right to Information (RTI) portals and open government data initiatives have made it easier for citizens to access government documents, request information, and monitor the implementation of public policies. Digitalizing government records and procedures reduces the likelihood of corruption by making processes more visible and traceable, thus ensuring that public policies are executed with integrity and fairness.

The role of IT in policy monitoring and evaluation is another area where technology has significantly transformed governance. Traditionally, policy implementation often lacked effective tracking mechanisms, leading to delays and inefficiencies in policy delivery. With the advent of IT, governments can now implement monitoring and evaluation (M&E) frameworks that use real-time data to assess the effectiveness of policies and programs. For instance, geo-spatial technologies like satellite imaging, combined with data analytics, enable governments to monitor environmental policies, track deforestation rates, and assess the impact of urban planning initiatives. Similarly, in sectors such as healthcare and education, IT tools can provide insights into the performance of public health schemes, educational reforms, and social welfare programs, allowing policymakers to make timely adjustments based on feedback and empirical evidence.

IT also facilitates cross-border policy coordination by enabling the digital sharing of information and best practices among countries. As globalization increases, nations must collaborate on various issues, including trade, climate change, security, and public health. IT allows governments to share policy frameworks, track global trends, and implement international agreements more effectively. For example, adopting blockchain technology for tracking the supply chain of goods can enhance the traceability and security of international trade, ensuring compliance with environmental standards and human rights regulations. Similarly, countries can cooperate on global challenges like climate change through digital platforms, using AI and machine learning to predict environmental patterns and share solutions for sustainable development.

Despite the immense potential of IT in policy transformation, several challenges need to be addressed to maximize its effectiveness. Digital literacy remains a significant barrier, particularly in rural areas and among marginalized groups, where access to technology and technical skills may be limited. For IT to serve as an effective tool for policy transformation, governments must prioritize digital inclusion by ensuring that all citizens have access to the infrastructure and the education necessary to navigate the digital landscape. Additionally, cybersecurity is an ongoing concern as governments collect sensitive data from citizens and businesses. Robust security frameworks must be in place to protect against cyber threats, data breaches, and identity theft. Finally, ensuring interoperability across diverse IT systems is crucial for seamless integration of policy-related data across multiple sectors, both at the national and international levels.

IT is reshaping thc landscape of policy transformation, offering governments powerful tools to improve governance, enhance transparency, foster citizen engagement, and promote economic and social development. In India, where governance challenges are immense and diverse, IT is proving to be a game-changer in transforming public administration and policy delivery. Through digital platforms, data analytics, and automated systems, IT simplifies processes and empowers citizens, making governance more participatory, efficient, and accountable. As the country moves towards a digitally enabled future, it is imperative for policymakers to continuously harness the potential of IT to build a more inclusive, transparent, and effective governance model. The convergence of technology and policy presents an unprecedented opportunity to foster sustainable development, enhance governance, and improve the quality of life for all citizens.

Present Challenges in Policy Formulation and Execution

The rapid integration of Information Technology (IT) tools into policy formulation and execution processes has significantly transformed governance. IT has the potential to enhance policy modelling, improve data collection, and facilitate data-driven decision-making. However, despite the undeniable promise, several challenges persist that hinder the smooth execution of policies and their ability to address the evolving needs of society. These challenges are primarily rooted in the limitations of current IT tools for policy modelling, the fragmentation of digital systems, inefficiencies in data-driven policymaking, and the gaps in policy implementation and monitoring. Understanding these challenges is essential for developing solutions to improve governance and policymaking in the digital age.

1. Current IT Tools for Policy Modeling, Data Collection, and Decision–Making

At the core of modern governance lies the use of data-driven policy modelling and evidence-based decision-making. IT tools such as Big Data, Artificial Intelligence (AI), Machine Learning (ML), and Geographical Information Systems (GIS) have been employed in various countries to enhance the quality and efficiency of policymaking. These tools allow for collecting and analysing vast amounts of data, which helps governments understand current trends, forecast future needs, and design policies more responsive to citizens' demands. For instance, data analytics is widely used to track the effectiveness of public policies, particularly in sectors such as healthcare, education, and agriculture. Governments now collect and analyze data related to healthcare services, educational performance, and crop yields, using this information to model future trends and policy outcomes. Through AI and ML algorithms, decision-makers can process vast datasets and extract patterns, which can help design predictive policies rather than reactive ones. Similarly, GIS tools enable governments to visualize and analyze spatial data, aiding urban planning, resource management, and disaster preparedness.

Despite these advancements, the effectiveness of these tools is contingent upon the quality of data and the accuracy of models used for decision-making. The reliance on historical data and limited access to real-time data often hampers the ability of policymakers to respond dynamically to fast-changing situations. In countries like India, where large-scale data collection is frequently inconsistent or of low quality, the efficacy of these tools can be severely limited. There is often a lack of interoperability between different IT systems, further complicating the gathering of accurate, up-to-date data for policy modelling.

2. Challenges in Policy Implementation Due to Fragmented Digital Systems One of the most pressing challenges in modern governance is the fragmentation of digital systems. While IT has revolutionized the policymaking process in many ways, the infrastructure supporting it often remains disconnected. For instance, government departments may use different software platforms for data collection, analysis, and service delivery. These systems may not communicate with each other effectively, resulting in data silos and inefficiencies. When digital systems are not fully integrated, policymaking becomes disjointed, and the information required to make informed decisions may not be readily accessible.

The fragmentation of digital systems also creates coordination challenges between different levels of government, agencies, and stakeholders. National-level policies may not be fully implemented locally due to differences in digital capabilities and infrastructure. For instance, policies aimed at financial inclusion may be delayed or disrupted if local branches of banks or post offices do not have access to reliable digital platforms or local government agencies cannot integrate national IT systems. This fragmentation complicates efforts to achieve uniform policy execution, leading to regional disparities in policy outcomes. The absence of standardized systems across government institutions can impede collaboration and data sharing. Government policies increasingly require inputs from various sectors (such as health, education, transport, and urban development), but without a common IT platform, these sectors cannot easily exchange information. This disconnect between policy design and its execution often results in delays, miscommunication, and inefficiencies.

3. Inefficiencies in Data-Driven Policymaking: The growing reliance on data-driven policymaking introduces opportunities and challenges. On the one hand, data analytics and real-time data collection offer the potential for more informed decision-making. On the other hand, several inefficiencies and pitfalls affect the effectiveness of data-driven policies. A key challenge is the inaccurate or incomplete data that can lead to poor policy design. Data collected may be outdated, fragmented, or poorly structured, which impacts its usefulness in policymaking. Inaccurate data inputs can result in misguided policy initiatives that do not align with citizens' needs. For example, data availability and quality in rural and underserved regions can be a significant issue in India. The digital divide exacerbates this problem, as large sections of the population lack reliable access to the internet, smartphones, or digital literacy, further limiting the scope and accuracy of data collection. This results in policy initiatives that may not accurately reflect the realities of the most vulnerable populations, undermining public service effectiveness.

Another major inefficiency in data-driven policymaking is the lack of timely data processing. Even when data is collected, its use in decision-making is often delayed due to lengthy processing times or the sheer complexity of analyzing vast amounts of information. Policies need to be adjusted and fine-tuned over time, but delays in data analysis hinder the ability of governments to adapt quickly. When data insights are processed and acted upon, the original issue may have changed or evolved, making the policy responses less relevant or effective. The issue of data overload is also a significant concern. With the proliferation of digital systems and sensors collecting vast amounts of data, governments often struggle to manage and make sense of the information they gather. While IT tools can help process large datasets, they cannot always provide clear insights without proper human oversight and interpretation. Policymakers may become overwhelmed by the sheer volume of data, making it harder to focus on the most important variables that influence policy decisions.

- **Gaps in Policy Implementation and Monitoring:** Even when policies are well-designed and data-driven, gaps in implementation and monitoring often hinder their success. A critical challenge lies in monitoring and evaluating the impact of policies in real-time. Governments often lack the necessary tools to track policy implementation on the ground, especially in large, diverse countries like India. Without continuous monitoring systems, it becomes difficult to identify bottlenecks or failures early on, resulting in corrective action delays. For instance, a policy aimed at improving access to healthcare may not be fully realized unless monitoring systems can track whether resources (such as healthcare workers or medicines) are being distributed as intended. Policy intent and implementation discrepancies can go unnoticed without continuous tracking.

Another problem arises in measuring policy effectiveness. Even when monitoring systems are in place, they often do not provide the necessary feedback to understand whether the policy is achieving its objectives. This gap in feedback loops makes it difficult for policymakers to understand what is working and what is not. In many instances, evaluations of policy outcomes occur only after the policy has been implemented for a long time, by which point corrective action may be difficult or costly. The political landscape plays a significant role in policy implementation and monitoring. Political interests, bureaucratic inertia, and the lack of accountability mechanisms often lead to suboptimal policy implementation. Even when policies are designed based on rigorous data and evidence, their execution may be compromised due to political will, resource constraints, or resistance from implementing agencies. For example, the National Rural Employment Guarantee Act (MGNREGA), which was designed to guarantee employment for rural workers, faced significant challenges in implementation due to administrative delays, corruption, and insufficient monitoring of work sites. This highlights the importance of having the right IT tools for monitoring and having effective governance structures in place to ensure corrective actions are taken when needed.

Inadequate training and capacity building within government agencies also contribute to gaps in policy execution. While IT tools may provide the infrastructure to track policies, a lack of skilled personnel to use these systems can severely limit their effectiveness. This challenge is exacerbated in lower-tier government offices, where technical skills are often underdeveloped, and the personnel are unfamiliar with advanced IT tools and data analysis methods. Consequently, even well-designed policies can fail to deliver their intended outcomes because of poor implementation practices or the inability to properly use the tools available for monitoring and evaluation.

IT as a Catalyst for Policy Change: Leveraging AI, Big Data, and Blockchain for Governance Transformation

Information Technology (IT) has become an indispensable driver of policy change, revolutionizing the processes of policy formulation, implementation, and evaluation. By harnessing cutting-edge technologies such as Artificial Intelligence (AI), Big Data, and Blockchain, governments are better equipped to develop evidence-based policies, ensure transparent execution, and enhance accountability. These innovations improve efficiency and foster inclusivity and responsiveness in governance. IT has emerged as a transformative force of rapidly evolving societal needs, enabling policymakers to address complex challenges with precision and agility. This chapter explores how AI, Big Data, and Blockchain catalyse policy change, focusing on their application in evidence-based policymaking and transparent execution.

AI and Big Data for Evidence-Based Policymaking: Integrating AI and Big Data in policymaking has introduced a paradigm shift in governance, enabling evidence-based decisions grounded in real-time data and advanced analytics. Traditional policymaking often relied on outdated data and anecdotal evidence, leading to suboptimal outcomes. AI and Big Data address these limitations by providing tools to process vast datasets, identify patterns, and generate actionable insights. This capability is particularly relevant for countries like India, where diverse and dynamic socioeconomic conditions necessitate adaptive and inclusive policies.

1. **Enhanced Data Collection and Analysis:** AI-powered tools facilitate collecting, integrating, and analysing data from diverse sources, such as government databases, social media platforms, IoT devices, and public surveys. For instance, machine learning algorithms can process unstructured data, such as text or images, to derive meaningful insights. This capability allows policymakers to identify emerging trends, assess public sentiment, and predict the potential impact of policy interventions. In public health, AI and Big Data have been instrumental in monitoring disease outbreaks, optimizing resource allocation, and improving healthcare delivery.

During the COVID-19 pandemic, governments worldwide utilized AI to predict infection hotspots, model the spread of the virus, and evaluate the effectiveness of containment measures. Similarly, AI-driven analysis of student performance data in education enables targeted interventions to address learning gaps, ensuring that resources are directed where they are most needed.

2. **Predictive and Prescriptive Analytics:** AI and Big Data also empower policymakers with predictive and prescriptive analytics, enabling proactive governance. Predictive models use historical and real-time data to forecast future scenarios like population growth, climate change impacts, or economic fluctuations. These insights help governments design policies that anticipate challenges rather than merely react to them. Prescriptive analytics, on the other hand, suggests optimal courses of action based on scenario analysis. For example, urban planning authorities can use AI to simulate the impact of various infrastructure projects on traffic patterns, environmental sustainability, and economic growth. Such data-driven approaches ensure that policy decisions are evidence-based and aligned with long-term objectives.

3. **Reducing Bias and Enhancing Inclusivity:** When designed responsibly, AI algorithms can mitigate biases often influencing human decision-making. By relying on objective data and standardized analysis, AI can promote fairness and inclusivity in policy design. For instance, algorithms can analyze demographic data to identify underserved communities and allocate resources equitably. However, ensuring that the data used for training AI models is representative and free from systemic biases is essential, as flawed datasets can perpetuate existing inequities.

4. **Defies in Implementation:** Despite their potential, adopting AI and Big Data in policymaking faces several challenges. These include the lack of high-quality data, concerns about privacy and security, and the need for skilled personnel to develop and deploy AI systems. Addressing these challenges requires investments in data infrastructure, robust regulatory frameworks, and capacity-building initiatives.

Blockchain for Transparent and Efficient Policy Execution

With its decentralized and immutable architecture, blockchain technology offers a robust solution for enhancing transparency, accountability, and efficiency in policy execution. By recording transactions on a distributed ledger, Blockchain ensures that data is secure, traceable, and tamper-proof. This characteristic makes it particularly valuable for public governance, where transparency and trust are critical.

1. **Enhancing Transparency in Public Administration:** Blockchain's transparency is a game-changer for governance. Recording every transaction or decision on a public ledger allows stakeholders to verify the authenticity and integrity of data. For example, in public procurement processes, Blockchain can track bids, contracts, and payments, ensuring that all transactions are fair and corruption-free. This transparency reduces the scope for malpractices and builds public trust in government institutions. One notable application of Blockchain in governance is the implementation of land registry systems. In many countries, including India, land disputes are significant due to unclear ownership records and fraudulent transactions. Blockchain-based land registries provide a secure and transparent platform for recording property transactions, ensuring that ownership records are immutable and accessible to all stakeholders.

2. **Streamlining Welfare Distribution:** Blockchain technology is also transforming the execution of welfare programs. By creating tamper-proof records of beneficiaries and transactions, Blockchain ensures that government subsidies, pensions, and benefits reach the intended recipients without leakages. For instance, Blockchain-enabled direct benefit transfer (DBT) systems eliminate intermediaries, reducing delays and minimizing the risk of corruption. In India, initiatives like the e-RUPI digital payment system are leveraging Blockchain principles to improve the efficiency of welfare distribution. These systems enable targeted and conditional transfers, ensuring that funds are used for their intended purpose, such as healthcare, education, or agricultural inputs.

3. **Securing Digital Identity:** Blockchain is critical in securing digital identities, which are essential for accessing government services. By creating decentralized identity systems, Blockchain eliminates identity theft risk and ensures that individuals have control over their data. These systems are particularly valuable for marginalized populations, who often lack formal identity documents and face barriers to accessing public services.

   For example, Blockchain-based identity platforms can integrate with Aadhaar, India's biometric-based digital identity system, to enhance its security and functionality. Such integration ensures that individuals can securely access banking, healthcare, and education services, promoting financial and social inclusion.

4. **Promoting Accountability Through Smart Contracts:** Smart contracts, a key feature of Blockchain technology, enable the automatic execution of agreements when predefined conditions are met. These contracts can be used to enforce policy compliance and reduce administrative overhead.

For instance, in environmental governance, smart contracts can monitor industrial emissions and impose penalties automatically if limits are exceeded. This automation ensures accountability and reduces the burden on regulatory authorities.

5. **Defies in Blockchain Adoption:** While Blockchain holds immense potential for policy execution, its adoption faces several challenges. These include the high computational cost of maintaining Blockchain networks, scalability concerns, and regulatory clarity. Integrating Blockchain with IT systems and processes requires significant technical expertise and infrastructure.

Synergy Between AI, Big Data, and Blockchain: The combined use of AI, Big Data, and Blockchain can create a powerful ecosystem for policy change. For example, AI can analyze data to identify policy priorities, Blockchain can ensure the transparency and security of data, and Big Data can provide the scale needed to address complex governance challenges. Together, these technologies enable a seamless flow of information across policy formulation, execution, and evaluation stages. Consider the case of **smart cities**, where these technologies converge to address urban challenges. AI-powered sensors collect real-time data on traffic, energy usage, and waste management, while Blockchain ensures the integrity of this data and facilitates secure transactions. Big Data analytics then provides insights to optimize urban planning and resource allocation. This integrated approach improves efficiency and enhances the quality of life for urban residents.

Current Landscape – Data-Driven Governance and Challenges in Policy Execution and Monitoring

The modern era has witnessed a profound shift in governance paradigms, driven by the exponential growth of data and its transformative potential in policymaking. Initiatives like Digital India epitomize the aspiration to harness technology for effective governance, citizen empowerment, and inclusive development. These efforts foster transparency, improve service delivery, and enhance accountability. However, despite the significant progress enabled by data-driven governance, gaps in policy execution and monitoring remain persistent challenges. These issues undermine the realization of policy objectives and highlight the need for systemic improvements.

Data-Driven Governance Initiatives – Role of Digital India: The **Digital India** program, launched in 2015, represents India's flagship initiative to build a digitally empowered society and knowledge economy. It encompasses multiple dimensions, including digital infrastructure, governance, and service delivery, to bridge the digital divide and foster inclusive development. By integrating Information and Communication Technology (ICT) into governance, the initiative seeks to create a framework for data-driven decision-making.

- **Digital Infrastructure for Data-Driven Governance:** Under Digital India, substantial investments have been made in building a robust digital infrastructure to facilitate data-driven governance. The Aadhaar system, with over 1.3 billion biometric-based identities, serves as a cornerstone for targeted delivery of government services. The BharatNet project, aiming to connect over 250,000-gram panchayats with high-speed broadband, enhances rural connectivity and enables data collection at the grassroots level. The IndiaStack, comprising digital public goods like e-KYC, e-Sign, and the Unified Payments Interface (UPI), has revolutionized service delivery by making processes seamless, paperless, and cashless. These platforms have empowered citizens while creating a vast reservoir of data for policymakers to analyze and optimize services.
- **Data-Driven Decision-Making:** The availability of real-time data has enabled governments to design and implement evidence-based policies. For instance, the Pradhan Mantri Fasal Bima Yojana (PMFBY) leverages remote sensing data to assess crop losses and provide timely insurance claims to farmers. Similarly, POSHAN Abhiyaan, India's flagship program for nutritional improvement, employs data analytics to identify high-burden districts and track progress in reducing malnutrition. Artificial Intelligence (AI) and Machine Learning (ML) have further augmented policymakers' ability to derive actionable insights. Predictive analytics tools help forecast trends in public health, education, and resource allocation, allowing governments to address challenges proactively.
- **Citizen-Centric Service Delivery:** Digital platforms have transformed the citizen-government interface, enabling faster, more transparent service delivery. Portals like DigiLocker provide secure access to important documents, while grievance redressal platforms such as CPGRAMS enable citizens to voice their concerns and receive timely resolutions. These platforms foster accountability and encourage participatory governance.

Gaps in Policy Execution: Despite the potential of data-driven governance, several gaps impede the seamless execution of policies. These gaps stem from challenges in digital infrastructure, data quality, and inter-agency coordination.

- **Uneven Digital Infrastructure:** While significant progress has been made in expanding digital connectivity, disparities persist, particularly in rural and remote regions. The digital divide exacerbates inequalities as citizens in interconnected areas struggle to access digital services. This divide undermines the inclusivity of programs like Digital India, limiting their effectiveness in bridging socioeconomic gaps. For instance, while urban areas have embraced digital payment systems, rural communities often rely on cash transactions due to limited internet penetration and digital literacy. Such disparities hinder the uniform execution of policies aimed at financial inclusion and other developmental goals.

- **Data Quality and Reliability:** Data-driven governance relies heavily on the availability of accurate, comprehensive, and up-to-date data. However, data quality remains a significant challenge. Data collection mechanisms often lack standardization, leading to inconsistencies and inaccuracies. Additionally, incomplete or outdated datasets compromise the ability to make informed policy decisions. The issue of real-time data availability is particularly critical in dynamic sectors like healthcare and disaster management. Delays in data collection and processing can result in delayed responses, reducing the effectiveness of interventions.
- **Fragmentation and Lack of Interoperability:** Policy execution is often hindered by the fragmentation of digital systems across government departments. Different agencies use disparate platforms and databases that are not fully interoperable, creating data silos. This fragmentation limits the seamless flow of information and complicates efforts to design holistic, multi-sectoral policies. For example, integrating health, education, and social welfare datasets to address child malnutrition holistically may be difficult if these sectors operate on separate digital systems. The lack of a unified data ecosystem impedes coordination and reduces the effectiveness of policy interventions.
- **Human Resource Challenges:** Effective policy execution requires robust technology and skilled personnel to manage and utilize these tools. A significant gap exists in the technical capacity of government officials, particularly at the grassroots level. Many government employees lack the training to analyze and interpret data or operate advanced IT systems, leading to inefficiencies in policy implementation. Capacity-building initiatives have been slow to keep pace with the rapid technological advancements, leaving a critical gap in human resources needed to leverage data-driven governance fully.

Citizen Feedback Systems for Inclusive Governance: Citizen feedback systems are pivotal in fostering participatory governance and ensuring policies align with public needs. By providing platforms for citizens to voice their opinions, these systems enhance accountability and responsiveness in governance.

- **Real-Time Feedback Mechanisms:** IT-enabled feedback systems like mobile apps and online portals allow citizens to provide real-time policy implementation and service delivery input. For example, platforms like MyGov in India enable citizens to share suggestions, report grievances, and participate in policy discussions. These inputs help policymakers identify gaps and refine their strategies.

- **Monitoring Public Sentiment:** AI-powered Sentiment analysis tools analyze feedback data to gauge public opinion on specific policies. This information helps governments understand citizen priorities and address concerns proactively. For instance, sentiment analysis can identify opposition or support among affected communities during large-scale infrastructure projects, guiding more inclusive decision-making.
- **Addressing Grievances Efficiently:** Grievance redressal platforms, such as the Centralized Public Grievance Redress and Monitoring System (CPGRAMS), use IT tools to track complaints, ensure timely responses, and monitor resolution rates. These systems enhance citizen trust by demonstrating accountability and responsiveness.

Geospatial Analytics for Spatially-Driven Policies: Geospatial analytics leverages spatial data from satellites, drones, and GIS systems to provide a geographical context to policymaking. This approach enables governments to address location-specific challenges and optimize resource management.

- **Mapping and Monitoring Resources:** Geospatial technologies help policymakers map natural resources, monitor land use changes, and track environmental degradation. For instance, satellite imagery monitors deforestation, urbanization, and water resource depletion. These insights guide policies on conservation, urban planning, and sustainable development.
- **Disaster Management:** Geospatial analytics is indispensable in disaster management, providing real-time data on weather patterns, flood zones, and evacuation routes. Governments can use this information to develop early warning systems, allocate resources effectively, and minimize disaster impacts.
- **Urban Planning and Infrastructure Development:** Spatial data is critical for urban planning, helping governments design efficient transportation networks, allocate land for housing, and ensure equitable access to amenities. For example, GIS-based tools identify underserved areas and prioritize infrastructure development projects.
- **Monitoring Policy Outcomes:** Geospatial analytics enables governments to monitor the spatial impact of policies, ensuring that resources are distributed equitably and outcomes are aligned with objectives. For instance,

spatial data can track the implementation of rural electrification programs, identifying regions that still lack access to electricity.

Synergy of IT Tools in Policy Enablement

Integrating AI, Blockchain, citizen feedback systems, and geospatial analytics creates a comprehensive ecosystem for policy enablement. These technologies complement each other, enhancing governance's efficiency, transparency, and inclusivity. For example, geospatial analytics maps urban challenges in smart city initiatives, AI analyzes data to propose solutions, Blockchain ensures transparent execution, and citizen feedback systems gather resident input. This synergy fosters holistic policymaking, ensuring interventions are data-driven, participatory, and impactful.

Case Studies – PM Gati Shakti and the Digital India Mission in Enabling Real-Time Policy Adaptation: The PM Gati Shakti initiative and the Digital India Mission are exemplary case studies that illustrate how Information Technology (IT) drives governance transformation in India. Both initiatives demonstrate the integration of digital tools in policy formulation, execution, and monitoring, enabling real-time adaptation and optimization. They highlight India's commitment to leveraging IT for inclusive growth, operational efficiency, and sustainable development. This section provides an in-depth analysis of these initiatives, focusing on their design, implementation, impact, and lessons learned.

PM Gati Shakti – Digital Platform for Multimodal Infrastructure Development: The PM Gati Shakti - National Master Plan, launched in 2021, is a landmark initiative to revolutionise infrastructure development in India. By integrating geospatial data, digital platforms, and inter-ministerial coordination, PM Gati Shakti seeks to optimize resource allocation, reduce project delays, and enhance the efficiency of multimodal logistics.

Objectives and Design: The PM Gati Shakti platform is designed to achieve three key objectives:

1. **Integrated Planning:** By creating a unified digital framework, the platform aligns infrastructure projects across sectors such as roads, railways, ports, and energy.
2. **Real-Time Monitoring:** Using geospatial technology provides real-time updates on project progress, enabling proactive decision-making.
3. **Efficiency and Sustainability:** It aims to reduce costs and promote sustainable development by minimising redundancies and delays.

Role of IT in Gati Shakti: The backbone of PM Gati Shakti is its GIS-based digital platform, which integrates data from over 16 ministries. This system enables:

- **Mapping of Existing Infrastructure:** The platform provides a comprehensive geospatial database of existing infrastructure assets, helping identify gaps and overlaps.
- **Scenario-Based Planning:** Through predictive analytics, planners can simulate various scenarios to optimize project design and execution.
- **Real-Time Collaboration:** Inter-ministerial coordination is facilitated by sharing data and insights on a unified platform.

Impact and Achievements: PM Gati Shakti has significantly improved infrastructure planning and execution. Key achievements include

- **Time Savings:** The digital platform has streamlined project approvals, reducing timelines by up to 50%.
- **Cost Efficiency:** Eliminating redundancies and optimizing resource allocation significantly reduce project costs.
- **Environmental Benefits:** The initiative minimizes ecological disruptions and reduces carbon footprints by enabling route optimisation.

Digital India Mission – Empowering Governance Through IT: The Digital India Mission, launched in 2015, is a comprehensive initiative to transform India into a digitally empowered society and knowledge economy. The mission focuses on three core areas:

1. Digital Infrastructure as a Utility to Citizens
2. Governance and Services on Demand
3. Digital Empowerment of Citizens

Objectives and Vision: Digital India envisions leveraging IT to ensure inclusive growth, enhance service delivery, and foster transparency. The mission aims to bridge the digital divide by providing universal access to digital infrastructure and promoting digital literacy.

Key Pillars and IT Integration: Digital India encompasses multiple programs that use IT to address diverse governance challenges:

- **Aadhaar:** The world's largest biometric-based digital identity system, Aadhaar enables targeted delivery of subsidies and services, reducing leakages.
- **UMANG App:** A unified platform offering access to over 1,200 government services across sectors.

- **BharatNet:** A project to provide high-speed internet to over 250,000 gram panchayats, fostering rural connectivity.
- **DigiLocker:** A platform for storing and sharing digital documents securely.

Role of IT in Real–Time Policy Adaptation: Digital India has enabled governments to adopt policies in real-time by leveraging data-driven insights and citizen feedback:

- **Data Analytics for Evidence-Based Policies:** Platforms like the COVID-19 Aarogya Setu app used real-time data to monitor infection trends, enabling dynamic containment strategies.
- **Feedback-Driven Adaptation:** Grievance redressal platforms such as CPGRAMS collect citizen feedback, providing valuable inputs for policy refinement.
- **Remote Monitoring of Programs:** IT tools allow real-time tracking of schemes like the PM Kisan Samman Nidhi, ensuring timely disbursal benefits to farmers.

Achievements and Impact: Digital India has made significant strides in enhancing governance outcomes:

- **Financial Inclusion:** Initiatives like Jan Dhan-Aadhaar-Mobile (JAM) trinity have brought millions into the formal banking system, ensuring direct benefit transfers to beneficiaries.
- **Enhanced Transparency:** By digitizing records and automating processes, Digital India has reduced opportunities for corruption.
- **Citizen Empowerment:** Digital literacy programs under Digital India have empowered rural communities, enabling them to access information, services, and opportunities.

The key lesson from Digital India is that citizen-centric design, coupled with robust IT systems, can address systemic inefficiencies and promote inclusive development.

Comparative Analysis: PM Gati Shakti and Digital India

While both PM Gati Shakti and Digital India leverage IT to transform governance, their focus areas differ:

- **PM Gati Shakti** emphasizes infrastructure development and uses geospatial analytics for integrated planning.
- **Digital India** focuses on service delivery and citizen empowerment through digital platforms.

Despite these differences, both initiatives demonstrate common themes:

1. **Data-Driven Governance:** Both programs rely on data analytics and real-time monitoring to adapt policies dynamically.
2. **IT as an Enabler:** The integration of advanced IT tools, such as GIS in PM Gati Shakti and mobile apps in Digital India, underscores the role of technology in bridging governance gaps.
3. **Collaborative Ecosystems:** Both initiatives highlight the importance of inter-agency coordination and public-private partnerships.

Future Scenario

The advent of predictive analytics, artificial intelligence (AI), and blockchain technology is revolutionizing policy design, implementation, and enforcement. As these technologies mature, their integration into governance frameworks will redefine how governments address complex challenges, anticipate future scenarios, and deliver transparent, efficient, and accountable services. This chapter explores the potential of these tools to shape the future of governance, focusing on their applications in dynamic policy formulation and real-time policy enforcement, and discusses the challenges and opportunities they present.

Predictive Analytics and AI in Dynamic Policy Formulation: Predictive analytics and AI have transformed decision-making processes in various domains, and their potential in policymaking is immense. By analyzing historical and real-time data, these tools enable governments to anticipate trends, identify risks, and design policies that adapt to changing circumstances.

Forecasting Socioeconomic Trends: Predictive analytics uses advanced algorithms to analyze large datasets and identify patterns that can forecast socioeconomic changes. For example:

- **Labor Market Trends:** AI models can analyze employment data, education statistics, and industry reports to predict unemployment rates and skill gaps, enabling targeted interventions.
- **Economic Forecasting:** Predictive tools assess financial indicators such as inflation, trade flows, and investment patterns, guiding monetary and fiscal policy decisions.

These capabilities allow policymakers to design proactive strategies that address potential crises before they materialize, fostering economic stability.

Optimizing Resource Allocation: AI-powered predictive analytics helps governments allocate resources more efficiently by identifying areas of greatest need. For instance:

- In healthcare, predictive models can forecast disease outbreaks and optimize the distribution of medical supplies and personnel.
- AI analyzes weather patterns, soil data, and crop health in agriculture to recommend resource allocation that maximizes yield and minimizes waste.

Enhancing Real-Time Policy Adaptation: Predictive tools enable dynamic policy formulation by providing real-time insights into changing conditions. For example:

- During the COVID-19 pandemic, predictive models analyzed infection rates and mobility data to inform containment strategies and healthcare resource deployment.
- In disaster management, predictive analytics forecasts the impact of natural calamities, guiding evacuation plans and relief efforts.

This ability to adapt policies in real-time ensures that governance remains responsive to emerging challenges.

Addressing Social Inequities: AI and predictive analytics can identify patterns of inequality and discrimination, enabling governments to design targeted policies that promote inclusivity. For example:

- Predictive models can analyze educational disparities and recommend investments in underserved regions.
- AI tools assess the impact of welfare programs on marginalized communities, helping refine their design and implementation.

Blockchain for Transparent, Real-Time Policy Enforcement: Blockchain technology is a game-changer for policy enforcement, offering unprecedented transparency, efficiency, and trust. Its decentralized, immutable ledger ensures that policy execution is tamper-proof, auditable, and real-time.

Ensuring Transparency in Policy Execution: Blockchain records every transaction on a distributed ledger, creating a transparent and traceable system for policy implementation. For example:

- **Public Procurement:** Blockchain ensures that contracts are awarded fairly and funds are utilized efficiently, reducing corruption in government procurement processes.
- **Welfare Programs:** Blockchain-based systems track the disbursement of benefits, ensuring that they reach intended recipients without leakages or delays.

Automating Compliance with Smart Contracts: Smart contracts, a feature of Blockchain, automate policy enforcement by executing predefined actions when specific conditions are met. For example:

- In environmental governance, smart contracts can monitor industrial emissions and impose penalties for violations automatically.
- In financial regulation, they ensure compliance with tax policies by automatically deducting applicable taxes from transactions.

Enabling Real-Time Monitoring and Auditing: Blockchain enables real-time monitoring of policy outcomes, allowing governments to identify and address implementation gaps promptly. For instance:

- In education, Blockchain tracks the distribution of scholarships, ensuring timely delivery and preventing misuse.
- In infrastructure development, it monitors project timelines and budgets, flagging delays and cost overruns.

Strengthening Citizen Engagement: Blockchain-based systems empower citizens by giving them access to tamper-proof records of policy implementation. For example:

- Inland governance, Blockchain creates immutable property records, reducing disputes and ensuring transaction transparency.
- Blockchain enables secure and transparent voting systems in electoral processes, enhancing voter confidence.

Integration of Predictive Analytics, AI, and Blockchain: The synergy between predictive analytics, AI, and Blockchain creates a powerful ecosystem for governance innovation. For example:

- **Predictive Monitoring:** AI-powered analytics identify risks and opportunities, while Blockchain ensures that policy responses are executed transparently and efficiently.
- **Dynamic Feedback Loops:** Real-time data from Blockchain systems feeds into predictive models, enabling continuous refinement of policies based on ground realities.
- **Collaborative Platforms:** Blockchain fosters inter-agency collaboration by providing a shared, tamper-proof record of policy implementation, while AI analyzes these records to generate insights.

Immediate Applications

- **Smart Cities and Urban Governance:** Integrating AI, predictive analytics, and Blockchain can transform urban governance. For instance:
    - Predictive models analyze traffic patterns to optimize public transportation.
    - Blockchain ensures the transparent allocation of resources for infrastructure projects.
    - AI-powered systems monitor air quality and recommend interventions to reduce pollution.
- **Healthcare and Pandemic Preparedness:** In healthcare, predictive analytics forecasts disease outbreaks, AI designs personalized treatment plans, and Blockchain ensures secure sharing of patient data. This integrated approach enhances the efficiency and equity of healthcare delivery.
- **Climate Action and Environmental Governance:** AI models predict the impact of climate change, guiding mitigation and adaptation strategies. Blockchain tracks carbon credits and ensures compliance with international environmental agreements, fostering accountability in climate action.
- **Education and Skill Development:** Predictive analytics identifies skill gaps, AI designs personalized learning programs, and Blockchain creates verifiable records of educational achievements. This ecosystem promotes lifelong learning and workforce readiness.

4

People-Centric IT Solutions for Empowering People

Introduction

Integrating information technology (IT) into governance and development frameworks has revolutionized how governments engage with citizens, driving empowerment across multiple dimensions. People-centric IT solutions focus on bridging the digital divide, promoting inclusivity, enhancing accessibility, and enabling active participation in governance and economic activities. These solutions leverage mobile technology, artificial intelligence (AI), cloud computing, and blockchain to create inclusive, transparent, and user-friendly platforms, ensuring equitable access to opportunities and resources.

Bridging the Digital Divide: One of the primary challenges in empowering people through IT is addressing the digital divide the gap between individuals with access to digital technologies and those without. In many developing nations, including India, large sections of the population remain excluded from the benefits of digital transformation due to socioeconomic, geographic, or infrastructural constraints.

- **Affordable Connectivity:** Initiatives such as BharatNet aim to provide high-speed internet to rural and remote areas, enabling millions to access digital services.
- **Mobile-First Solutions:** With mobile phones being more accessible than other digital devices, governments and organizations design applications that operate on low-cost smartphones, ensuring broader reach.
- **Digital Literacy Programs:** Schemes like Pradhan Mantri Gramin Digital Saksharta Abhiyan (PMGDISHA) focus on imparting basic digital skills to rural populations, enabling them to leverage IT tools for financial transactions, e-learning, and accessing government services.

Enhancing Accessibility to Public Services: IT solutions have significantly transformed the delivery of public services, making them more accessible, efficient, and citizen-centric.

- **Unified Digital Platforms:** Applications like UMANG (Unified Mobile Application for New-age Governance) integrate multiple government services into a single platform, allowing citizens to access benefits related to healthcare, education, and welfare programs seamlessly.
- **Aadhaar Integration:** The biometric-based Aadhaar system has enabled targeted delivery of subsidies, pensions, and scholarships, eliminating leakages and ensuring that benefits reach the intended recipients.
- **E-Governance Portals:** Digital platforms such as DigiLocker simplify document verification processes by providing citizens with secure online storage for certificates and records.

Promoting Financial Inclusion: Financial inclusion is a critical empowerment aspect, particularly for marginalized communities. IT solutions have enabled individuals to access formal banking systems, credit facilities, and digital payment mechanisms.

- **Jan Dhan Yojana and JAM Trinity:** The combination of Jan Dhan accounts, Aadhaar, and mobile connectivity has brought millions into the formal financial system, enabling direct benefit transfers and reducing dependency on intermediaries.
- **Mobile Payment Solutions:** Platforms like BHIM (Bharat Interface for Money) and UPI (Unified Payments Interface) have made digital transactions simple and accessible, fostering a cashless economy.
- **Microfinance and Credit Access:** AI-driven platforms analyze transaction data to assess creditworthiness, enabling small-scale entrepreneurs and farmers to access microloans.

Fostering Citizen Participation in Governance: People-centric IT solutions empower individuals to play an active role in governance by enhancing transparency, accountability, and public participation.

- **Grievance Redressal Platforms:** Digital tools like CPGRAMS (Centralized Public Grievance Redress and Monitoring System) allow citizens to report issues and provide feedback on government services, ensuring their voices are heard.
- **Crowdsourcing Policy Inputs:** Platforms like MyGov enable citizens to contribute ideas and suggestions for policy formulation, fostering collaborative governance.
- **Open Data Portals:** These platforms promote transparency and empower citizens to analyze and monitor public projects by making government data publicly available.

Revolutionizing Education and Skill Development\: IT-driven educational tools are transforming learning experiences, enabling personalized, accessible, and scalable education for individuals across all demographics.

- **E-Learning Platforms:** Initiatives like SWAYAM and DIKSHA offer free online courses catering to diverse learning needs, from primary education to advanced professional skills.
- **AI-Powered Personalization:** AI-driven learning platforms adapt content to individual learners' needs, enhancing engagement and retention.
- **Remote and Blended Learning:** Digital tools enable education in remote areas, addressing teacher shortages and infrastructural gaps.
- **Skill Development:** Programs like Skill India leverage IT to provide vocational training through interactive modules and virtual simulations.

Empowering Women and Vulnerable Populations: People-centric IT solutions are pivotal in addressing women and marginalized groups' unique challenges.

- **Social Security Access:** Digital platforms streamline the delivery of social welfare benefits to women, senior citizens, and individuals with disabilities, ensuring timely and transparent disbursement.
- **Digital Marketplaces:** Platforms like Mahila e-Haat enable women entrepreneurs to showcase and sell their products online, fostering economic empowerment.
- **Health and Safety Applications:** Mobile apps provide critical information on maternal health, nutrition, and safety, addressing gender-specific needs.

Supporting Sustainable Development: People-centric IT solutions align with global sustainability goals by addressing environmental, social, and economic dimensions.

- **Environmental Conservation:** IT tools monitor natural resources and promote sustainable practices through applications such as geo-tagging of afforestation projects.
- **Smart Cities:** Digital solutions optimize energy usage, waste management, and urban mobility, improving quality of life while minimizing ecological footprints.
- **Agricultural Innovations:** IT-enabled platforms provide farmers with real-time weather forecasts, market prices, and best practices, enhancing productivity and resilience.

Approach Forward: While people-centric IT solutions offer immense potential, their effective implementation faces several challenges:

- **Digital Literacy Gaps:** Ensuring widespread adoption requires continuous efforts in educating citizens about digital tools.
- **Infrastructure Deficits:** Many rural and remote areas still lack reliable internet connectivity and digital infrastructure.
- **Data Privacy and Security:** Protecting citizens' data from breaches and misuse is essential for maintaining trust.
- **Customizing Solutions:** IT tools must address the diverse needs of India's population, taking into account linguistic, cultural, and economic diversity.

Current Scenario – IT in Education, Healthcare, and Public Welfare

Integrating **Information Technology (IT)** into education, healthcare, and public welfare has transformed these critical sectors, making them more accessible, efficient, and inclusive. Across the globe, particularly in developing nations like India, IT plays a pivotal role in bridging gaps, democratizing access to essential services, and fostering sustainable development. This section explores the current landscape of IT applications in these domains, highlighting advancements, challenges, and potential pathways for further integration.

IT in Education – Revolutionizing Learning Paradigms: The education sector has experienced a paradigm shift by incorporating IT, making learning more personalized, accessible, and scalable. IT tools and platforms address systemic challenges, such as inadequate teacher-student ratios, outdated curricula, and geographic educational barriers, particularly in underserved regions.

• **E-Learning Platforms and Digital Classrooms:** Digital platforms have emerged as powerful tools for education delivery:

- **Massive Open Online Courses (MOOCs):** Platforms like SWAYAM and edX provide free or affordable courses on diverse subjects, democratizing access to quality education.
- **Smart Classrooms:** Interactive whiteboards, projectors, and e-content enhance traditional teaching methods, fostering engagement and better comprehension.
- **Digital Libraries:** Resources such as National Digital Library of India (NDLI) offer access to millions of books, journals, and research articles, catering to students and researchers alike.

• **AI-Driven Personalization:** AI tools enable tailored learning experiences by analyzing students' performance and adapting content to suit individual needs:

- **Adaptive Learning Systems:** Platforms like BYJU's use AI to modify lesson plans and exercises based on a learner's progress.
- **Intelligent Tutoring Systems:** These systems simulate a one-on-one tutor, providing real-time feedback and addressing knowledge gaps.

• **Overcoming Barriers to Access:** IT helps bridge educational divides:

- **Remote Learning:** Tools like Zoom, Google Classroom, and Microsoft Teams became essential during the COVID-19 pandemic, ensuring the continuity of education.
- **Multilingual and Inclusive Content:** IT platforms offer resources in regional languages and assistive technologies for students with disabilities.

IT in Healthcare – Transforming Service Delivery: Healthcare has been one of the most transformative domains for IT, enabling advancements in diagnostics, treatment, and patient care. IT solutions are bridging gaps in healthcare infrastructure and addressing challenges such as inequitable access, inefficiencies, and high costs.

• **Telemedicine and Remote Healthcare:** Telemedicine platforms connect patients with healthcare providers in real-time, overcoming geographic barriers:

- **National Telemedicine Service:** Initiatives like eSanjeevani have facilitated millions of consultations, particularly in rural and underserved areas.
- **Mobile Health Apps:** Applications such as Practo and mHealth enable remote consultations, appointment scheduling, and health monitoring.

• **Electronic Health Records (EHRs):** EHRs have streamlined patient data management:

- **Unified Patient Records:** Systems like Ayushman Bharat Digital Mission aim to create integrated health IDs for citizens, facilitating seamless access to medical histories.
- **Data-Driven Insights:** EHRs enable healthcare providers to analyze trends, improving diagnostics and personalized treatment plans.

• **AI and Big Data in Healthcare:** AI-powered systems and Big Data analytics are driving breakthroughs in healthcare:

- **Diagnostics:** Tools like IBM Watson Health analyze medical data to assist in diagnosing diseases such as cancer.
- **Predictive Analytics:** AI predicts disease outbreaks by analyzing environmental and demographic data, enabling proactive interventions.

- **Robotic Assistance:** Surgical robots enhance precision and reduce recovery times.

• **Public Health Campaigns:** IT facilitates large-scale health campaigns:

- **Vaccination Drives:** Digital systems track vaccine distribution and monitor coverage, as demonstrated during India's COVID-19 vaccination campaign.
- **Awareness Programs:** IT platforms disseminate health information and promote preventive care, reaching millions through social media and SMS campaigns.

IT in Public Welfare: Driving Inclusive Development: Public welfare programs aim to improve the quality of life for citizens, particularly the marginalized. IT has redefined how welfare schemes are designed, implemented, and monitored, ensuring transparency and accountability.

• **Direct Benefit Transfers (DBTs):** DBTs eliminate intermediaries in welfare disbursement:

- **Aadhaar Integration:** Linking Aadhaar to welfare schemes ensures targeted delivery of benefits, such as subsidies and pensions.
- **Real-Time Monitoring:** Systems like Public Financial Management System (PFMS) track fund transfers, reducing leakages and delays.

• **E-Governance Portals:** Digital platforms enhance citizen engagement and service delivery:

- **UMANG App:** Integrates over 1,000 services from central, state, and local governments, providing a one-stop solution for citizens.
- **Grievance Redressal Systems:** Portals like CPGRAMS empower citizens to report issues and receive timely resolutions.

• **Social Inclusion Through IT:** IT enables inclusivity in welfare programs:

- **Skill Development:** Platforms like Skill India Digital Portal provide vocational training to marginalized communities, enhancing employability.
- **Digital Marketplaces:** Initiatives such as Government e-Marketplace (GeM) support artisans and small businesses by connecting them directly to buyers.
- **Women Empowerment:** Applications like Mahila e-Haat promote women entrepreneurs, fostering financial independence.

- **Disaster Management and Relief:** IT plays a critical role in managing crises:
    - **Real-Time Monitoring:** Geospatial and IoT technologies track natural disasters, enabling timely relief measures.
    - **Relief Distribution:** Blockchain ensures transparent aid distribution, reducing fraud and inefficiencies.

Interconnected Challenges and Opportunities: While IT has revolutionized education, healthcare, and public welfare, certain challenges persist across all three domains:

- **Digital Infrastructure:** Expanding broadband connectivity and ensuring affordable device access is essential.
- **Capacity Building:** Training stakeholders, including teachers, healthcare providers, and government officials, is critical to fully leveraging IT tools.
- **Data Protection:** Robust frameworks must be established to safeguard sensitive information.

Opportunities for synergies among these domains are significant. For example, a unified digital identity system like Aadhaar can integrate welfare, healthcare, and education services, enabling holistic development.

Addressing India's Digital Divide: Urban-Rural Disparities and Gender-Socioeconomic Barriers

India's rapid digital transformation has been marked by unprecedented growth in internet penetration, mobile connectivity, and access to digital services. However, this progress has also highlighted significant inequalities in digital access, known as the digital divide. This divide is most pronounced across urban-rural lines, gender dynamics, and socioeconomic disparities, creating barriers to inclusive development. Addressing these challenges is imperative for India to harness the full potential of the digital revolution and achieve its socioeconomic goals.

Urban-Rural Disparities in Digital Access: Urban India has benefitted disproportionately from the digital revolution, while rural areas struggle with inadequate infrastructure, affordability issues, and limited digital literacy.

- **Infrastructure Deficits**
    - **Connectivity Challenges:** Due to superior infrastructure, urban regions enjoy better internet speeds and coverage. In contrast, rural areas face inconsistent broadband access, slow speeds, and network outages. For example, as of 2023, rural internet penetration stood at approximately 38% compared to 67% in urban areas.

- **Electricity Supply:** Reliable electricity is a prerequisite for digital devices and internet usage. Many rural areas still face power shortages, affecting digital access.
- **Mobile Network Coverage:** Urban areas benefit from dense mobile towers and 4G/5G coverage, whereas rural regions often rely on 2G/3G networks, limiting access to high-speed internet services.

- **Cost and Affordability**
  - The affordability of digital devices and internet plans is a significant barrier in rural areas, where incomes are often lower. Urban households are more likely to afford smartphones, laptops, and broadband services.
- **Digital Literacy Gaps**
  - Many rural citizens are unfamiliar with navigating digital platforms, a challenge compounded by low literacy levels and the absence of local-language content. Urban populations with better education levels are more adept at leveraging digital technologies for education, healthcare, and financial services.
- **Government Initiatives:** The Indian government has launched several initiatives to address urban-rural digital disparities:
  - **BharatNet Project:** This initiative aims to provide high-speed broadband connectivity to 250,000 gram panchayats across India, bringing the internet to rural areas.
  - **Digital Villages:** Programs like DigiGaon aim to transform villages into digitally empowered hubs with telemedicine, skill development, and e-governance services.
  - **Low-Cost Devices:** Partnerships with private companies to introduce affordable smartphones and tablets have improved device accessibility in rural areas.

Gender Barriers in Digital Access: The digital divide in India is not just geographical but also deeply gendered, with women significantly less likely to access digital tools and services compared to men.

- **Gender Gaps in Internet Usage**
  - According to a 2022 report by the Internet and Mobile Association of India (IAMAI), only about 33% of internet users in rural areas are women, compared to 51% in urban areas. This disparity highlights persistent socio-cultural norms that limit women's digital participation.

- **Device Ownership:** Women are less likely to own smartphones or other digital devices. Many families prioritize men's access to technology, leaving women dependent on shared or borrowed devices.

- **Socio-Cultural Norms**
  - In many parts of India, societal expectations restrict women's mobility and access to education, hindering their ability to engage with digital tools.
  - **Digital Literacy:** Women often lack the basic digital skills to navigate the internet or use mobile applications effectively, exacerbating their exclusion from digital ecosystems.

- **Empowering Women Through IT:** Several initiatives have been introduced to promote digital inclusion for women:
  - **Internet Saathi Program:** A joint initiative by Google and Tata Trusts that trains rural women to use the internet, empowering them to become digital ambassadors in their communities.
  - **Mahila E-Haat:** An online platform under the Ministry of Women and Child Development that helps women entrepreneurs showcase and sell their products.
  - **Digital Literacy Campaigns:** Programs like PMGDISHA (Pradhan Mantri Gramin Digital Saksharta Abhiyan) prioritize women participants, providing them with essential digital skills.

Socioeconomic Barriers to Digital Inclusion: Economic inequalities and societal stratifications further widen the digital divide, creating distinct challenges for marginalized groups.

- **Affordability Issues**
  - **Device Costs:** The high cost of smartphones, computers, and internet services is a barrier for economically weaker sections.
  - **Data Costs:** Despite India having some of the lowest mobile data rates globally, the cost remains prohibitive for many low-income households.

- **Educational Inequalities**
  - Access to digital education is heavily skewed in favor of affluent families. During the COVID-19 pandemic, urban and wealthy households seamlessly transitioned to online learning, while rural and underprivileged students struggled due to a lack of devices and connectivity.

- **Exclusion of Marginalized Communities**
  - **Caste and Tribal Disparities:** Scheduled Castes (SC) and Scheduled Tribes (ST) face systemic barriers to education and economic opportunities, limiting their digital adoption.
  - **Urban Slums:** Even within urban areas, slum dwellers often lack access to the infrastructure and services needed for digital inclusion.
- **Government Interventions**
  - **Affordable Internet Plans:** Initiatives like Digital India aim to make Internet services more affordable for economically weaker sections.
  - **Subsidized Devices:** Programs providing subsidized tablets or laptops to students from low-income families help bridge socioeconomic divides.
  - **Local Content Development:** Encouraging digital content creation in regional languages ensures inclusivity for linguistically diverse populations.

Role of Technology in Bridging the Digital Divide: Technology itself can be leveraged to overcome many barriers to digital inclusion.

- **Mobile-First Solutions**
  - With over **1.1 billion mobile connections** in India, mobile technology has become a critical enabler of digital access. Apps designed for low-bandwidth environments and compatibility with basic smartphones can significantly enhance rural and marginalized communities' digital engagement.
- **Artificial Intelligence and Machine Learning**
  - AI-driven platforms can provide personalized digital literacy training, adapting content to users' needs and skill levels.
  - **Language Translation Tools:** AI can bridge linguistic barriers, making digital platforms accessible to India's linguistically diverse population.
- **Solar-Powered Connectivity Solutions**
  - Solar-powered Wi-Fi hotspots can connect internet to areas without reliable electricity, addressing infrastructure deficits in rural regions.
- **Community Networks**
  - Localized community networks, operated by residents, provide affordable and sustainable internet solutions in underserved areas.

Future Directions and Recommendations: To effectively bridge the digital divide, a multi-pronged approach is required:

- **Investment in Infrastructure:** Expanding broadband and mobile networks to rural and remote areas should remain a top priority.
- **Affordable Devices and Services:** Collaborations between governments and private companies can reduce the cost of devices and internet plans.
- **Targeted Policies for Women and Marginalized Groups:** Tailored programs addressing specific barriers faced by women, SC/ST communities, and economically disadvantaged groups are essential.
- **Promoting Digital Literacy:** Comprehensive digital literacy campaigns, delivered through schools, community centers, and online platforms, can empower citizens to leverage technology effectively.
- **Monitoring and Evaluation:** Regular assessment of digital inclusion programs ensures accountability and identifies areas for improvement.

IT Initiatives for Public Welfare – eGovernance and AI-Based Solutions

Integrating Information Technology (IT) into public welfare has revolutionized the design, implementation, and delivery of services in India, promoting transparency, efficiency, and inclusivity. IT initiatives such as Aadhaar and Direct Benefit Transfer (DBT) have redefined e-governance, while AI-based solutions in education and healthcare address underserved communities' needs. This section explores the current landscape of IT-driven public welfare, highlighting the transformative potential of these initiatives and the challenges that remain.

eGovernance Initiatives: egovernance refers to using IT to improve government services, engage citizens, and enhance accountability. In India, flagship initiatives like Aadhaar and DBT exemplify how IT can empower citizens and streamline welfare programs.

• Aadhaar-Unique Identity for Every Citizen:

Launched by the **Unique Identification Authority of India (UIDAI)** in 2009, Aadhaar has become the cornerstone of India's e-governance framework. As of 2023, over 1.3 billion people have been enrolled in the system.

- **Biometric and Demographic Data:** Aadhaar uses biometrics (fingerprints and iris scans) and demographic information to give each citizen a unique 12-digit identification number.
- **Integration with Services:** Aadhaar is linked to a wide array of government and private services, such as opening bank accounts, obtaining subsidies, and accessing healthcare.

- **Benefits of Aadhaar**
  - **Targeted Delivery:** By verifying identities, Aadhaar ensures that benefits reach the intended recipients, reducing leakages in welfare programs.
  - **Financial Inclusion:** Aadhaar-enabled Payment Systems (AePS) allow rural populations to access banking services without traditional infrastructure.
  - **Unified Identity:** Aadhaar eliminates the need for multiple identification documents, simplifying interactions with government agencies.
- **Direct Benefit Transfer (DBT): Revolutionizing Subsidy Delivery:** The DBT program, introduced in 2013, leverages Aadhaar to transfer subsidies and benefits directly to beneficiaries' bank accounts.
  - **Elimination of Middlemen:** DBT reduces corruption by bypassing intermediaries.
  - **Real-Time Transfers:** Beneficiaries receive funds directly and promptly, enhancing trust in government programs.
  - **Applications:** DBT is used for subsidies on LPG (Pratyaksh Hanstantrit Labh or PAHAL scheme), scholarships, pensions, and employment guarantees under MGNREGA.

AI-Based Solutions for Underserved Communities: Artificial Intelligence (AI) is increasingly being leveraged to address systemic challenges in education and healthcare, particularly for underserved communities in rural and remote areas.

- **AI in Education:** AI has the potential to revolutionize education by enabling personalized learning, addressing teacher shortages, and making quality education accessible to all.
- **Personalized Learning**
  - **Adaptive Learning Platforms:** AI-powered platforms like BYJU's and Khan Academy tailor content to individual learning needs, helping students grasp concepts at their own pace.
  - **Real-Time Feedback:** AI systems analyze student performance and provide instant feedback, enabling timely intervention.
- **Bridging Gaps in Teacher Availability**
  - **Virtual Tutors:** AI-powered chatbots and voice assistants can simulate one-on-one teaching experiences, addressing teacher shortages in rural areas.
  - **Multilingual Support:** AI facilitates learning in regional languages, catering to India's diverse linguistic landscape.

- **Inclusive Education**
  - **Assistive Technologies:** AI tools, such as speech-to-text systems and braille translators, support students with disabilities, ensuring inclusivity.
  - **AI-Powered Content Creation:** Automated tools generate textbooks and learning materials in multiple languages, addressing content shortages in underserved regions.
- **AI in Healthcare:** AI-driven solutions are transforming healthcare by enhancing diagnostics, treatment, and accessibility for underserved populations.
- **Diagnostics and Early Detection**
  - **Disease Prediction Models:** AI systems analyze patient data to predict diabetes, cancer, and cardiovascular conditions. For example, Wadhwani AI uses AI to detect tuberculosis in rural areas.
  - **Imaging Analysis:** AI-powered tools interpret X-rays, MRIs, and CT scans with high accuracy, reducing diagnostic errors.
- **Telemedicine and Remote Care**
  - **AI-Driven Chatbots:** Virtual health assistants provide preliminary diagnoses, suggest treatments, and schedule appointments, reducing the burden on healthcare providers.
  - **Teleconsultations:** AI supports remote consultations by analyzing patient symptoms and assisting doctors in providing accurate diagnoses.
- **Public Health Interventions**
  - **Epidemic Monitoring:** AI systems track disease outbreaks by analyzing environmental and demographic data, enabling timely interventions.
  - **Vaccination Campaigns:** AI helps optimize vaccination schedules and monitor coverage, as demonstrated during India's COVID-19 vaccination drive.
- **Impact on Public Welfare:** The convergence of e-governance and AI-driven solutions has had a transformative impact on public welfare in India:
  - **Transparency and Accountability:** IT tools have reduced corruption and improved trust in government programs.
  - **Inclusivity:** Digital platforms ensure that marginalized groups have access to essential services.
  - **Efficiency:** Automation and real-time data analysis enhance the speed and effectiveness of welfare delivery.

Recommendations for Enhancing IT-Driven Welfare: To maximize the potential of IT in public welfare, several measures are needed:

- **Strengthening Digital Infrastructure:** Expanding broadband connectivity and mobile networks to rural areas is critical.
- **Promoting Digital Literacy:** Comprehensive training programs should equip citizens with the skills to use digital platforms effectively.
- **Ensuring Data Privacy:** Robust legal frameworks must protect sensitive data from misuse.
- **Scaling AI Solutions:** Partnerships between governments, private sector players, and academia can accelerate the development and deployment of AI-based tools.

India's IT-driven initiatives in e-governance and AI-based education and healthcare are redefining public welfare by making services more transparent, efficient, and inclusive. While significant progress has been made, addressing challenges such as privacy concerns, data quality, and the digital divide is essential to ensure equitable access to these innovations. By building on the successes of Aadhaar, DBT, and AI-driven solutions, India can create a resilient, technology-enabled framework for public welfare, empowering citizens and driving sustainable development.

Future Vision for Universal Digital Literacy by 2047

As India approaches its centenary of independence in 2047, achieving universal digital literacy will be one of the most transformative milestones in its journey toward inclusive growth and global competitiveness. With digital technology increasingly becoming the backbone of economic, social, and cultural progress, ensuring equitable access to digital tools and skills cannot be overstated. Universal digital literacy, combined with personalized learning systems powered by Artificial Intelligence (AI), can fundamentally reshape India's education landscape, bridge gaps in accessibility and affordability, and empower marginalized communities to participate fully in the digital economy. By 2047, India envisions a digitally literate population capable of leveraging technology for personal growth, economic empowerment, and social well-being, driving the nation's advancement across sectors while ensuring no one is left behind.

Achieving Universal Digital Literacy: Universal digital literacy is using digital devices to navigate and effectively engage with digital content, tools, and systems. It encompasses many competencies, from basic skills such as operating smartphones and using the internet to more advanced capabilities like coding, digital communication, and data analysis. For India, a country with over a billion people, including a large proportion of rural, elderly, and underprivileged populations, achieving digital literacy by 2047 presents both challenges and immense opportunities.

India's path to universal digital literacy must be inclusive, targeting diverse groups across age, gender, socioeconomic status, and geographical location. Ensuring equitable access to digital infrastructure such as reliable internet connectivity, smartphones, and affordable devices will be the first step in achieving widespread digital literacy. Government initiatives like the Digital India program have significantly improved digital access and connectivity in remote regions, but the accessibility challenge remains a key hurdle. To overcome this, India must expand its physical infrastructure and address affordability issues through subsidies, public-private partnerships, and community-based digital literacy programs. Local languages and culturally relevant content will be essential to making digital literacy accessible to all, particularly for those in rural areas where the linguistic divide poses a significant barrier to engagement with digital content. In addition to infrastructure, there must be a concerted push for digital education and skill-building programs. These should be available in schools, universities, and community centers, targeting youth and adults, especially in rural areas. Partnerships with technology companies, academic institutions, and non-governmental organizations will be critical in expanding access to digital learning materials and courses. Online platforms and mobile apps, leveraging India's vast mobile network, can provide accessible, self-paced learning, overcoming the barriers of physical distance and resource constraints. Over the coming decades, offline and online learning tools will upskill the population, making digital literacy a universal feature of India's social and educational fabric.

Personalized Learning Systems Using AI: One of the most promising prospects in the vision of universal digital literacy is the potential for personalized learning systems powered by Artificial Intelligence (AI). AI can transform the educational landscape by adapting learning materials to suit individual learners' needs, strengths, and weaknesses. This personalized approach offers a solution to many of the challenges faced by India's education system, such as overcrowded classrooms, one-size-fits-all teaching methods, and unequal access to quality education.

By 2047, AI-based personalized learning platforms will be central to promoting digital literacy and enhancing educational outcomes across all levels. These platforms, already in their nascent stages globally, will evolve to become increasingly sophisticated, offering tailored learning experiences that adapt in real-time to students' progress and learning styles. For example, AI can identify patterns in how students engage with material, automatically adjusting the complexity of lessons, offering additional resources for areas where students are struggling, or accelerating content for those progressing faster. In India's diverse educational ecosystem, this can significantly improve the quality of education by ensuring that every student, irrespective of their background, can receive personalized attention and support. Also, AI will allow for integrating various forms of learning from video lectures to interactive simulations and quizzes creating a rich and diverse learning environment. AI-powered educational tools like intelligent tutoring systems (ITS) can mimic one-on-one tutoring, allowing students to engage with content at their own pace and according to their unique needs. These systems will use natural language processing (NLP) and machine learning algorithms to understand student queries and provide real-time feedback, making learning more interactive, engaging, and effective.

For marginalized communities, personalized AI-driven learning tools can overcome significant barriers, such as limited access to qualified teachers and educational resources. For example, students in rural or remote areas may not have access to a highly trained math tutor, but AI tools can help fill that gap by providing 24/7 access to educational content and tailored support. Similarly, AI systems can help learners with disabilities, such as visual or hearing impairments, by offering customized formats for learning content or using speech recognition and text-to-speech technologies to assist comprehension. The deployment of AI in education will also extend beyond academic learning, facilitating skill development in areas like coding, digital marketing, graphic design, and other vocational skills crucial for future employment. This can be especially valuable for India's large informal workforce, enabling individuals to acquire new skills and improve their employability without needing traditional, costly educational pathways.

Bridging Gaps in Accessibility and Affordability: Achieving universal digital literacy by 2047 also requires addressing significant gaps in accessibility and affordability. Digital literacy cannot be achieved without the broader context of socioeconomic inequalities. In India, the digital divide is pronounced, with urban areas enjoying better access to the internet and technology compared to rural areas, where a significant portion of the population still lacks reliable internet connectivity, smartphones, or computer access. One of the most critical aspects of this future vision is ensuring that technology is accessible to all, irrespective of socioeconomic status, geographical location, or educational background. Governments and private enterprises must collaborate to design inclusive IT systems that account for the socioeconomic realities of various populations. For example, affordable smartphones, subsidized internet plans, and free public Wi-Fi in underserved areas can greatly enhance access. Additionally, local community centers, libraries, and schools can serve as hubs for digital learning, offering free access to digital tools and support for individuals who might not have the means to own personal devices. Further, digital content must be made inclusive by being available in various languages, dialects, and formats. India's linguistic diversity uniquely challenges digital literacy, with over 22 official languages and hundreds of regional dialects. Educational content must cater to this diversity, whether delivered through videos, e-books, or online courses. Text-to-speech technology, automatic translation tools, and multilingual interfaces can ensure that digital resources are accessible to all, regardless of the language spoken.

Affordability is another significant barrier to achieving universal digital literacy. While the cost of technology has been decreasing globally, many families in rural and economically disadvantaged areas still struggle to afford the necessary tools for digital engagement. Innovative financing models such as pay-per-use schemes, government subsidies, or low-cost financing options for devices and data packages can help address this issue. Public-private partnerships can also play a role in ensuring that digital tools are available at affordable rates, particularly through community-based digital learning initiatives.

Looking Forward – Connected, Educated, and Empowered India

By 2047, India aims to be a nation where digital literacy is not a privilege but a fundamental skill required for personal growth and social participation. This vision extends beyond just education; it seeks to ensure digital literacy becomes a core part of India's broader development agenda. Digital literacy will enable citizens to access various services, from government schemes and healthcare services to job opportunities and social empowerment initiatives. It will also give them the tools to actively participate in the democratic process, express their views, and engage with global markets. As the digital economy continues to grow, the benefits of digital literacy will cascade through all sectors of society. Workers with digital skills will be better equipped to compete in the knowledge economy, and small businesses will be able to expand their reach through e-commerce platforms and digital marketing. Digital literacy will also drive innovation as more people from diverse backgrounds contribute to creating digital tools, apps, and platforms that meet the needs of India's people.

5

Transforming Processes Through IT

Introduction

In the contemporary digital era, Information Technology (IT) has emerged as a pivotal tool in transforming business and governance processes, enabling greater efficiency, transparency, and innovation. The process transformation facilitated by IT is not merely a technical upgrade but a fundamental rethinking of how organizations operate, deliver services, and interact with stakeholders. This transformation affects every facet of an organization, from internal workflows to customer engagement, and extends to how governments interact with citizens. Integrating IT into processes fosters greater flexibility, scalability, and adaptability, allowing organisations to respond swiftly to changing environments and demands.

Process Transformation: Process transformation refers to the redesign or improvement of existing processes using IT systems to increase efficiency, reduce costs, enhance customer satisfaction, and improve overall performance. This transformation is rooted in re-engineering workflows, leveraging IT tools and platforms to automate tasks, eliminate redundancies, and streamline operations. While traditional processes were managed manually or through isolated legacy systems, IT systems provide a cohesive framework that integrates diverse functions, breaking down silos, and enabling seamless communication across different departments or entities. The scope of process transformation is vast, covering various domains:

- **Business Operations:** IT enables companies to optimize processes across multiple functions from supply chain management to customer service.
- **Governance:** E-governance initiatives have transformed public administration, simplifying processes like licensing, registration, and service delivery.
- **Healthcare:** IT solutions, such as Electronic Health Records (EHR), have revolutionized patient care by enabling seamless data exchange and improving operational efficiency.
- **Education:** Online learning management systems and administrative automation have transformed traditional educational processes, making them more flexible and scalable.

Key Drivers of Process Transformation through IT: The adoption of IT to transform processes is driven by several factors, which include the desire for greater operational efficiency, cost reduction, enhanced customer satisfaction, and the competitive pressures of the digital age. Key drivers include:

- **Technological Advancements:** Technological innovations like cloud computing, big data, artificial intelligence (AI), machine learning (ML), and Internet of Things (IoT) have accelerated process transformation. These technologies allow businesses to process vast amounts of data, automate decision-making, and create intelligent, self-optimizing systems that can continuously improve processes. For example, AI-powered chatbots can automate customer service tasks, reducing operational costs and improving response times.
- **Data-Driven Decision Making:** The shift towards data-driven decision-making has significantly influenced process transformation. Organizations can optimise processes by collecting, analyzing, and using large volumes of real-time data. For example, predictive analytics can be used to anticipate customer demand, allowing businesses to adjust production schedules accordingly. This leads to better resource management, waste reduction, and increased agility.
- **Globalization and Market Competition:** Globalization and the rise of digital platforms have heightened competition, forcing organizations to rethink their processes for greater responsiveness and efficiency. IT systems provide the scalability and flexibility needed to cater to global markets and diverse customer demands. Automation tools help businesses reduce turnaround times, improve service delivery, and enhance customer satisfaction.
- **Customer Expectations:** With the advent of digital technologies, customers now expect faster, more personalized, and seamless business interactions. IT enables businesses to align their internal processes with these heightened expectations by automating routine tasks, providing real-time data, and offering personalized services. This is particularly evident in sectors like retail, where omnichannel customer service platforms allow businesses to interact with customers across various digital touchpoints.

Key Areas of Process Transformation Through IT: The application of IT for process transformation spans across various domains, with several key areas seeing profound changes:

- **Workflow Automation:** IT enables the automation of repetitive tasks, which traditionally require human intervention. Software tools such as Enterprise

Resource Planning (ERP) systems and Robotic Process Automation (RPA) platforms allow for the streamlining of business workflows. For instance, in manufacturing, IT systems monitor and adjust production lines in real-time, reducing downtime and optimizing efficiency. Similarly, in finance, RPA automates tasks such as invoice processing, reducing errors and speeding up turnaround times.

- **Data Integration and Interoperability:** In the absence of integrated systems, organizations often operate in silos, resulting in inefficiencies and data duplication. IT enables integration disparate systems into unified platforms, ensuring seamless communication and data sharing. For example, integrating electronic health records (EHR) across multiple healthcare providers ensures patient data is accessible in real-time, improving decision-making and reducing errors. This interconnectedness of systems, driven by IT, enables smoother workflows and more accurate data usage.
- **Real-Time Monitoring and Control:** The ability to monitor processes in real-time is another transformative effect of IT. With IoT sensors, RFID tags, and cloud-based dashboards, businesses can track and monitor processes across the value chain in real-time. In logistics, for example, tracking shipments with GPS and sensor data enables businesses to optimize routes, predict delays, and improve customer satisfaction. In manufacturing, real-time monitoring helps identify bottlenecks and quality issues, allowing for immediate corrective actions.
- **Decision-Making and Predictive Analytics:** IT systems are increasingly used to enhance decision-making by providing data-driven insights. Predictive analytics, powered by machine learning algorithms, enables businesses to anticipate market trends, customer behavior, and potential risks. In retail, for instance, data analytics can forecast customer buying patterns, allowing businesses to adjust inventory levels and pricing strategies accordingly. This proactive decision-making leads to more efficient processes, as organizations can adjust operations in anticipation of demand rather than reacting after the fact.

Benefits of Transforming Processes Through IT: The integration of IT into organizational processes brings several critical benefits:

- **Enhanced Efficiency:** Automating processes through IT reduces the need for manual interventions, leading to faster completion of tasks, fewer errors, and optimized resource utilization. Automated systems streamline repetitive workflows, reduce delays, and save human resources for more value-added tasks. For example, an automated customer support system using AI chatbots can handle basic queries, leaving customer service representatives to address more complex issues.

- **Cost Reduction:** Organizations can lower operational costs by optimizing processes and reducing the need for human intervention in routine tasks. Cloud computing reduces the need for expensive on-site infrastructure, while automation decreases labor costs. IT-driven systems can improve energy efficiency and reduce waste, contributing to cost savings.
- **Improved Transparency and Accountability:** IT systems provide greater visibility into operations, making tracking and monitoring performance easier. In government and public administration, IT has enhanced transparency through e-governance platforms that provide real-time updates on the status of services and requests. In business, project management tools allow managers to track the progress of initiatives, monitor team performance, and hold employees accountable.
- **Agility and Scalability:** Organizations that leverage IT for process transformation are better positioned to respond to market changes and scale operations. Cloud-based systems, for example, allow businesses to rapidly scale their IT infrastructure without significant upfront investment, enabling them to expand operations quickly. This agility is crucial in an era of constant technological change and shifting market demands.
- **Enhanced Customer Experience:** IT-driven process transformation ultimately leads to improved customer experiences. Businesses can offer their customers more responsive and tailored solutions through automation, personalised services, and streamlined communication channels. For example, AI-based recommendation engines provide personalized suggestions to customers based on their browsing history, leading to higher satisfaction and increased sales.

Transforming processes through IT is about implementing new technology and reimagining how organizations operate, deliver services, and interact with customers. Integrating IT systems into business operations, governance, and various sectors like healthcare and education can revolutionize how societies function, driving efficiency, cost savings, and customer satisfaction. However, the challenges of cybersecurity, resistance to change, and skills gaps must be addressed for successful transformation. As we move forward, the evolution of IT technologies will further refine and optimize these processes, enabling more adaptive, transparent, and resilient systems to meet the needs of an increasingly complex and interconnected world.

Current Scenario

In recent years, business process automation (BPA) has emerged as a transformative force across various industries, promising substantial improvements in operational efficiency, cost reduction, and accuracy. Organizations can streamline workflows by automating repetitive, time-consuming tasks, freeing up valuable human resources for strategic tasks, and creating more scalable and agile business operations. The integration of technologies such as Robotic Process Automation (RPA), Artificial Intelligence (AI), and Machine Learning (ML) into business processes has enabled significant advancements in industries like finance, healthcare, manufacturing, and supply chain management. However, as businesses increasingly embrace automation, a complex landscape of challenges related to sustainability and the limitations of traditional workflows emerges, particularly when striving to align business operations with sustainable development goals (SDGs). This detailed exploration addresses the current scenario of BPA, its impact on operational efficiency, the limitations of sustainability-driven process designs, and the inefficiencies inherent in traditional workflows across key sectors.

Role and Impact of Business Process Automation on Efficiency: At its core, business process automation is designed to eliminate manual intervention, streamline workflows, and increase the overall efficiency of an organization. In sectors like finance, automation tools can handle tasks such as transaction processing, compliance checks, and report generation, allowing human employees to focus on higher-value decision-making tasks. Automation solutions enable faster inventory management, order fulfilment, and logistics coordination in supply chain management. By reducing the complexity and time required for these functions, BPA significantly boosts operational efficiency.

• **Automation Technologies Driving Efficiency:** Robotic Process Automation (RPA), AI, and cloud computing are pivotal in automating business processes. RPA is particularly valuable in automating rule-based tasks that require little human judgment, such as data entry, invoicing, and account reconciliation. On the other hand, AI and machine learning algorithms enable the automation of more complex decision-making processes by learning from historical data and providing real-time insights.

- **Robotic Process Automation (RPA):** RPA bots mimic human actions in repetitive tasks, significantly reducing processing time and minimizing human error. In banking, RPA is used to automate tasks like customer onboarding and transaction verification, which were traditionally time-consuming and prone to error.
- **Artificial Intelligence (AI) and Machine Learning (ML):** AI-driven automation enables organizations to optimize decision-making, forecasting, and personalization at scale. In marketing, AI-based algorithms tailor

advertisements and promotional content to specific customer segments, enhancing engagement and conversion rates.

- **Cloud Computing:** Cloud-based automation solutions provide scalability and flexibility, enabling businesses to adjust their processes based on demand. By centralizing data storage and management, cloud computing makes it easier for businesses to collaborate across departments and locations, improving operational efficiency.

Tangible Benefits of Automation: Increased Speed and Accuracy: Automation leads to faster processing times with fewer errors. For example, AI-driven diagnostics and automated patient scheduling systems reduce wait times and improve patient care in the healthcare sector.

- **Cost Reduction:** By automating labour-intensive processes, organizations can reduce overhead costs and minimize the need for manual intervention. In manufacturing, automation technologies such as robotic arms have significantly reduced production costs and minimized human error, leading to higher product quality and more efficient operations.
- **Scalability and Flexibility:** Automated systems can handle large volumes of work without additional human resources, making it easier to scale operations. Automated warehouses and AI-driven route optimization tools enable companies to handle more orders efficiently, even as demand fluctuates.

Limitations in Sustainability-Driven Process Designs: While business process automation has proven to be highly effective in improving operational efficiency, there are several challenges when it comes to designing processes that are also sustainable. Sustainability-driven process designs are those that not only focus on efficiency but also consider long-term environmental, economic, and social impacts. The growing pressure to integrate sustainability into business processes has led to the realization that, though efficient, traditional automated systems may not always align with sustainability goals.

- **Environmental Impact of Automation:** One of the primary concerns regarding sustainability is its environmental impact. Automation technologies, particularly those involving high energy consumption or non-renewable resources in hardware infrastructure, can inadvertently contribute to environmental degradation. For instance, data centers that house cloud computing infrastructure often require massive amounts of electricity to operate and cool the systems. While cloud computing provides efficiency benefits, it generates a significant carbon footprint if powered by non-renewable energy sources.

- **Energy Consumption:** The growing reliance on automated systems increases the need for energy to power data centers and manufacturing equipment. As automation becomes more pervasive, it is critical to design energy-efficient systems that rely on renewable energy sources to reduce their environmental impact.
- **Electronic Waste:** The proliferation of automation technologies and the constant upgrading of systems lead to electronic waste (e-waste) generation. Many automated systems require specialized hardware, which, when obsolete, contributes to the growing global e-waste crisis.

• **Social Impacts and Labor Market Disruptions:** Another limitation of automation in sustainability is the social impact on the workforce. While automation can significantly improve operational efficiency, it can also result in job displacement, particularly in industries where machines replace routine, manual tasks. In regions with limited access to reskilling opportunities, this can exacerbate inequalities and create social instability.

- **Job Displacement:** As more tasks are automated, the demand for low-skilled labor diminishes. This can contribute to unemployment or underemployment, particularly in manufacturing, retail, and customer service sectors.
- **Inequality in Access to Technology:** The adoption of advanced automation tools is often concentrated in certain regions or industries, leading to disparities in access to technological advancements. This inequality can prevent marginalized communities from benefiting from automation's efficiencies and opportunities.

Economic Considerations: While automation can lead to cost savings, it introduces initial investments and long-term sustainability complexities. Small and medium-sized enterprises (SMEs) may find investing in expensive automation technologies challenging, particularly in developing economies. The efficiency gains achieved through automation may not always translate into equitable economic growth if the benefits are not widely distributed.

- **Capital Costs:** The initial investment in automation technologies, including hardware, software, and infrastructure, can be prohibitive for many businesses. Even if the long-term savings from automation are significant, the upfront costs may be a barrier for smaller businesses or those in emerging markets.
- **Concentration of Wealth:** Automation can exacerbate wealth inequality, as larger corporations are more likely to benefit from automation due to their financial resources, while smaller businesses struggle to keep up. This

disparity can result in a concentration of economic power and a widening wealth gap.

Inefficiencies in Traditional Workflows Across Key Sectors: Despite the undeniable benefits of BPA, many sectors still face inefficiencies due to outdated workflows, manual processes, and fragmented systems. These inefficiencies manifest in various ways, from slower response times and higher operational costs to errors and poor customer service. The continued reliance on traditional workflows limits the full potential of automation technologies and impedes progress toward a more streamlined, efficient, and sustainable model.

- **Healthcare Sector Inefficiencies:** In healthcare, traditional workflows often involve paper-based systems, manual record-keeping, and fragmented patient information across multiple departments. These inefficiencies can result in delays, misdiagnoses, and poor patient outcomes. While Electronic Health Records (EHRs) and telemedicine have improved efficiency, many hospitals and clinics still struggle with interoperability between systems, leading to incomplete or inconsistent patient data.

  - **Manual Administrative Tasks:** In many healthcare facilities, administrative tasks such as patient registration, billing, and insurance verification are performed manually. This wastes time and increases the likelihood of errors, compromising patient care.
  - **Data Silos:** Fragmented health information systems make it difficult for healthcare providers to access comprehensive patient histories, leading to inefficiencies and suboptimal care. Integrating automated systems into electronic health record management is necessary to ensure that data flows seamlessly between departments.

Manufacturing and Supply Chain Challenges: In manufacturing and supply chains, traditional workflows are often hindered by outdated systems that do not fully integrate with modern digital technologies. Although automation has significantly advanced in industries like automotive manufacturing, many sectors still rely on legacy systems that hinder efficiency and flexibility.

- **Manual Inventory Management:** In many manufacturing environments, inventory management is still done manually or using outdated systems, leading to stockouts, overstocking, and increased costs.
- **Slow Production Cycles:** Traditional production processes are often rigid and unable to adapt quickly to changing market demands. The lack of real-time monitoring and predictive analytics prevents manufacturers from responding quickly to shifts in demand, resulting in inefficiencies and missed opportunities.

Public Sector and Governance: Government agencies often operate using a patchwork of outdated systems, which limits the efficiency of services offered to citizens. Manual processes like taxation, licensing, and social welfare programs create delays and inaccuracies, reducing public trust and undermining service delivery.

- **Bureaucratic Red Tape:** Many public sector processes are bogged down by bureaucratic inefficiencies, where multiple layers of approval and paper-based processes slow down decision-making and service delivery.
- **Lack of Data Integration:** Governments often struggle with siloed data systems that prevent departments from sharing critical information in real-time. This results in inefficiencies and a lack of coordinated action across different branches of government.

Moving Toward Sustainable and Efficient Business Processes: While business process automation presents a substantial opportunity to improve efficiency and reduce costs, the path to fully realizing these benefits is fraught with challenges. Sustainability-driven process designs must account for automation technologies' environmental, social, and economic implications, particularly regarding energy consumption, waste generation, and job displacement. Additionally, inefficiencies in traditional workflows across key sectors highlight the need for comprehensive digital transformation, where automation technologies integrate seamlessly with existing systems to optimize performance. The future of process automation must be centered around improving operational efficiency and ensuring that these improvements are aligned with sustainable development goals. By addressing the limitations of current processes and leveraging the power of emerging technologies such as AI, blockchain, and IoT, businesses and governments can create smarter, more sustainable systems that benefit all stakeholders.

Process Inefficiencies in Key Sectors

In the quest for sustainable development and economic growth, addressing inefficiencies in key sectors is paramount. Agriculture, manufacturing, and urban planning are three pillars of any nation's economy, yet they face significant process inefficiencies that hinder overall productivity, sustainability, and economic growth. These inefficiencies limit the growth potential and exacerbate environmental degradation, social inequalities, and economic disparities. Understanding the root causes of process inefficiencies in these sectors and the potential for reform is crucial for the holistic advancement of economies and societies. This section delves into the specific process inefficiencies in agriculture, manufacturing, and urban planning and explores how they affect productivity, sustainability, and the broader goals of equitable growth.

Agricultural Process Inefficiencies: Agriculture remains a cornerstone of many developing economies, including India, where it is a livelihood for millions of people and a critical part of the national economy. However, the agricultural sector continues to be plagued by inefficiencies in several key processes. These inefficiencies stem from outdated farming practices, inadequate infrastructure, limited access to technology, and insufficient integration of data-driven insights into farming operations.

Traditional Farming Methods and Low Productivity: Despite advancements in agricultural technologies and research, much of the agricultural sector in developing countries still relies heavily on traditional farming methods. These methods often involve labor-intensive practices, limited mechanization, and an over-reliance on rain-fed irrigation, leading to low productivity. Traditional practices, while sometimes culturally significant, are often inefficient when compared to modern techniques that integrate advanced farming tools, such as precision agriculture.

- **Limited Mechanization:** Many small and medium-sized farms lack the resources to invest in modern machinery. Without access to modern tractors, plows, or automated irrigation systems, farmers are forced to rely on manual labor, which reduces production efficiency and increases costs.
- **Low Yield Crops:** Traditional agricultural methods do not always consider the latest advancements in crop breeding, pest management, or irrigation technology, which can limit crop yields. This inefficiency directly impacts food production and the economy, especially in countries where agriculture is the primary source of income.

Fragmented Supply Chains and Post-Harvest Losses: The agricultural supply chain in many regions is highly fragmented, with smallholders, local markets, and intermediaries all playing a role. This fragmentation leads to storage, transportation, and distribution inefficiencies, often resulting in post-harvest losses. The lack of proper storage facilities and modern logistics infrastructure also exacerbates these losses, which can be as high as 30-40% in certain developing countries.

- **Lack of Cold Storage:** Many farmers do not have access to cold storage facilities, leading to spoilage of perishable crops such as fruits and vegetables. This problem is particularly prevalent in tropical and sub-tropical regions where high temperatures accelerate decay.
- **Inefficient Distribution Channels:** The reliance on informal intermediaries in the agricultural supply chain often means that products are sold to farmers and consumers at lower prices. This process increases costs and delays in getting products to market, reducing the agricultural sector's efficiency and profitability.

Limited Use of Data and Technology: While technology has the potential to revolutionize agriculture, the use of modern tools such as satellite imaging, Internet of Things (IoT) devices, and big data analytics remains limited in many rural areas. These tools can enhance decision-making by providing real-time data on weather patterns, soil conditions, and pest infestations, allowing farmers to make more informed choices. However, the adoption of these technologies remains slow due to factors such as high initial costs, lack of digital literacy, and insufficient infrastructure.

- **Precision Agriculture:** Technologies like GPS-guided tractors, soil sensors, and drones can optimize resource use, reducing the waste of water, fertilizer, and pesticides. However, the adoption rate of these technologies is still low among smallholders due to cost barriers and lack of awareness.
- **Climate Resilience:** Climate change poses a significant challenge for agriculture, and there is a need for better data on weather patterns and environmental conditions to make farming more resilient. However, the lack of real-time data integration into farm management systems prevents farmers from responding effectively to climate fluctuations.

Manufacturing Process Inefficiencies: Manufacturing is another sector where process inefficiencies impact productivity and sustainability. Manufacturers often face challenges in developing and developed economies due to outdated equipment, inefficient workflows, poor supply chain management, and a lack of technological integration. These inefficiencies hinder the sector's ability to compete globally, increase operational costs, and strain environmental resources.

Outdated Technology and Equipment: One of the key contributors to inefficiency in the manufacturing sector is the use of outdated technology and equipment. Many factories, especially in developing economies, still rely on legacy machines and manual processes, which limit their ability to produce at scale and with high precision. This inefficiency results in higher labor costs, slower production cycles, and frequent breakdowns.

- **Low Automation:** The lack of automation in manufacturing processes increases the time and labor required to produce goods, reduces output consistency, and raises production costs. Automated systems, such as robotic arms, 3D printing, and intelligent manufacturing systems, can enhance efficiency, but adopting such technologies is often slow in traditional manufacturing industries.
- **Downtime and Maintenance:** Outdated machinery often leads to unplanned downtime, halting production and reducing manufacturing systems' overall efficiency. Maintenance costs for older equipment are also higher, further increasing operational expenses.

Inefficient Supply Chain Management: The manufacturing sector depends on a complex network of suppliers, distributors, and logistics systems. Inefficiencies in supply chain management, such as delays in procurement, stockouts, overstocking, and poor coordination, increase costs and reduce the overall efficiency of manufacturing operations.

- **Inventory Management:** In many manufacturing sectors, inventory management remains a manual process, resulting in overstocking, stockouts, or the holding of obsolete stock. This inefficiency increases working capital costs and reduces the ability of manufacturers to meet demand on time.
- **Logistics Delays:** Poor logistics and transportation systems contribute to delays in the delivery of raw materials, components, and finished goods. Inefficiencies in transportation and warehousing result in longer lead times, reduced production cycles, and higher transportation costs.

Environmental Impacts and Sustainability Concerns: Manufacturing processes often consume vast amounts of natural resources and generate significant waste, contributing to environmental degradation. While automation and optimization of workflows can improve efficiency, many manufacturing sectors still have a long way to go in integrating sustainability into their processes.

- **Energy Consumption:** Manufacturing industries are major energy consumers, with much of their energy coming from non-renewable sources. The adoption of energy-efficient technologies and the shift to renewable energy sources are critical to reducing the environmental impact of manufacturing processes.
- **Waste Generation:** Inefficient manufacturing processes often lead to high waste levels in raw materials and finished products. The lack of circular economy principles in many industries means that much of the waste generated is not reused or recycled, contributing to environmental pollution.

Urban Planning Process Inefficiencies: Urban planning is critical to managing the growth of cities and ensuring that urban areas can support their populations sustainably. However, inefficiencies in urban planning hinder the development of livable and sustainable cities. These inefficiencies often stem from outdated planning methods, lack of integration between various city departments, and failure to adopt technology and data-driven decision-making.

Fragmented Urban Planning and Lack of Integration: In many urban areas, the planning process is fragmented, with various departments handling different aspects of urban development without sufficient coordination. This fragmentation leads to inefficient allocation of resources, inconsistent policies, and overlapping initiatives that fail to address the holistic needs of the city.

- **Siloed Departments:** Urban planning involves multiple sectors, including housing, transportation, healthcare, and environmental management. However, these departments often operate in silos, with limited communication and data sharing. As a result, urban development plans are frequently disjointed, leading to inefficiencies and missed opportunities for synergy.
- **Lack of Data Integration:** Urban areas generate vast data, from traffic patterns and energy consumption to waste generation and air quality. However, this data is often not integrated into planning and decision-making processes, making it difficult for city officials to optimize urban systems for sustainability, efficiency, and quality of life.

Inefficient Land Use and Infrastructure Development: One of the most significant inefficiencies in urban planning is the underutilization of land and poorly planned infrastructure development. In many rapidly growing cities, urban sprawl leads to inefficient land use, longer commute times, and increased infrastructure costs.

- **Urban Sprawl:** Unplanned or poorly managed urban sprawl increases the distance between residential, commercial, and industrial zones, leading to traffic congestion, environmental degradation, and higher infrastructure costs. Compact city designs and better zoning regulations can reduce urban sprawl and enhance infrastructure efficiency.
- **Inadequate Infrastructure:** Many urban areas suffer from underdeveloped infrastructure, such as inefficient public transportation systems, insufficient waste management, and inadequate water and energy distribution networks. The lack of integrated, modern infrastructure systems leads to inefficiencies and higher operational costs.

Sustainability Challenges in Urban Planning: As cities grow, sustainable urban planning becomes more urgent. Urban areas consume a disproportionate share of global resources and produce significant greenhouse gas emissions. Inefficiencies in urban planning, such as the over-reliance on private transportation or failing to incorporate green spaces and renewable energy systems, exacerbate these environmental challenges.

- **Transportation Systems:** Inefficient public transportation systems and heavy reliance on private vehicles increase traffic congestion, fuel consumption, and air pollution. Developing integrated, multimodal public transportation systems can reduce the environmental impact of urban transportation.

- **Waste and Energy Management:** Urban areas generate vast amounts of waste and consume large quantities of energy. Efficient waste management systems, including recycling and composting programs, and the adoption of renewable energy sources, can significantly reduce the environmental footprint of urban areas.

Addressing Process Inefficiencies for a Sustainable Future

The process inefficiencies in agriculture, manufacturing, and urban planning are significant barriers to sustainable development and economic growth. These inefficiencies, whether rooted in outdated practices, inadequate infrastructure, or fragmented systems, prevent these sectors from realizing their full potential. However, substantial improvements can be made in these key areas by integrating technology, improving supply chain coordination, adopting sustainable practices, and focusing on data-driven decision-making. Governments, industries, and communities must work together to optimize processes and drive innovation for a more sustainable future. By addressing inefficiencies, these sectors can contribute more effectively to national economic growth, environmental sustainability, and social equity.

IT-Driven Process Automation

In recent years, technological advancements have catalyzed the evolution of industries toward more sustainable, efficient, and optimized processes. One of the most compelling areas of innovation is in IT-driven process automation, which has proven to be a key enabler in the transition toward zero-waste systems. Leveraging technologies such as robotics, the Internet of Things (IoT), and artificial intelligence (AI), organizations are improving their operational efficiency and driving sustainability by minimizing waste, conserving resources, and reducing carbon footprints. Zero-waste systems, which focus on maximizing resource use, reducing waste generation, and promoting recycling and reuse, are now increasingly viable due to the synergy between these technologies.

Role of Robotics in Process Automation: Robotics, a core element of process automation, has significantly transformed industries by introducing more precise, efficient, and scalable manufacturing processes. Robotics plays a pivotal role in enhancing operational efficiency, reducing waste generation, and optimizing the recycling and reuse of materials.

– **Robotics for Precision and Efficiency:** One of the most significant advantages of robotics in process automation is its precision and ability to perform repetitive tasks accurately. Robots can be programmed to handle complex, intricate tasks such as sorting materials, assembling products, and conducting quality control checks. This precision is crucial for implementing zero-waste principles in manufacturing, as it minimizes the likelihood of defects and material waste. By ensuring that every action is executed with high accuracy, robotic systems reduce the chances of overproduction, misallocation of resources, and errors that lead to waste. For example, in a recycling facility, robotic arms equipped with sensors and AI algorithms can sort waste materials (such as plastic, metal, and paper) more efficiently than human workers, ensuring that the maximum amount of recyclable materials is extracted. This automated sorting process increases the efficiency of recycling operations and enhances the quality of the recycled materials, contributing to a circular economy where waste is minimized and resources are reused.

– **Collaborative Robots (Cobots) in Manufacturing:** In manufacturing environments, collaborative robots (cobots) work alongside human workers to enhance productivity while minimizing waste. Unlike traditional robots that work in isolation, cobots are designed to interact safely with humans, making them ideal for environments where human oversight is necessary. Cobots can assist in material handling, assembly, and packaging, reducing the need for excess materials and labour-intensive tasks. By automating manual processes, cobots enable manufacturers to reduce operational costs, avoid overproduction, and streamline workflowsall of which contribute to the goal of zero-waste manufacturing. Cobots can be integrated into existing production lines, enabling manufacturers to scale their operations without significant infrastructure overhauls. This flexibility helps businesses reduce waste while maintaining high levels of productivity and efficiency.

The Internet of Things (IoT) and Data-Driven Waste Management: The IoT represents a network of connected devices that can collect, exchange, and analyze data in real-time. This interconnectivity is particularly transformative for zero-waste initiatives, as it enables businesses to monitor and optimize resource use, track waste generation, and adjust processes to minimize environmental impact.

– **Real-Time Monitoring of Resource Consumption:** One of the primary challenges in achieving zero-waste systems is the lack of real-time data on resource consumption and waste generation. IoT sensors embedded within production equipment, supply chains, and waste management systems enable businesses to monitor resource use in real-time, providing invaluable insights into inefficiencies and areas for improvement. For instance, IoT-enabled smart meters can track energy consumption across manufacturing facilities, while machine sensors can monitor water usage and material waste. By analyzing this data, organizations can identify patterns of resource wastage and take corrective actions to optimize their operations. Real-time monitoring also allows businesses to detect equipment malfunctions or inefficiencies that may lead to excess resource consumption, enabling faster response times and minimizing downtime.

– **Optimizing Waste Collection and Sorting:** The IoT also plays a key role in waste management by optimizing the collection and sorting of waste materials. IoT sensors can be deployed in waste bins, smart trash compactors, and recycling stations to monitor the level of waste and determine the optimal time for collection. This eliminates the need for waste collection based on fixed schedules, reducing fuel consumption and emissions associated with unnecessary waste collection trips. IoT devices can help track the types and quantities of discarded materials, providing data to optimize waste sorting processes and improve recycling rates. By integrating IoT technology into the waste management process, municipalities and businesses can create a more efficient and sustainable system that ensures the maximum amount of materials are recycled, reused, or composted, minimising waste.

– **Data-Driven Decision-Making for Circular Economy:** The wealth of data generated by IoT devices can also drive smarter, more sustainable decision-making. By analyzing data collected from across supply chains, manufacturing processes, and waste management systems, companies can identify opportunities to adopt circular economy principles. For example, IoT data can reveal opportunities to reuse waste materials as raw inputs in other processes, leading to reduced waste generation and a more sustainable manufacturing model. This data-driven approach enhances operational efficiency and facilitates the implementation of zero-waste strategies in industries that traditionally have high waste outputs.

Artificial Intelligence (AI) for Predictive Analytics and Optimization: AI is critical in IT-driven process automation, optimizing resource use, minimizing waste, and enhancing efficiency. By leveraging machine learning algorithms, deep learning models, and advanced data analytics, AI systems can predict future trends, optimize workflows, and improve waste management strategies.

– **Predictive Analytics for Waste Reduction:** One of the most powerful applications of AI in zero-waste systems is predictive analytics. AI models can analyze historical data on resource consumption, production patterns, and waste generation to forecast future demand, optimize inventory levels, and predict when equipment will need maintenance. This predictive capability allows businesses to minimize waste by aligning production schedules with demand, reducing overproduction and excess inventory. For example, in a manufacturing setting, AI-powered predictive models can optimize the use of raw materials, ensuring that only the required amount is ordered based on real-time demand forecasts. This reduces the chances of surplus materials that may ultimately go to waste. Predictive maintenance algorithms can also forecast when machinery will require repairs, preventing equipment failures that could result in resource wastage or production downtime.

– **AI for Waste Stream Optimization:** AI algorithms can also optimize waste streams by identifying the most effective methods for sorting, processing, and recycling waste materials. By analyzing waste composition data and identifying patterns in waste generation, AI systems can recommend the best sorting strategies to maximize recycling rates and minimize landfill contributions. AI-powered robotics can also assist in automating waste sorting processes, using advanced machine learning techniques to recognize and categorize materials for recycling with high accuracy. AI can facilitate the development of new waste-to-resource technologies. For example, machine learning models can identify novel ways to repurpose industrial byproducts or find innovative uses for waste materials in other sectors, thus contributing to a circular economy and zero-waste manufacturing.

– **Smart Manufacturing Systems:** In smart manufacturing, AI and machine learning algorithms are essential in real-time process optimization. AI systems can dynamically adjust production processes by analyzing data from various sourcesIoT sensors, robotics, and production linesto minimize waste and improve efficiency. For instance, AI can optimize the raw material used in producing goods, ensuring no excess or shortfall. Additionally, AI systems can optimize energy consumption during production, reducing the carbon footprint of manufacturing operations.

Considerations in Implementing IT-Driven Automation: Despite the significant potential of IT-driven process automation through robotics, IoT, and AI, several challenges must be addressed to realize the full benefits of zero-waste systems.

– **High Initial Investment and Technological Infrastructure:** Adopting robotics, IoT, and AI technologies often requires substantial upfront investment in hardware, software, and infrastructure. Many businesses, tiny and medium-sized enterprises (SMEs), may struggle to afford the initial costs associated with implementing these technologies. Existing systems may need to be upgraded or replaced to accommodate new technologies, further increasing the financial burden.

– **Data Privacy and Security Concerns:** Data privacy and security concerns grow as organizations increasingly rely on IoT devices and AI algorithms. The interconnected nature of IoT devices creates potential vulnerabilities that malicious actors could exploit to gain access to sensitive data or disrupt operations. Businesses must invest in robust cybersecurity measures to safeguard their systems and ensure the integrity of the data they collect and analyze.

– **Skills Gap and Workforce Transformation:** The successful implementation of IT-driven process automation requires a skilled workforce in robotics, IoT, and AI technologies. However, many industries have a skills gap, as existing employees may not have the necessary expertise to operate, maintain, and optimize these advanced systems. Training and upskilling initiatives are essential to bridge this gap and ensure the workforce is prepared to handle the demands of an automated, zero-waste economy.

IT-driven process automation, mainly through the integration of robotics, IoT, and AI, offers immense potential to create zero-waste systems that are both efficient and sustainable. By enhancing precision, optimizing resource use, and improving waste management, these technologies contribute to realising a circular economy where waste is minimized, resources are reused, and environmental impact is reduced. While challenges such as high initial investment, data security, and workforce transformation must be addressed, the long-term benefits of these technologies in achieving sustainability goals are undeniable. As industries continue to evolve, the combination of robotics, IoT, and AI will play a pivotal role in reshaping the future of waste management, manufacturing, and overall resource optimization, driving economies toward a more sustainable and resilient future.

Case Study

Digitization has become a transformative force in reshaping industries, governance, and infrastructure in rapid technological evolution. As one of the world's largest and most diverse economies, India is witnessing an accelerated move towards digitalization in various sectors. This shift has significantly impacted the country's supply chain systems, the development of smart cities, and the redesign of processes across industries. Through the strategic application of information technology, India aims to enhance operational efficiency, promote sustainability, and improve overall governance. This case study explores the role of digitization in India's supply chain systems, the development of smart cities, and the redesign of business processes, focusing on the effectiveness, challenges, and outcomes of these digital transformations.

Digitization in India's Supply Chain Systems: India's supply chain systems, traditionally characterized by inefficiencies, fragmentation, and a lack of integration, have faced a substantial transformation in recent years due to digitization. With an economy that relies heavily on sectors like agriculture, manufacturing, and retail, efficient supply chain management is crucial to sustaining growth and meeting the demands of a fast-growing population.

– **Role of Technology in Supply Chain Optimization:** One of the key advancements in India's supply chain systems is the integration of digital technologies like the Internet of Things (IoT), blockchain, and Artificial Intelligence (AI). These technologies address critical challenges such as inventory management, transportation inefficiencies, and demand forecasting. For instance, IoT sensors are being used to track goods in real-time, providing accurate data on the location, condition, and expected delivery time of products. This real-time tracking reduces delays, improves delivery timelines, and enhances customer satisfaction.

With its transparency and traceability, blockchain technology is gaining traction in India's supply chain systems. By recording every transaction in a decentralized ledger, blockchain enhances trust among stakeholders and reduces fraud. In sectors like agriculture, where supply chains are often complex and involve multiple intermediaries, blockchain can provide an immutable record of transactions, ensuring that goods are sourced ethically and delivered most efficiently. AI and machine learning algorithms are also being employed to optimize inventory management. By predicting demand patterns using historical data, AI tools enable businesses to maintain optimal stock levels, reducing excess inventory or stockouts. These technologies significantly reduce the cost of inventory management while also contributing to sustainability by minimizing waste and preventing overproduction.

E-NAM (National Agriculture Market): A notable example of digital transformation in India's supply chain is implementing the National Agriculture Market (e-NAM) initiative. Launched by the Government of India, e-NAM is a pan-India digital platform that connects farmers with wholesale buyers, helping streamline the agricultural supply chain. By using this platform, farmers can access a wider market and receive real-time price information, enabling them to make better decisions about when and where to sell their produce.

e-NAM has helped mitigate the issues of middlemen exploitation, delayed payments, and poor pricing transparency. It provides a nationwide centralized database of agricultural produce, allowing farmers to access information about various crops' quality, quantity, and pricing. The digital platform integrates payment systems, ensuring prompt and transparent transactions. By digitizing the agricultural supply chain, e-NAM has significantly improved the livelihood of farmers while promoting sustainable farming practices and reducing food waste.

Smart Cities Integrating Digital Technologies for Urban Transformation: India's ambitious Smart Cities Mission, launched in 2015, aims to create 100 cities that use digital technology to improve the quality of life for citizens. This initiative seeks to integrate smart infrastructure, digital governance, and sustainable development practices, enhancing urban living through innovative solutions.

Smart Infrastructure: IoT, Sensors, and Automation: Smart cities rely heavily on IoT technologies to connect and automate various urban systems. For instance, smart meters and sensors are deployed in critical infrastructure such as water supply, electricity grids, and waste management systems to monitor usage and optimize resource allocation. IoT-enabled streetlights adjust their brightness based on the time of day or the presence of pedestrians, saving energy while enhancing public safety. Similarly, smart waste management systems use sensors in trash bins to monitor fill levels and schedule collection only when necessary, reducing the number of unnecessary collection trips and fuel consumption. AI and machine learning algorithms further optimize urban processes by analyzing data collected from IoT devices. In traffic management, AI models predict traffic patterns and adjust traffic signals in real-time to ease congestion. This improves traffic flow and reduces fuel consumption and air pollution. In energy management, AI systems predict peak consumption times and adjust energy supply accordingly, ensuring a reliable power supply while minimizing wastage.

Case Study

One of the successful examples of India's Smart Cities Mission is Ahmedabad, which has implemented several smart technologies to improve the quality of life for its citizens. The city has focused on smart traffic management, waste management, and water conservation to address its growing urban challenges. Regarding smart traffic management, Ahmedabad has deployed a network of IoT sensors and cameras at key intersections to monitor traffic flow in real-time. AI-powered traffic control systems analyze this data to adjust signal timings, preventing bottlenecks and reducing congestion. The city has also implemented an intelligent parking management system, helping citizens find available parking spots through a mobile app. This reduces the time spent searching for parking, reducing fuel consumption and traffic congestion.

Ahmedabad has also introduced smart waste management solutions, including smart bins that alert authorities when they are full. The city's water supply system has been enhanced through IoT sensors, which monitor water levels and detect leaks, preventing water wastage. These initiatives have improved operational efficiency and contributed to sustainability by optimizing resource use and reducing environmental impact.

Process Redesign – Leveraging Digital Technologies for Efficiency: The digitization of processes extends beyond supply chains and urban infrastructure to redesign business processes across sectors. Process redesign focuses on automating, streamlining, and improving workflows to reduce inefficiencies, cut costs, and enhance service delivery. In India, various sectors such as healthcare, education, manufacturing, and governance have embraced IT-driven process redesign to improve service delivery and customer satisfaction.

Digital Governance: E-governance initiatives in India, such as the Digital India program, have transformed public sector service delivery. By digitizing key government services and processes, the government has improved accessibility, efficiency, and transparency. For example, implementing the Goods and Services Tax (GST) system has streamlined tax collection, making it easier for businesses to comply with tax regulations. The system uses digital platforms for invoicing, payment, and filing, reducing delays and fraud while improving revenue generation.

Another significant e-governance initiative is the Public Distribution System (PDS) transformation, where ration cards are linked to biometric data through Aadhaar, ensuring that benefits reach the intended recipients. This digitization reduces inefficiencies, curtails corruption, and ensures that food security programs are more effective in assisting vulnerable populations.

Case Study: The implementation of the Goods and Services Tax (GST) is an excellent case study of process redesign through digitization in India. GST replaced multiple indirect taxes with a single, unified tax structure, streamlining tax collection and reducing the burden on businesses. Introducing a nationwide digital platform for tax filing and payment significantly reduced administrative costs, minimized errors in tax assessments, and improved compliance. Through the GST system, businesses can upload invoices, track their tax obligations, and receive refunds efficiently, ensuring a seamless flow of goods and services across the country. The GST system's digital nature makes it easier to monitor cross-border transactions, reducing tax evasion and improving government revenue collection.

Opportunities in Digitization: While India's digitization initiatives in supply chains, smart cities, and process redesign have demonstrated substantial progress, they are not without challenges. One of the primary hurdles is the digital divide, with significant disparities in internet access and digital literacy between urban and rural areas. In rural India, where a large portion of the population remains disconnected from digital services, the full benefits of these digital initiatives may not be realized. Data privacy and security concerns are central to successfully implementing these digital systems. As more data is collected and shared across platforms, robust cybersecurity measures are crucial to protect sensitive information and prevent breaches. However, these challenges also present opportunities for innovation. The government and private sector are working to improve internet connectivity, particularly in rural areas, through initiatives like BharatNet, which aims to provide high-speed broadband to remote areas. Additionally, a growing emphasis is on improving digital literacy so citizens can fully engage with and benefit from digital platforms.

Digitization reshapes India's supply chain systems, smart cities, and process workflows, creating a more efficient, transparent, and sustainable future. Through technologies like IoT, AI, blockchain, and robotics, India is moving toward a digital ecosystem that enhances productivity, optimizes resource use, and improves governance. The case studies of e-NAM, Ahmedabad Smart City, and GST implementation illustrate the potential of these digital initiatives to revolutionize industries and public services. As India continues to embrace digital transformation, overcoming challenges such as the digital divide, data security, and infrastructure limitations will ensure that these innovations benefit all sectors of society. By continuing to invest in digital infrastructure, India has the opportunity to build a more resilient, sustainable, and equitable future.

Future Scenario

As the world faces the pressing challenges of resource depletion, environmental degradation, and rising waste generation, industries and economies increasingly turn to technology-driven solutions to address these issues. India's aspirations for sustainable development, the future will see the integration of advanced technologies like Artificial Intelligence (AI), robotics, and Digital Twin technology to revolutionize industries, businesses, and environmental systems. Combined with principles of the circular economy and zero-waste systems, these technologies offer transformative opportunities to optimize resource use, minimize waste, and create sustainable, resilient systems. This detailed exploration envisions how autonomous processes powered by AI and robotics and circular economy processes enabled by digital twin technology will shape India's future landscape for sustainability, efficiency, and economic growth.

Autonomous Processes Driven by AI and Robotics for Zero-Waste Systems: Zero-waste systems aim to eliminate waste by ensuring every product is reused, repaired, or recycled into new products, significantly reducing the environmental impact of industries and human activities. Achieving such systems will require integrating cutting-edge technologies that allow for autonomous processessystems that operate without human intervention. These processes will not only optimize the use of resources but also ensure that production, consumption, and disposal happen sustainably.

Role of AI and Robotics in Automating Waste Management: Artificial Intelligence, combined with robotics, will enable smart waste management systems capable of automatically sorting, recycling, and processing waste in real-time. In India's context, where waste management remains one of the most significant urban challenges, integrating AI and robotics into waste management practices can lead to highly efficient and sustainable outcomes. AI algorithms can be used to analyze vast amounts of data from sensors and cameras embedded in waste management systems to determine the composition of waste materials. This enables automated waste sorting, ensuring that recyclable materials such as plastics, metals, and paper are efficiently separated from non-recyclable waste.

Robotics will further enhance this process by performing physical tasks such as separating, compacting, and even autonomously processing waste. For instance, robotic arms can be deployed in recycling facilities to sort waste at high speeds, reducing the need for manual labor and human involvement in dangerous conditions. With AI systems continuously learning from waste data, these systems can improve over time, adapting to new materials and changing waste patterns, optimizing both cost and environmental impact. In addition to recycling, robotics will play a role in the production and manufacturing processes, enabling industries to operate in a zero-waste manner. AI-powered robots will analyze production data to identify inefficiencies, minimize waste generation, and enhance resource utilization in real-time. By integrating AI into production systems, businesses can ensure that raw materials are used to their maximum potential while waste generated in the manufacturing process is minimized or repurposed.

Autonomous Waste-to-Energy Systems: Beyond waste sorting and recycling, autonomous waste-to-energy (WTE) technologies will play a crucial role in zero-waste systems. These processes involve converting non-recyclable waste into energy, reducing waste in landfills while simultaneously generating power. Robotics and AI will optimise WTE systems by continuously monitoring, adjusting, and improving the process to ensure the most efficient waste conversion into energy. AI algorithms will optimize combustion or gasification processes, while robotics can automate the maintenance and operation of complex WTE facilities. Likewise, AI-powered autonomous systems will enable decentralized energy generation from waste. Small-scale, autonomous waste-to-energy plants can be deployed in rural or remote areas where centralized waste management infrastructure is often lacking. These plants can be fully monitored and managed through AI-based systems that adjust to changing waste inputs, ensuring optimal energy production and reducing reliance on traditional, non-renewable energy sources.

Circular Economy Processes Enabled by Digital Twin Technology: A circular economy, where resources are kept in use for as long as possible, is central to long-term sustainability. Digital Twin technology, which involves creating digital replicas of physical assets, processes, or systems, holds significant potential to facilitate the shift toward a circular economy by enabling real-time monitoring, predictive analytics, and optimization of resource flows. By replicating the entire lifecycle of products and materials, digital twins allow for a comprehensive understanding and management of resource use, waste production, and the potential for recycling or reusing materials.

Digital Twin Technology for Sustainable Product Design: One of the key challenges in the traditional linear economy is the disposal of products at the end of their life cycle. In a circular economy, products are designed for longevity, repairability, and eventual recycling. Digital Twin technology is poised to play a transformative role in product lifecycle management by enabling the design and development of products that can be easily disassembled, repaired, or upcycled. By creating digital replicas of products and simulating their behavior in different usage conditions, companies can optimize designs for durability and minimize resource waste during production. For example, in the automotive industry, Digital Twin technology can simulate the performance of a vehicle's components and predict wear and tear, enabling manufacturers to design cars that require fewer repairs and are more easily recycled at the end of their life. This would reduce the need for raw materials, lower emissions, and help minimize the environmental footprint of the automotive sector. Similarly, in electronics, digital twins can be used to simulate and test how devices degrade over time, enabling manufacturers to create products that are more energy-efficient and easier to recycle.

Digital Twin Technology for Supply Chain and Resource Management: Digital Twin technology can also be applied to the supply chain, enabling businesses to create virtual representations of their entire resource flow. This allows companies to simulate how raw materials move through the supply chain, track resource usage, and identify inefficiencies. By utilising digital twins, businesses can optimize resource allocation, reduce waste, and improve material recycling. For instance, by simulating the logistics of transporting goods, digital twins can predict delays, optimize routes, and enhance the efficiency of delivery systems, reducing fuel consumption and emissions. Digital twins enable the tracing of materials throughout their entire lifecycle, from extraction to disposal, recycling, or repurposing for circular economy. By creating a closed-loop system, businesses can monitor when and how materials can be reused or recycled, reducing waste and promoting the efficient use of resources. For instance, a product's digital twin might track its use and predict when it will need maintenance or replacement, allowing for timely interventions to prolong its lifespan. This data can be used to design better recycling programs, where products or components are collected and sent for remanufacturing or reuse.

Real-Time Monitoring and Feedback Loops: Digital Twin technology, when combined with IoT sensors and AI, enables real-time monitoring of resource flows in circular economy systems. By continuously collecting data from physical assets, digital twins can provide up-to-date insights into how resources are used and whether any inefficiencies or waste are occurring. This data can be fed back into the system to trigger automatic adjustments in the manufacturing process, logistics, or waste management systems, ensuring that resources are used as efficiently as possible. Industries can create more agile and responsive circular economy systems that adapt quickly to changing conditions. For instance, in the energy sector, a digital twin of an energy grid can monitor energy consumption and flow from renewable sources, ensuring that the grid optimizes clean energy and minimizes reliance on non-renewable sources. Similarly, in the waste management sector, a digital twin can track the flow of materials through recycling centers, optimizing sorting processes and reducing contamination rates to improve the efficiency of recycling operations.

Case Study

Smart cities, as part of India's ongoing urban transformation, are ideal environments for the application of digital twin technology. Cities like Chandigarh, Pune, and Bengaluru have already started to incorporate IoT-based smart solutions for urban management, and the next step is to integrate these systems into a comprehensive digital twin framework. In the case of waste management, for example, integrating IoT sensors, AI, and digital twins would enable real-time monitoring of waste generation, collection, and disposal, ensuring that resources are used effectively and that waste is minimized. These data-driven insights could allow municipalities to optimize collection routes, improve recycling rates, and create better feedback loops for citizen engagement. Digital twins can also enhance the circular economy in smart cities by facilitating the sharing of resources. Platforms powered by digital twins can track the availability and usage of shared resources such as energy, water, and public transport, allowing cities to optimize their infrastructure while ensuring equitable access. This would align with the broader goals of sustainable urban development by reducing waste and improving citizens' overall quality of life.

The future scenario of autonomous processes driven by AI and robotics for zero-waste systems and circular economy processes enabled by digital twin technology presents a promising vision for India's sustainable development. By leveraging these technologies, industries, municipalities, and governments can optimise resource use and promote environmental stewardship and economic resilience. While challenges such as infrastructure gaps, digital literacy, and the need for regulatory frameworks remain, the potential of these advanced technologies to drive significant change in waste management, production processes, and urban development cannot be overstated. As India continues to embrace digital transformation, integrating AI, robotics, and digital twin technology will play a crucial role in creating a more sustainable, circular economy that benefits businesses, consumers, and the environment.

6

IT-Enhanced Sustainable Products

Introduction

In global environmental challenges, sustainable products have gained considerable prominence. Sustainable products are those designed, manufactured, and disposed of in ways that minimize environmental impact, promote resource efficiency, and contribute to a circular economy. Integrating information technology (IT) into sustainable products' design, production, and lifecycle management has emerged as a crucial enabler of innovation and efficiency in achieving environmental sustainability. IT-enhanced sustainable products leverage technologies such as Artificial Intelligence (AI), the Internet of Things (IoT), blockchain, and data analytics to reduce waste, optimize resource use, extend product lifecycles, and enhance transparency. By incorporating IT into the lifecycle of products, manufacturers and designers can make smarter decisions that have far-reaching benefits for both the environment and the economy. This detailed analysis will explore how IT contributes to creating sustainable products, focusing on product design, resource optimization, manufacturing processes, consumer engagement, and end-of-life management.

IT in Product Design: Creating Sustainable Products from the Start: The lifecycle of a product begins at the design stage, where decisions related to materials, energy consumption, and overall environmental impact are made. In traditional product design, considerations often focus solely on functionality and cost-efficiency, with limited regard for environmental consequences. However, with IT tools such as AI and Computer-Aided Design (CAD) software, manufacturers can now design products with sustainability in mind from the beginning. AI-powered design tools can analyze vast datasets, simulate environmental impacts, and recommend materials and processes that minimize energy consumption and waste generation throughout the product's life cycle.

AI and machine learning algorithms can help designers identify patterns in material use and predict the environmental impact of different design choices. By incorporating sustainability criteria such as energy efficiency, recyclability, and renewable materials into the design phase, AI tools allow for optimising product performance with minimal environmental costs. For example, in the apparel industry, AI is being used to design clothing that requires fewer resources to produce, incorporates sustainable fibers, and is easily recyclable. Similarly, in the automotive industry, AI is utilized to optimize car designs, reducing weight and enhancing fuel efficiency while minimizing the environmental impact of manufacturing.

Resource Optimization with IoT and Data Analytics: The Internet of Things (IoT) and data analytics provide manufacturers with real-time insights into how resources are used in the production and use phases of a product's life cycle. IoT sensors embedded in products or manufacturing processes can track variables such as energy consumption, material usage, and emissions, providing data that can be analyzed to identify inefficiencies and areas for improvement. By collecting and analyzing this data, companies can optimize the use of raw materials, minimize waste, and improve energy efficiency in real-time.

For instance, in smart appliances, IoT sensors allow products like refrigerators, washing machines, and HVAC systems to monitor energy consumption and optimize their performance to reduce electricity usage. These products can also communicate with other devices, enabling users to adjust settings remotely for maximum efficiency. Data analytics platforms can aggregate the data from these devices and provide manufacturers and consumers with actionable insights to reduce environmental impact and lower costs. IoT and data analytics can be applied in supply chain management to track the movement of materials and goods, ensuring that resources are used efficiently and waste is minimized throughout the entire supply chain.

Manufacturing Processes – Optimizing Efficiency and Reducing Waste: IT-enhanced manufacturing processes, often called Industry 4.0, play a critical role in developing sustainable products by optimizing production processes to minimize waste and energy consumption. Technologies like AI, robotics, and 3D printing are revolutionizing manufacturing by improving precision, reducing resource consumption, and minimizing the environmental footprint of industrial operations.

AI-driven automation in manufacturing systems allows for more efficient and precise production runs, where materials are used optimally and waste is minimized. AI algorithms can predict the optimal quantity of raw materials required for each production cycle, ensuring that excess waste is avoided. Robotics further enhances this process by performing tasks with high precision and speed, reducing errors and material waste. For example, in electronics manufacturing, robots can assemble components with high precision, producing less defective output and reducing the wastage of materials like soldering components and semiconductors. Additive manufacturing, or 3D printing, has become a powerful tool for creating sustainable products. 3D printing uses only the material necessary to create an object, eliminating the waste typically generated through traditional subtractive manufacturing processes. This technology allows for the creation of complex geometries that would be difficult or impossible to achieve through conventional manufacturing methods, reducing material usage while enhancing product design flexibility. Industries such as aerospace, automotive, and healthcare are already adopting 3D printing to manufacture lightweight, durable parts with less material waste, ultimately contributing to more sustainable production methods.

Consumer Engagement and Smart Products: IT is also increasingly important in promoting consumer engagement and encouraging sustainable behaviors. Smart products incorporating IoT and AI provide consumers with real-time data on their usage patterns, allowing them to make informed decisions about optimising their energy consumption and reducing waste. For instance, smart thermostats can adjust heating or cooling based on a user's preferences and habits, ensuring energy is used efficiently while maintaining comfort. Similarly, smart lighting systems can adjust brightness based on the time of day or occupancy, reducing energy consumption.

The integration of sustainability information into consumer-facing products has the potential to drive widespread adoption of green products. By leveraging digital platforms, companies can educate consumers about the environmental benefits of their products and empower them to make sustainable choices. For instance, mobile apps that track the carbon footprint of a product or suggest environmentally-friendly alternatives are gaining popularity. Similarly, blockchain technology is being utilized to create transparent supply chains, allowing consumers to verify the sustainability credentials of the products they purchase, such as whether they were made using renewable energy or are recyclable at the end of their life cycle.

End-of-Life Management and Circular Economy: The concept of circularity, which emphasizes the reuse, recycling, and repurposing of products and materials, is a core aspect of sustainable product design. IT plays a crucial role in managing the end-of-life phase of products by enabling manufacturers and consumers to track products throughout their life cycle. Digital technologies such as blockchain, IoT, and AI facilitate efficient collection, sorting, and recycling of products once they end their useful life. For example, blockchain technology allows for the creating transparent and traceable systems for managing end-of-life products. By embedding blockchain in product packaging or tagging systems, manufacturers can track products and materials as they are returned for recycling or remanufacturing, ensuring that circular economy principles process them. Additionally, IoT sensors embedded in products enable tracking of wear and tear, helping consumers and manufacturers assess when products should be repaired, reused, or recycled, further promoting circularity.

In electronics industry, where e-waste is a significant concern, IT-driven solutions for end-of-life management are particularly important. AI algorithms are being developed to optimize the dismantling and recycling of electronic products, ensuring that valuable materials such as precious metals and rare earth elements are recovered and reused. Similarly, IoT and data analytics can track the quantity and composition of e-waste, enabling more efficient and effective recycling programs.

Case Studies: Several companies have already adopted IT-driven solutions to create sustainable products, demonstrating the effectiveness of these technologies in driving environmental and economic sustainability. One example is the use of AI and IoT in producing smart home appliances. Companies like Bosch and Philips are incorporating AI into home appliances such as washing machines and dishwashers, enabling them to optimize water and energy usage based on real-time data from sensors. These appliances help consumers reduce their environmental impact and enhance product longevity through predictive maintenance, which allows consumers to know when a product needs repair or replacement. In the automotive industry, companies like Tesla are leveraging AI and robotics to design electric vehicles that are more sustainable than traditional gasoline-powered cars and optimized for performance, energy efficiency, and recyclability. Tesla's AI-powered manufacturing system helps streamline production processes, ensuring minimal waste generation and reducing the carbon footprint of manufacturing. The company's vehicles also integrate IoT and AI to enhance energy efficiency, providing drivers with insights into their energy consumption patterns and enabling optimal battery management for longer vehicle lifespans.

IT-enhanced sustainable products represent a critical innovation in transitioning to a more sustainable, circular economy. By leveraging AI, IoT, blockchain, and data analytics, businesses can design, manufacture, and manage products that optimize resource use, reduce waste, and empower consumers to make more sustainable choices. These technologies contribute to environmental sustainability and create economic opportunities by improving efficiency, reducing costs, and opening new markets for sustainable products. As India and other countries prioritize sustainable development, integrating IT into product design, manufacturing, and lifecycle management will be essential to achieving long-term environmental goals. The future of sustainable products lies in the ability to harness technology to create more efficient, transparent, and circular systems that minimize waste, reduce carbon footprints, and enhance resource utilization.

Current Scenario

In the evolving landscape of sustainable manufacturing, eco-design tools and lifecycle analysis (LCA) software have emerged as key technological solutions that enable businesses to reduce their environmental impact. The primary objective of eco-design is to optimize the entire product lifecycle from raw material extraction to product use and eventual disposal by reducing resource consumption, minimizing waste, and improving energy efficiency. These tools allow companies to make more informed decisions during the design phase and ensure that sustainability is integrated into the core of their manufacturing processes. Similarly, lifecycle analysis (LCA) software comprehensively evaluates a product's environmental performance by quantifying the ecological footprint over its entire lifecycle. However, while these tools hold tremendous potential for transforming manufacturing practices towards sustainability, their adoption faces numerous challenges. These challenges range from technical and financial barriers to organizational resistance and a lack of standardized methodologies. This section explores the role of eco-design tools and LCA software in sustainable manufacturing, highlighting the barriers preventing widespread adoption.

Eco-Design Tools – Enabling Sustainable Product Development: Eco-design tools are crucial in guiding designers to create resource-efficient, recyclable, and less polluting products. These tools provide decision-makers with data that help identify opportunities to reduce energy use, emissions, and material waste in the design phase. Software such as *GaBi*, *SimaPro*, and *SolidWorks Sustainability* allows designers to simulate different design options and assess their environmental impact before production begins. By providing quantitative insights into material selection, energy consumption, and waste generation, eco-design tools enable designers to make decisions that significantly reduce a product's carbon footprint.

Eco-design tools are particularly valuable for industries that rely heavily on resource-intensive processes, such as electronics, automotive, and consumer goods. For example, in the electronics industry, eco-design tools can be used to select materials that are not only low-impact but also recyclable, helping reduce the environmental consequences of e-waste. In the automotive sector, these tools enable the design of vehicles that optimize material use, reduce fuel consumption, and incorporate renewable energy sources. Through detailed simulations, eco-design tools also allow designers to identify potential problems with product durability and performance, which could lead to early failure and increased waste. These tools are increasingly integrated with other IT solutions, such as Product Lifecycle Management (PLM) systems, which help streamline product development. By integrating eco-design tools with PLM, manufacturers can align sustainability goals with operational efficiency and ensure that environmental considerations are woven into every production stage. Eco-design tools are not just about minimizing negative environmental impacts but are also integral to creating value through innovation and market differentiation.

Lifecycle Analysis Software: Lifecycle analysis (LCA) is another pivotal approach to sustainability that involves assessing the environmental impacts associated with all stages of a product's life, from raw material extraction to disposal. LCA software, such as *GaBi*, *SimaPro*, and *OpenLCA*, is designed to help businesses measure and reduce the environmental footprint of their products. These software tools evaluate multiple environmental factors, including energy consumption, greenhouse gas emissions, water usage, waste generation, and resource depletion, by compiling data from each phase of the product lifecycle.

LCA software helps companies identify "hotspots" areas of the production process where the greatest environmental impact occurs. For example, in the production of a smartphone, an LCA might reveal that the extraction of rare earth metals, such as lithium and cobalt, has a disproportionate environmental impact compared to other stages of the product's lifecycle, like assembly or distribution. With this knowledge, companies can focus their sustainability efforts on these critical points, such as responsibly sourcing raw materials or developing recycling programs to recover valuable materials. Additionally, LCA software helps companies meet industry standards and regulations by providing transparent, quantifiable data that can be used in environmental certification schemes, such as ISO 14001 or the EU Ecolabel. LCA software helps businesses communicate their sustainability efforts to consumers, investors, and regulatory bodies by providing verifiable data on a product's environmental performance. As global demand for transparency and sustainability in product manufacturing rises, companies that leverage LCA tools gain a competitive edge by demonstrating their commitment to minimizing their ecological footprint.

Defies in Adopting IT-Driven Sustainable Manufacturing: While eco-design tools and LCA software can significantly enhance manufacturing processes' sustainability, several challenges hinder their widespread adoption. These barriers span technical, financial, organizational, and regulatory domains, and overcoming them is crucial for advancing the transition toward more sustainable manufacturing practices.

- **High Initial Costs:** The financial barrier is one of the most significant obstacles to adopting IT-driven sustainable manufacturing tools. Acquiring eco-design and LCA software can be prohibitive, especially for small and medium-sized enterprises (SMEs) that may not have the capital to invest in high-end tools. Integrating these tools into existing workflows and systems often requires additional training, infrastructure, and ongoing support investment. For many companies, particularly those operating on tight margins, the upfront cost of implementing these technologies outweighs the perceived benefits, leading them to delay or avoid adoption. However, the costs associated with conducting lifecycle assessments can be exceptionally high due to the need for detailed, accurate data across the entire product lifecycle. Gathering data on everything from raw material sourcing to end-of-life disposal requires significant time and resources, which can deter businesses from investing in LCA analysis. While some software solutions offer more affordable options for smaller enterprises, the overall financial burden remains a substantial barrier to widespread adoption.
- **Complexity and Expertise Requirements:** The complexity of eco-design tools and LCA software presents another challenge. These tools require

specialized knowledge and expertise, and the learning curve for mastering these technologies can be steep. For instance, LCA software involves complex calculations, data interpretation, and modeling that necessitate expertise in environmental science, engineering, and sustainability practices. Additionally, eco-design tools often require a deep understanding of product manufacturing processes, material properties, and the environmental impact of various production methods.

For many organizations, especially SMEs or those in emerging markets, the lack of skilled personnel trained in sustainability tools and techniques is a significant barrier to implementation. The need for specialized expertise also translates into increased operational costs, as companies may need to hire consultants or invest in internal training programs to utilize these technologies effectively. This complexity can discourage companies from adopting eco-design and LCA software, particularly when the immediate return on investment is unclear or difficult to quantify.

- **Fragmentation of Digital Systems:** The integration of eco-design and LCA tools into existing manufacturing processes is often hindered by the fragmentation of digital systems within organizations. Many manufacturers use legacy systems incompatible with modern sustainability tools, creating technical data sharing and integration challenges. For example, integrating eco-design tools with existing Product Lifecycle Management (PLM) systems or Enterprise Resource Planning (ERP) software can be difficult, as these systems often use different formats or databases. This lack of interoperability increases the complexity of adopting IT-driven sustainability practices and slows the transition toward more sustainable manufacturing. Companies frequently lack standardized data collection and reporting processes, making integrating sustainability data across various departments and functions challenging. To fully realize the potential of eco-design and LCA tools, businesses must invest in IT infrastructure that facilitates seamless data sharing and real-time monitoring across different product lifecycle stages This infrastructure, however, can be costly to develop and maintain, particularly for small enterprises with limited resources.

- **Data Quality and Availability:** One of the core challenges in implementing effective LCA and eco-design processes is the lack of high-quality, accurate, and accessible data. The effectiveness of these tools depends heavily on the availability of detailed data on raw materials, energy consumption, emissions, and waste generation across all stages of the product lifecycle. Datatal impacts are incomplete, and inconsis in many industries. Gathering the necessary data often requires companies to collaborate with suppliers, contractors, and other stakeholders, which can be complex and time-consuming. The absence of standardized data formats and reporting frameworks further exacerbates this issue. For instance, manufacturers may have access to sustainability data from their suppliers but struggle to integrate it into their systems due to differences in data formats or measurement criteria. This lack of reliable data creates uncertainties in decision-making, making it difficult for companies to fully evaluate their products' environmental impacts and identify improvement opportunities.

- **Regulatory and Market Barriers:** The regulatory landscape surrounding sustainability practices remains fragmented and inconsistent across regions, adding another layer of complexity to adopting IT-driven sustainability tools. While some countries have implemented regulations and standards to promote sustainable manufacturing practices, these standards vary significantly in scope, enforcement, and requirements. The lack of a unified global framework for sustainability in manufacturing can create confusion for businesses operating in multiple markets, complicating efforts to adopt standardized eco-design and LCA practices. Also, market demand for sustainable products often remains insufficient to justify the significant investments required for adopting these technologies. While there is growing consumer interest in environmentally responsible products, the willingness to pay a premium for such products is still limited in many markets. This can make it challenging for businesses to justify the financial costs of adopting eco-design tools and LCA software, particularly when the return on investment is not immediate or tangible.

Eco-Design and Product Lifecycle Analysis*:* Tools for Designing Sustainable Products

In recent decades, the demand for sustainable products has grown exponentially, driven by increasing consumer awareness, regulatory pressures, and the urgency to address environmental challenges such as climate change, resource depletion, and pollution. In response, eco-design and product lifecycle analysis (LCA) have emerged as fundamental methodologies to guide the design and manufacturing of products that minimize their environmental impact over their entire lifecycle. Eco-design involves integrating environmental considerations into the product design process, aiming to reduce the ecological footprint of products from raw material extraction to disposal. Product lifecycle analysis, on the other hand, provides a comprehensive tool for evaluating the environmental performance of a product throughout its lifecycle.

Eco-Design – Holistic Approach to Sustainability: Eco-design encourages designers and engineers to consider the environmental impacts of a product from the very beginning of the design process rather than as an afterthought. The goal is to create environmentally friendly, cost-effective, and socially responsible products throughout their lifecycle. The principles of eco-design focus on optimizing resource use, minimizing waste, reducing energy consumption, and enhancing recyclability. By applying these principles, eco-design reduces the environmental footprint of products and contributes to a circular economy, where products and materials are reused, refurbished, and recycled rather than disposed of. Eco-design tools are integral to this process, providing designers with the data and capabilities needed to make informed decisions about product design. These tools range from simple calculators for assessing material choices to complex software integrating environmental criteria into the design process. Some of the most widely used eco-design tools include the *Design for Environment (DFE)*, *SolidWorks Sustainability*, and *Eco-Design Studio*. These tools allow designers to simulate different product configurations, evaluate material choices, assess energy use, and estimate the product's overall environmental impact.

One of the key aspects of eco-design is the focus on energy efficiency. For example, in designing electronic products, eco-design tools can help identify low-energy components, such as energy-efficient processors or LED screens, to reduce the product's energy consumption during its use phase. Similarly, these tools can assist in identifying materials that are both sustainable and economically viable, such as biodegradable plastics or recycled metals. Eco-design tools can help manufacturers minimize waste by recommending easier disassembling product designs for recycling or reuse. Eco-design also encourages the use of renewable materials and the reduction of hazardous substances in products. By integrating eco-design principles into the design phase, manufacturers can significantly reduce the environmental impact of their products, creating a competitive advantage in an increasingly environmentally-conscious market. In addition to reducing negative environmental impacts, eco-design helps organizations comply with sustainability standards, certifications, and regulations, enhancing their reputation and marketability.

Product Lifecycle Analysis (LCA – Comprehensive Assessment Tool: Product lifecycle analysis (LCA) is a quantitative tool used to evaluate the environmental impacts of a product throughout its entire lifecycle from raw material extraction, manufacturing, and distribution to use and end-of-life disposal or recycling. LCA considers various environmental factors, including resource depletion, energy consumption, greenhouse gas emissions, water usage, and waste generation. LCA provides a comprehensive understanding of a product's environmental footprint, identifying hotspots or areas where significant environmental impacts occur and offering insights into how these impacts can be mitigated.

LCA can be applied at various stages of product development, from early-stage design to post-production, to help companies assess potential environmental trade-offs and optimize product designs. This tool allows manufacturers to evaluate multiple product design alternatives, assessing the most environmentally friendly design while maintaining performance, durability, and cost-effectiveness. By providing an evidence-based evaluation of a product's environmental impact, LCA also helps companies make more sustainable material, energy, and manufacturing decisions. LCA software, such as *SimaPro*, *GaBi*, and *OpenLCA*, has made it easier for manufacturers to conduct detailed assessments. These software platforms can analyse a product's environmental impact using life cycle inventory (LCI) data, including resource usage, emissions, and waste generation information. They also enable companies to model the environmental performance of products under different scenarios and compare the outcomes to inform decision-making.

LCA helps manufacturers understand the cumulative environmental impact of a product, identifying inefficiencies and potential improvements. For example, LCA could show that the manufacturing phase of a product generates significant emissions, while transportation or end-of-life disposal may have a relatively lower impact. With this knowledge, manufacturers can prioritize strategies to reduce emissions during manufacturing or implement more sustainable packaging to minimize the transportation footprint. In some cases, the analysis may also reveal that the use phase of a product has the most substantial environmental impact, prompting designers to focus on energy efficiency or extend product durability. The results of an LCA are typically presented in the form of an environmental impact assessment report, which helps companies meet regulatory requirements and obtain certifications such as ISO 14001 or the EU Ecolabel. LCA helps organizations communicate their sustainability efforts transparently to stakeholders, customers, and investors, providing credibility to their claims of environmental responsibility.

Tasks in Implementing Eco-Design and LCA in Product Development: While eco-design and LCA tools offer valuable support for creating sustainable products, their widespread adoption faces several challenges. The first challenge is related to these tools' complexity and data requirements. Effective LCA requires comprehensive and accurate data on materials, energy consumption, and emissions, which can be difficult to obtain. In many cases, the data required for LCA may not be available, especially when sourcing materials from a global supply chain with limited transparency. This data gap can hinder the accurate assessment of environmental impacts, leading to incomplete or inaccurate results.

The second task is the cost of implementing eco-design and LCA tools. For small and medium-sized enterprises (SMEs), the upfront costs associated with acquiring eco-design and LCA software, training personnel, and implementing these tools across the product development process can be significant. Many SMEs may also lack the technical expertise to interpret LCA results effectively or integrate them into their design processes. Larger companies with more resources may have an easier time adopting these tools, but SMEs may struggle to justify the costs if the environmental benefits are not immediately apparent or financially quantifiable. Integrating eco-design and LCA into existing product development processes can also be challenging. Many traditional manufacturing systems are not designed with sustainability in mind, and it can be difficult to shift organizational practices toward sustainability without a fundamental change in company culture. Employees may resist adopting new tools or workflows, and the lack of sustainability-focused leadership within an organization can lead to fragmented efforts that fail to create meaningful impact.

Another task is the complexity of managing trade-offs between sustainability and other product attributes, such as cost, performance, and marketability. While eco-design and LCA tools provide valuable insights into environmental impacts, they may also reveal trade-offs that could affect product quality, cost, or consumer acceptance. For instance, selecting a more sustainable material may increase production costs, while improving recyclability might compromise a product's durability. Balancing these competing priorities requires careful decision-making, and in some cases, the environmental benefits of eco-design and LCA may not outweigh the financial or functional challenges. Finally, the lack of industry-wide standardization in eco-design and LCA methodologies can complicate comparisons between different products or designs. While several standards, such as ISO 14040 and ISO 14044, provide guidelines for conducting LCA, these standards are not always universally applied, leading to inconsistencies in how environmental impacts are calculated. The lack of standardization in eco-design approaches and LCA methods makes comparing products across or even within the same industry difficult.

Opportunities and Way Forward: Despite the challenges, significant opportunities exist to enhance the adoption of eco-design tools and LCA in sustainable product development. One promising avenue is the increasing availability of cloud-based software solutions, which lower the barriers to entry by providing cost-effective, scalable tools for small businesses and startups. These tools offer affordable subscription models and enable businesses to access powerful eco-design and LCA capabilities without requiring significant upfront investment. These cloud-based tools often feature user-friendly interfaces that reduce the complexity of implementing sustainability practices.

Another opportunity lies in the growing adoption of Industry 4.0 technologies, including the Internet of Things (IoT), big data analytics, and artificial intelligence (AI). These technologies can help overcome some of the data challenges associated with LCA by enabling real-time data collection and analysis throughout a product's lifecycle. For example, IoT-enabled sensors can track a product's performance during use, providing valuable data on energy consumption, wear and tear, and end-of-life disposal. By integrating these technologies with eco-design and LCA tools, manufacturers can gain more accurate and dynamic insights into the environmental impact of their products. There is a growing recognition of the importance of circular economy principles in promoting sustainable product design. Eco-design tools emphasising recyclability, remanufacturing, and reuse are increasingly integrated into circular economy frameworks, helping companies close the loop on product lifecycles. Collaboration across industries and supply chains is also key to overcoming data gaps and improving the transparency of environmental performance. Developing global sustainability standards and databases could further promote consistency in eco-design and LCA practices and enable more meaningful comparisons between products and industries.

Emerging Technologies in Sustainable Manufacturing – 3D Printing and Biodegradable Materials

In recent years, sustainable manufacturing has witnessed transformative advancements driven by the growing need to reduce environmental impacts while meeting the increasing demand for customized, high-performance products. The most promising innovations in this area are 3D printing (additive manufacturing) and biodegradable materials. These emerging technologies offer substantial opportunities for reducing waste, conserving resources, and minimizing carbon footprints but also enable the creation of products with optimized efficiency, sustainability, and recyclability throughout their lifecycle. This detailed exploration of these technologies highlights their role in advancing sustainable manufacturing practices, their current applications, and the challenges and opportunities associated with their adoption across industries.

3D Printing: 3D printing is a transformative manufacturing process that involves creating objects layer by layer based on digital models. Unlike traditional subtractive manufacturing, where the material is cut from a larger block, 3D printing uses only the material necessary to build the object, significantly reducing waste. This additive process is particularly beneficial for sustainability because it minimizes excess material consumption, lowers energy usage, and facilitates the production of highly complex and customized products with minimal material waste. 3D printing allows for on-demand, localized production, reducing the need for large inventories and the environmental impact of transportation and logistics.

One of the key benefits of 3D printing for sustainable manufacturing is the ability to use a wide range of materials, including biodegradable, recyclable, and renewable feedstocks. Traditional manufacturing methods often rely on resource-intensive processes that require energy and water, leading to the depletion of natural resources and higher emissions. In contrast, 3D printing can utilize recycled materials or bio-based polymers, reducing the demand for virgin resources and minimizing the overall ecological footprint of the manufacturing process. For instance, materials such as polylactic acid (PLA), a biodegradable plastic derived from renewable plant sources, can be used in 3D printing, providing an environmentally friendly alternative to petroleum-based plastics. Additionally, 3D printing enables greater design flexibility, allowing manufacturers to produce highly customized and lightweight products. This flexibility can contribute to sustainability by reducing the need for excess material, enhancing the efficiency of product designs, and minimizing transportation costs. Lightweight products, for example, require less energy to transport, reducing fuel consumption and greenhouse gas emissions. The precision of 3D printing allows for the production of complex geometries that were previously impossible or economically unfeasible with traditional manufacturing techniques. This can result in better utilization of materials, as intricate designs can reduce the overall material needed for the product without compromising functionality or durability.

3D printing is particularly advantageous in aerospace, automotive, and healthcare industries, where material efficiency and customization are critical. In the aerospace sector, 3D printing has been used to manufacture lightweight components that help reduce fuel consumption and emissions. For example, GE Aviation has implemented 3D printing to produce turbine blades, which are lighter and more energy-efficient than their traditionally manufactured counterparts. Similarly, in the automotive industry, companies like BMW and Volkswagen have leveraged 3D printing to create lighter vehicle parts, optimizing fuel efficiency while reducing material waste. In healthcare, 3D printing enables the production of custom prosthetics and implants tailored to individual patients, enhancing treatment outcomes and reducing the need for mass-produced, standardized products.

Despite its promising sustainability potential, 3D printing also faces challenges. One of the primary concerns is the energy consumption associated with 3D printing machines, especially those that operate at high temperatures, such as metal printers. While the additive nature of the process minimizes material waste, the energy intensity of the printing process can offset some of its environmental benefits. Additionally, the adoption of 3D printing is still limited by the availability of eco-friendly materials and the need for further innovation in material science to create more sustainable feedstocks that do not compromise product performance. The widespread adoption of 3D printing for large-scale manufacturing requires infrastructure development, such as establishing local production hubs, which could involve significant investments in energy and resources.

Biodegradable Materials: The growing demand for environmentally friendly products has shifted toward developing and using biodegradable materials in manufacturing. Biodegradable materials break down naturally in the environment through microorganisms, thereby reducing the accumulation of plastic waste and minimizing the long-term environmental impact. These materials are derived from renewable resources such as plant-based starches, proteins, cellulose, algae, and fungi, making them more sustainable than traditional petroleum-based plastics. Integrating biodegradable materials into manufacturing processes offers a promising solution for mitigating plastic pollution, conserving natural resources, and reducing the carbon footprint of products.

One of the most widely recognized biodegradable materials is polylactic acid (PLA), derived from fermented plant sugars such as corn or sugarcane. Due to its biodegradability, compostability, and favourable mechanical properties, PLA is used in various applications, including packaging, consumer goods, and textiles. PLA breaks down in composting environments within a few months, offering a significant environmental advantage over traditional plastics, which can persist in landfills for hundreds of years. PLA is produced from renewable resources, which reduces reliance on fossil fuels and supports the development of a circular economy by promoting bio-based feedstocks.

Another emerging biodegradable material is polyhydroxyalkanoates (PHA), a family of bioplastics produced by microorganisms through the fermentation of plant-based feedstocks. PHAs are fully biodegradable and can be used in various applications, from packaging and agricultural films to medical devices and implantable materials. Unlike PLA, which requires specific industrial composting conditions to biodegrade, PHAs break down in various environments, including soil, water, and marine ecosystems. This makes PHAs a lovely alternative to conventional plastics in terms of their environmental impact, as they offer a potential solution to the growing problem of plastic pollution in the oceans.

Using biodegradable materials in manufacturing processes, especially with 3D printing, can create more sustainable products with reduced environmental impact. For example, combining PLA with 3D printing can produce custom-made, biodegradable products for various industries, such as packaging, automotive, and healthcare. This synergy between 3D printing and biodegradable materials allows for producing lightweight, low-waste products that are both environmentally friendly and cost-effective. Additionally, using biodegradable materials in additive manufacturing could address concerns related to the disposal of waste material generated by traditional manufacturing methods.

Despite the promising potential of biodegradable materials, several challenges remain in their widespread adoption. One of the primary limitations is the performance of biodegradable plastics compared to conventional polymers. Biodegradable materials may not always possess the same mechanical strength, durability, or thermal resistance as traditional plastics, which can limit their use in certain applications. The biodegradation process of these materials depends on environmental factors such as temperature, humidity, and the presence of microorganisms, which can vary widely depending on geographic location and disposal conditions. As a result, the effective degradation of biodegradable materials may not always occur in the intended timeframe or environment, leading to potential environmental concerns if these materials accumulate in landfills or natural ecosystems. Another challenge is the scalability and cost of biodegradable materials. While bioplastics like PLA and PHA have gained significant traction in niche markets, they are still relatively expensive compared to petroleum-based plastics, primarily due to the high cost of production and the need for specialized feedstocks. This makes it difficult for manufacturers to transition fully to biodegradable materials without increasing production costs, which may ultimately be passed on to consumers. Additionally, the production of biodegradable materials requires significant land and resource inputs, which could compete with food production or lead to deforestation if not managed responsibly. As a result, the sustainable sourcing and large-scale production of biodegradable materials remain a key challenge for the industry.

Synergies Between 3D Printing and Biodegradable Materials: Integrating 3D printing and biodegradable materials presents a unique opportunity to create highly efficient, customized products with reduced environmental impact. This synergy allows for more precise control over material usage, reducing waste during manufacturing and enabling the creation of products tailored to individual needs. Using biodegradable materials in 3D printing can enhance the sustainability of the process by ensuring that the end products can be composted or degraded without leaving behind harmful waste. For instance, 3D-printed biodegradable packaging made from PLA can replace traditional plastic packaging, reducing plastic waste in landfills and oceans. The combination of these technologies is particularly valuable in packaging, medical devices, and consumer goods sectors, where product customization and sustainability are increasingly important. In the medical field, for example, 3D printing with biodegradable materials can be used to create custom implants or scaffolds for tissue regeneration, which are eventually absorbed by the body, eliminating the need for surgical removal. In the packaging industry, biodegradable 3D-printed containers can offer a sustainable alternative to conventional plastic packaging, often ending up in landfills or the ocean.

Case Study – India's Electric Vehicle Ecosystem: India's electric vehicle (EV) ecosystem has been growing rapidly, driven by environmental concerns, technological advancements, and policy interventions aimed at reducing the country's dependence on fossil fuels and addressing its pollution crisis. The electric vehicle market in India is still in its nascent stages compared to global leaders like China and Europe, but significant steps have been taken to create a conducive environment for the growth of electric mobility. India's efforts to transition towards EVs are embedded within the broader context of sustainable development, economic growth, and environmental conservation. This case study examines the various dimensions of India's electric vehicle ecosystem, highlighting key challenges and successes and the role of policies, technology, and market dynamics in shaping the future of electric mobility in India.

Need for Electric Vehicles in India: India faces significant environmental challenges, including high levels of air pollution, exacerbated by transport sector emissions. According to the World Health Organization (WHO), India's air quality is among the worst in the world, with several cities ranking in the top 10 for high levels of particulate matter (PM2.5). The transport sector is a major contributor to air pollution, accounting for nearly 20% of total carbon emissions. In addition to environmental concerns, India is heavily dependent on imports for its oil and gas needs. This dependency creates energy security risks and adds to the country's fiscal deficit.

Promoting electric vehicles is a potential solution to both of these issues. EVs promise cleaner air by reducing tailpipe emissions and providing a pathway for the country to reduce its carbon footprint. The adoption of EVs is aligned with India's energy transition goals, which include increasing the share of renewable energy in the power mix and reducing the country's reliance on fossil fuels. The growing adoption of electric vehicles is also seen as an opportunity to reduce oil imports, thereby improving energy security and contributing to the nation's economic resilience.

Government Policies and Incentives: The Indian government has recognized the importance of EVs in achieving sustainability goals and has taken significant steps to promote electric mobility through a combination of policy incentives, subsidies, and infrastructure development programs. In 2015, the government launched the **National Electric Mobility Mission Plan (NEMMP) 2020**, which aims to promote the adoption of electric vehicles and associated infrastructure in the country. The plan aimed to support the development of EVs in India by incentivizing manufacturers, promoting research and development, and creating a policy environment that facilitates the growth of electric mobility.

The **Faster Adoption and Manufacturing of Hybrid and Electric Vehicles (FAME) scheme**, launched in 2015 and extended in phases, has been one of the primary policy instruments driving the adoption of EVs. The FAME initiative offers financial incentives for purchasing electric vehicles and support for charging infrastructure. The first phase of FAME focused on providing subsidies for electric two-wheelers, electric buses, and electric cars, while the second phase, launched in 2019, placed a greater emphasis on building the necessary charging infrastructure. Under the FAME II scheme, the government allocated INR10,000 crore (approximately USD 1.4 billion) to support EV adoption, focusing on offering incentives for adopting electric buses, two-wheelers, and cars. Several state governments have also introduced policies and incentives to promote EV adoption. For example, Delhi's Electric Vehicle Policy 2020 offers exemptions from road tax, registration fees, and incentives for electric two-wheelers, autos, and buses. Similarly, Maharashtra, Tamil Nadu, and other states have introduced their own EV policies, focusing on subsidies, tax exemptions, and the development of charging infrastructure. These state-level policies play a crucial role in driving EV adoption, as they provide local incentives that make electric vehicles more affordable and accessible to consumers.

One of the most critical aspects of the Indian government's push for electric vehicles is the emphasis on building a robust charging infrastructure. Developing charging stations is key to overcoming consumers' range anxiety when considering an electric vehicle. The Indian government and private players have been working to establish a widespread network of charging stations across urban and rural areas. The FAME II scheme allocated INR1,000 crore (approximately USD 137 million) for establishing charging stations to set up at least 2,700 charging stations across the country. This infrastructure development is crucial to ensuring EV owners can easily access charging points, helping to boost consumer confidence and adoption.

Technological Advancements and Industry Players: India's electric vehicle ecosystem is being shaped by the convergence of multiple technological advancements, including improvements in battery technology, charging infrastructure, and electric drivetrains. The core component of an EV is its battery, and the efficiency, cost, and lifespan of batteries are crucial determinants of the vehicle's overall performance and affordability. India is seeing significant investments in the research and development of lithium-ion batteries, which are currently the most widely used in EVs. Several companies, including Reliance Industries and Tata Group, are exploring opportunities in the battery manufacturing space in line with the government's goal to develop indigenous battery manufacturing capabilities. The development of battery-swapping technology is another promising avenue for overcoming the challenges of long charging times. Companies such as Bajaj Auto, Ola Electric, and Lithium Urban Technologies are exploring battery-swapping stations, where users can exchange a depleted battery for a fully charged one, thereby reducing downtime and making EVs more convenient for consumers, particularly for commercial use in sectors like logistics and ride-sharing.

The EV industry in India is also witnessing the emergence of a diverse range of players, from traditional automakers to startups. Established companies such as Tata Motors and Mahindra Electric are making significant strides in the electric vehicle market. Tata Motors, for example, launched the Nexon EV, an electric SUV that has garnered significant attention due to its performance, affordability, and range. Mahindra Electric, part of the Mahindra Group, has been a key player in electric mobility, offering electric cars such as the e2o Plus and eVerito. Other players, including Ola Electric, Ather Energy, and Revolt Motors, focus on electric two-wheelers, which have become a major segment in India's EV market due to their affordability and fuel efficiency.

The advent of new business models and service providers has further accelerated the growth of India's EV ecosystem. Ola Electric, for instance, has announced plans to establish one of the world's largest electric vehicle manufacturing plants in Tamil Nadu. This facility will produce two-wheelers and battery packs, helping scale production while reducing costs. The company has also launched its charging network, Ola Hypercharger, which aims to address the infrastructure gap by providing fast-charging stations across urban centers.

Market Dynamics and Consumer Adoption: While the policy environment and technological advancements have provided significant support to the electric vehicle ecosystem in India, the pace of adoption is still relatively slow, with several challenges hindering mass-market penetration. One of the primary barriers to EV adoption is the high upfront cost of electric vehicles compared to traditional internal combustion engine (ICE) vehicles. Even though the FAME scheme and state-level incentives have significantly reduced the cost gap, EVs remain expensive for the average Indian consumer, especially when compared to gasoline or diesel vehicles, which are often more affordable and widely available.

Range anxiety is another significant challenge for potential EV buyers. Although the range of electric vehicles has improved in recent years, many consumers remain sceptical about whether an EV will be able to meet their daily transportation needs without the risk of running out of battery. This challenge is exacerbated by the insufficient charging infrastructure in rural and semi-urban areas, where charging stations are still limited. In addition to these concerns, the lack of awareness and understanding about electric vehicles among Indian consumers remains a significant barrier. While the demand for EVs has been growing, it is still concentrated mainly in urban centers, where consumers are more likely to be environmentally conscious and have better access to charging infrastructure. Rural and semi-urban markets comprise a significant portion of India's population and remain underpenetrated by electric vehicles. Expanding the reach of EVs to these markets requires targeted awareness campaigns and efforts to address the unique challenges faced by consumers in these regions.

Future Scenario

As the global demand for sustainable production and consumption intensifies, industries are increasingly exploring cutting-edge technologies to reduce waste, improve efficiency, and minimize environmental impact. The future of manufacturing and consumer goods is poised for transformation, particularly with the integration of smart materials, 3D printing, and blockchain technology. These innovations, when harnessed effectively, have the potential to create zero-waste production systems, promote sustainable consumption, and offer unprecedented transparency in product lifecycle management. This section delves into the potential role of these technologies in reshaping manufacturing and consumption practices, focusing on their contribution to a sustainable future.

Smart Materials and 3D Printing for Zero-Waste Production: Zero-waste production has emerged as a critical goal for industries aiming to minimize their environmental footprint. Traditional manufacturing processes are often associated with significant material waste, energy consumption, and inefficiency. However, smart materials those that respond dynamically to environmental stimuli and 3D printing technologies offer promising solutions to these challenges, leading to more efficient, adaptive, and sustainable production methods.

Smart Materials: Definition and Applications: Smart materials are engineered materials that can change their properties in response to external stimuli such as temperature, pressure, light, or magnetic fields. These materials can be designed to perform specific functions, such as self-healing, energy absorption, or shape-memory effects, which makes them particularly suitable for sustainable manufacturing processes. By incorporating smart materials into product designs, manufacturers can reduce waste, extend the lifespan of products, and improve recyclability, all of which are fundamental components of a zero-waste manufacturing system.

For instance, self-healing materials, which can repair themselves when damaged, have the potential to extend the lifespan of products and reduce the need for frequent replacements. This capability reduces waste and conserves resources by minimizing the frequency of new materials being sourced and processed. Similarly, shape-memory alloys (SMAs) and polymers, which can return to their original form when exposed to specific environmental conditions, enable more efficient and adaptive designs, minimizing waste and energy usage during production. In sustainable packaging, smart materials can help reduce the environmental impact of single-use plastics. For example, biodegradable smart materials that respond to environmental conditions, such as moisture or temperature, can be used for packaging that decomposes naturally after use. This advancement could significantly reduce plastic waste's burden, a major global environmental challenge. In a circular economy, such materials can be integrated into closed-loop systems, where products are designed to be disassembled, reused, or recycled without contributing to landfills.

3D Printing and Additive Manufacturing for Waste Reduction: 3D printing, or additive manufacturing, represents a paradigm shift in how products are designed, prototyped, and produced. Unlike traditional subtractive manufacturing methods, which often involve cutting, milling, or carving materials from a larger block (leading to significant material waste), 3D printing builds products layer by layer based on a digital model. This method allows for precise control over material usage, reducing waste generation and enabling more efficient use of resources.

One of the most significant advantages of 3D printing in zero-waste production is its ability to create complex, customized geometries that are often impossible to achieve through traditional manufacturing methods. This precision minimizes material waste because the product is built exactly to specification, with no excess material required. The ability to print only the necessary components means that production can be more localized, reducing the need for transportation and decreasing the carbon footprint associated with traditional manufacturing processes.

Using recycled materials in 3D printing is another promising avenue for sustainability. 3D printers can use various waste materialssuch as plastic bottles, industrial byproducts, or discarded electronicsas feedstock for producing new objects. This not only diverts waste from landfills but also reduces the need to source new raw materials, a critical aspect of the circular economy. Advances in material science have led to the development of biodegradable filaments for 3D printing, further enhancing the sustainability of this technology. These bioplastics, made from renewable resources, can be used to create eco-friendly products that are easier to recycle or compost at the end of their life cycle.

In the manufacturing sector, 3D printing also has the potential to disrupt supply chains by enabling on-demand production. Instead of manufacturing products in bulk and storing them in warehouses, which can lead to overproduction and surplus waste, 3D printing allows for the production of items as needed. This reduces inventory waste, lowers the carbon footprint of transportation, and ensures that resources are not overutilized in the production process. The flexibility of 3D printing also facilitates product repairs and modifications, eliminating the need to replace entire products and reducing overall material consumption.

Blockchain Technology for Sustainable Consumption: While smart materials and 3D printing are revolutionizing production processes, blockchain technology enables transparent, efficient, and sustainable consumption. Blockchain, a decentralized and immutable digital ledger, allows for the transparent tracking and verification of transactions, assets, and processes across a supply chain. This characteristic is particularly beneficial in ensuring that products are sourced, produced, and distributed sustainably, fostering consumer trust and encouraging responsible consumption patterns.

Blockchain for Transparent Product Lifecycles: Blockchain provides the infrastructure needed for full product lifecycle traceability. Consumers are increasingly concerned about their purchases' environmental and social impact, and blockchain technology enables companies to provide transparent, verifiable information about the sourcing, production, and disposal of their products. By recording every stage of a product's journeyfrom raw material extraction to manufacturing, distribution, and disposalblockchain can assure consumers that their products were made with sustainable practices. For instance, blockchain can track the provenance of materials, such as rare metals or palm oil, ensuring that they were sourced sustainably and without contributing to deforestation or human rights violations. It can also monitor the carbon footprint of manufacturing processes, helping consumers make informed decisions about the environmental impact of their purchases. This level of transparency is critical in sustainable consumption, as it empowers consumers to choose products that align with their environmental values.

The **fashion industry**, for example, has increasingly adopted blockchain to verify the sustainability of its supply chains. Through blockchain, brands can track the sourcing of materials like cotton, wool, and leather, ensuring that they are produced ethically and sustainably. Similarly, blockchain in the food industry allows consumers to verify the origin of their food, from farm to table, ensuring that products are grown or produced using environmentally friendly methods.

Blockchain for Circular Economy and Recycling: Blockchain's potential extends to developing circular economy models where products are reused, repaired, or recycled instead of discarded. In a circular economy, products are designed to be returned to the supply chain at the end of their lifecycle for recycling or remanufacturing. Blockchain can help facilitate this process by tracking the ownership and condition of products, ensuring that materials are returned to the production system and reused efficiently.

For example, electronics recycling could be greatly improved through the use of blockchain. By tracking electronic product ownership and recycling history, blockchain ensures that valuable components, such as rare earth metals, are recovered and reused, reducing the demand for new raw materials and minimizing e-waste. Blockchain also provides a way to verify the quality of recycled materials, ensuring that they meet the necessary standards for reuse in the production of new products.

Opportunities for the Future: Governments, corporations, and research institutions have a chance to collaborate on creating scalable, cost-effective solutions for integrating these technologies into existing systems. As technology costs decrease and infrastructure improves, the widespread adoption of these innovations will contribute significantly to achieving zero-waste production and fostering sustainable consumption patterns. The future scenario for sustainable production and consumption increasingly depends on the synergy between smart materials, 3D printing, and blockchain technology. These technologies, each transformative in its own right, can contribute to creating closed-loop systems that minimize waste, enhance resource efficiency, and promote transparency in product lifecycles. By integrating these innovations, industries can move toward more sustainable practices while consumers gain the tools to make informed, environmentally responsible choices. As the global push for sustainability intensifies, these technologies will play a pivotal role in shaping a circular economy where waste is minimized, resources are conserved, and the environmental footprint of production and consumption is significantly reduced.

7

Platforms as Enablers of Sustainability

Introduction

In the modern age of digital transformation, platforms have become critical enablers of sustainability. These digital infrastructures, which facilitate connectivity, resource sharing, and collaboration, are pivotal in accelerating the adoption of sustainable practices across various sectors. Platforms enable the effective integration of resources, stakeholders, and technologies, offering scalable solutions that can help achieve long-term environmental, economic, and social sustainability goals. Whether through promoting circular economy models, enabling efficient resource use, fostering innovation in green technologies, or driving policy changes, platforms serve as central hubs for fostering sustainable development.

Types of Platforms: At its core, a platform is a digital infrastructure or service that connects multiple stakeholders, facilitating the exchange of goods, services, information, or energy. These platforms can vary in scope and function, ranging from e-commerce platforms, which enable the trade of products, to energy platforms, which manage the distribution of renewable energy. Platforms are often designed to leverage the network effect, where the platform's value increases as more users participate and share resources. The term "platform" encompasses a variety of systems designed to optimize resource efficiency, promote sustainable practices, and accelerate the transition to low-carbon economies. Several types of platforms play a crucial role in enabling sustainability. These include:

- **Collaborative Platforms:** These platforms foster collaboration between different stakeholders, such as individuals, organizations, and governments. They facilitate knowledge sharing, joint problem-solving, and collective action on sustainability issues. Examples include open-source platforms for environmental data sharing or collaborative platforms for sustainable development projects.

- **Marketplaces for Sustainable Products and Services:** These digital marketplaces connect consumers with sustainable goods and services, ranging from renewable energy options to eco-friendly products. They provide a venue for businesses that adopt sustainable practices to reach environmentally-conscious consumers, directly linking sustainability and consumer demand.

- **Energy Platforms:** These platforms manage renewable energy's generation, distribution, and consumption. They integrate energy from diverse sources, such as solar, wind, and hydro, and optimize usage through advanced analytics, artificial intelligence (AI), and blockchain technology. By facilitating energy trading, peer-to-peer energy exchanges, and real-time energy consumption monitoring, these platforms help optimize energy use and reduce carbon footprints.
- **Circular Economy Platforms:** Circular economy platforms enable businesses and consumers to participate in a closed-loop economic system where products and materials are reused, refurbished, and recycled. These platforms track the lifecycle of products, promote resource efficiency, and incentivize the reduction of waste.
- **Sustainability Reporting Platforms:** These platforms collect, analyze, and report data related to sustainability performance. They assist organizations in tracking key sustainability metrics, such as carbon emissions, water usage, and waste production. These platforms facilitate informed decision-making for businesses and consumers by providing transparent sustainability reports.

Role of Platforms in Enabling Sustainable Development: The role of platforms in enabling sustainability is multifaceted. These digital infrastructures are enablers of resource optimization, innovation, collaboration, and transparency all of which are central to achieving the United Nations' Sustainable Development Goals (SDGs).

Resource Optimization: One of the key ways platforms contribute to sustainability is by optimizing the use of resources. Resource inefficiency, overproduction, and waste are common problems in traditional systems. Platforms enable more efficient use of materials, energy, and labor by streamlining operations and facilitating data-driven decision-making. For example, smart grids in energy platforms enable the efficient distribution of electricity. They use real-time data and predictive analytics to match energy supply with demand, reducing energy wastage and optimizing renewable energy sources. Similarly, in manufacturing, platforms that facilitate Industry 4.0 technologies such as IoT, AI, and machine learning can help optimize production lines, reduce material waste, and improve energy efficiency. By ensuring that resources are used only when needed and in optimal quantities, platforms contribute to reducing overall environmental impacts.

In **agriculture**, precision farming platforms allow farmers to optimize irrigation, fertilizer, and pesticide use, significantly reducing water waste, chemical runoff, and energy consumption. Using platforms in agriculture helps ensure that sustainable farming practices are adopted and scaled to meet the growing demands of global food systems.

Fostering Innovation for Sustainability: Platforms play a significant role in fostering innovation, particularly in developing and adopting green technologies. Through collaborative spaces, stakeholders such as businesses, universities, research institutions, and governments can collectively share knowledge, pool resources, and innovate. This is particularly important in addressing complex environmental challenges like climate change, biodiversity loss, and resource depletion. For instance, crowdsourcing platforms enable innovators and entrepreneurs to test and launch sustainable products and services. These platforms offer an avenue for funding, collaboration, and exposure, allowing sustainable innovations to gain traction in the market. A prominent example is Kickstarter, where sustainable startups in renewable energy, eco-friendly fashion, and sustainable agriculture find support from investors and customers.

Also, The adoption of platforms in industries such as electric vehicles (EVs), solar energy, and clean technology has accelerated the pace of innovation. EV charging infrastructure, for example, is being optimized through digital platforms that manage and monitor charging stations, reducing wait times and enabling the efficient use of renewable energy.

Promoting Collaboration and Knowledge Sharing: Collaboration is essential for addressing global sustainability challenges. Platforms facilitate collaboration between governments, businesses, civil society, and consumers, ensuring all stakeholders are aligned in their sustainability efforts. By fostering dialogue, knowledge sharing, and joint action, platforms create ecosystems where diverse perspectives are integrated into decision-making processes. For example, UN Global Compact's sustainability platforms allow companies to share best practices, benchmark sustainability progress, and develop new partnerships to achieve the SDGs. Platforms like these help build cross-sector collaborations, where private, public, and nonprofit entities can come together to solve complex sustainability problems. Similarly, platforms like EcoPost, which facilitates plastic waste recycling, engage local communities, waste pickers, and businesses in creating a sustainable waste management system. These platforms promote **inclusive collaboration**, ensuring that all sectors of society participate in the transition to a sustainable future.

Enhancing Transparency and Accountability: Sustainability requires transparency, particularly when it comes to the tracking of environmental, social, and governance (ESG) metrics. Platforms provide data collection, monitoring, and reporting tools, allowing organizations to share their sustainability performance with stakeholders, including consumers, investors, and regulatory bodies.

Blockchain technology ensures that sustainability data is immutable and transparent when integrated into platforms. Platforms leveraging blockchain enable the secure, transparent tracking of supply chains, ensuring products are sustainably sourced and ethically produced. For example, FairChain is a platform that allows coffee farmers to track the entire coffee supply chain, ensuring that sustainability standards are met and that the economic benefits of coffee production are fairly distributed among producers. Similarly, platforms for sustainable finance enable investors to assess the sustainability impact of their portfolios and make informed decisions that align with their values. By increasing transparency, these platforms hold businesses accountable for their sustainability claims and drive more responsible corporate behavior.

Defies and Opportunities in Platform-Driven Sustainability: While platforms offer vast potential in enabling sustainability, several challenges must be addressed. One significant barrier is data privacy and security. Protecting sensitive information is crucial as platforms rely heavily on data collection, particularly in industries like energy and agriculture. The digital divide remains an issue, as not all regions or populations have access to the necessary technology or digital literacy to engage with sustainability platforms effectively. Another is the lack of standardization in sustainability reporting and practices across industries. Without universally agreed-upon metrics and guidelines, the effectiveness of platforms in driving sustainability can be undermined. To overcome this challenge, international organizations and governments must work together to establish consistent frameworks for measuring and reporting sustainability efforts.

However, challenges present opportunities for further development and collaboration. These barriers can be mitigated by improving data governance and ensuring inclusivity in platform design. As platforms evolve, the opportunities for scaling sustainable practices and driving systemic change will only increase. Platforms serve as key enablers of sustainability by optimizing resource use, fostering innovation, promoting collaboration, and enhancing transparency across various sectors. In the future, platforms will continue to play a pivotal role in transitioning the global economy toward sustainability. Their ability to connect stakeholders, integrate technologies, and scale solutions will be essential in addressing the complex and interconnected challenges of environmental, social, and economic sustainability. As digital infrastructure improves and adoption expands, platforms will be integral to shaping a more sustainable and equitable world for future generations.

Current Scenario

In the current era, the role of digital platforms, specifically cloud computing and shared economy platforms, in reducing carbon footprints is increasingly significant. As the world grapples with the challenges of climate change, integrating these platforms into various sectors offers innovative solutions to reduce energy consumption, optimize resource use, and lower greenhouse gas emissions. The application of these technologies has the potential to improve the efficiency of industries and accelerate the global transition to a more sustainable, low-carbon economy. Cloud computing and shared economy platforms are key enablers of sustainable development, providing businesses and individuals with scalable, cost-effective, and environmentally conscious solutions.

Cloud Computing – Catalyst for Energy Efficiency and Carbon Reduction: Cloud computing refers to delivering computing servicesincluding storage, processing power, and softwareover the internet. Instead of maintaining physical infrastructure, businesses and individuals can access computing resources remotely, which brings significant sustainability benefits. One of the primary environmental advantages of cloud computing is its ability to optimise resources and energy efficiency, directly contributing to reducing carbon footprints.

Centralized Infrastructure and Energy Efficiency Traditional data centers, which were widely used before the advent of cloud computing, require massive amounts of energy for operation and cooling. These data centers often operate inefficiently, with significant energy consumption for servers that are underutilized or idle, leading to higher carbon emissions. Cloud computing addresses this inefficiency by offering centralized infrastructure that optimizes resource usage. Cloud service providers such as Amazon Web Services (AWS), Google Cloud, and Microsoft Azure have invested heavily in green technologies, such as energy-efficient data centers and renewable energy sourcing, to minimize the environmental impact of their operations. For example, these providers now use advanced cooling techniques and artificial intelligence (AI) to optimize the power usage of their data centers, reducing the energy required for cooling by adjusting airflow, temperature, and humidity in real-time. Consolidating resources in cloud data centers, as opposed to the distributed approach of traditional data centers, also leads to better utilization of physical hardware. As a result, cloud computing has a lower energy consumption per user than maintaining individual on-site infrastructure, thus reducing overall carbon footprints.

Scalability and Resource Efficiency: Cloud computing also helps reduce carbon emissions through scalability. In the past, businesses had to over-provision their computing resources to accommodate peak demand, often wasting energy when the resources were not in use. Cloud services allocate resources dynamically, allowing users to pay only for their needs. Cloud platforms can scale up or down in response to demand, ensuring that energy and resources are utilized efficiently and only when necessary. This on-demand nature of cloud computing allows organizations to reduce unnecessary energy use drastically and, therefore, their carbon footprints. Cloud computing enables organizations to access powerful data analytics tools that track and analyze energy consumption, optimize energy use, and minimize waste. For instance, businesses can leverage cloud-based energy management systems to monitor real-time energy usage across their operations, identify inefficiencies, and make informed decisions on energy-saving measures.

Renewable Energy Integration: Leading cloud providers increasingly integrate renewable energy sources into their operations. Companies like Google, for example, have committed to running their global data centers entirely on renewable energy and are working toward achieving carbon neutrality. By investing in renewable energy, cloud service providers reduce their dependence on fossil fuels, directly contributing to lowering carbon emissions in the sector.

Shared Economy Platforms – Facilitating Sustainability Through Resource Sharing: In addition to cloud computing, shared economy platforms play an equally vital role in reducing carbon footprints. The shared economy, called the collaborative or peer-to-peer economy, refers to systems where individuals share access to goods, services, or resources via digital platforms. Examples of shared economy platforms include services like Uber for transportation, Airbnb for lodging, WeWork for office space, and platforms for sharing tools, appliances, and even renewable energy resources. The central idea behind shared economy platforms is to reduce overconsumption and optimize the use of resources by facilitating the sharing and reuse of existing products. By eliminating the need for individuals and businesses to own and maintain underutilized assets, these platforms help reduce waste, lower production demands and minimize environmental impacts.

Transportation and Mobility Solutions: One of the most prominent shared economy sectors contributing to carbon footprint reduction is shared transportation. Traditional vehicle ownership leads to inefficient use of resources, with most vehicles idle for significant portions of the day. Shared mobility platforms such as Uber, Lyft, and BlaBlaCar have enabled individuals to access transportation without needing private car ownership. This shift reduces the number of cars on the road and lowers fuel consumption and emissions. Further, shared mobility platforms often facilitate using electric vehicles (EVs). Services like Uber have begun incorporating EVs into their fleets, which can significantly reduce emissions compared to traditional gasoline-powered vehicles. The widespread adoption of shared electric vehicles is a promising development, particularly in urban areas where air quality is a major concern. Additionally, ride-sharing platforms encourage autonomous vehicles, which, when integrated with renewable energy sources, can further reduce emissions and optimize the use of existing infrastructure.

Shared Housing and Space Utilization: The shared economy also extends to housing and accommodation. Platforms like Airbnb and HomeAway allow homeowners to rent out unused space, ensuring that existing resources are used efficiently. This reduces the demand for new construction and the associated environmental costs of building materials, energy use, and waste generation. By maximizing the utilization of available space, these platforms contribute to reducing urban sprawl, which is a major contributor to environmental degradation and resource consumption. Co-working spaces provided by platforms like WeWork allow businesses to share office infrastructure, reducing the need for multiple companies to own and operate separate office buildings. This model optimizes the use of office space, reduces energy consumption, and leads to cost savings, all while minimizing the carbon footprint associated with building maintenance and construction.

Waste Reduction and Circular Economy: Another vital aspect of shared economy platforms is their role in promoting the circular economy, which emphasizes the reuse, recycling, and repurposing of materials and goods to minimize waste. Platforms such as Rent the Runway, which allows users to rent clothing, and Tool Libraries, which facilitate the shared use of tools, contribute to a circular economy by ensuring that products are reused rather than disposed of after a single use. By extending the lifecycle of products and services, these platforms reduce the demand for raw materials, energy consumption, and waste generation, ultimately reducing carbon footprints. Additionally, peer-to-peer (P2P) energy sharing platforms enable individuals and businesses to share excess energy from renewable sources like solar panels. Platforms such as Power Ledger and Lo3 Energy are pioneering efforts in enabling decentralized energy trading. By allowing individuals and businesses to sell surplus energy, these platforms reduce reliance on centralized energy grids, which are often powered by fossil fuels. This decentralized approach promotes renewable energy, directly reducing greenhouse gas emissions and supporting a more sustainable energy system.

Synergy Between Cloud Computing and Shared Economy Platforms for Sustainability: The synergy between cloud computing and shared economy platforms presents a powerful tool for reducing carbon footprints and fostering a more sustainable future. Cloud computing drives energy efficiency, enables scalability, and integrates renewable energy sources, while shared economy platforms facilitate resource sharing, reduce waste, and optimize consumption patterns. These platforms help businesses and individuals make more informed, sustainable decisions and reduce their environmental impact. These platforms' continued development and adoption will be crucial in meeting global sustainability targets, such as those outlined in the Paris Agreement and the UN Sustainable Development Goals (SDGs). By leveraging the capabilities of cloud computing and the shared economy, society can accelerate its efforts to reduce carbon emissions, optimize resource use, and transition toward a low-carbon, circular economy. However, for these platforms to realize their full potential, issues related to data privacy, accessibility, and regulation must be addressed. Governments and industry stakeholders must collaborate to create policies and standards that support the growth of these platforms while ensuring that sustainability benefits are maximized. As the adoption of cloud computing and shared economy platforms continues to grow, their role in reducing carbon footprints will be increasingly central to the global effort to combat climate change and build a more sustainable world.

Importance of Platforms in Viksit Bharat

In the journey toward a Viksit Bharat (Developed India) by 2047, digital platforms have emerged as a pivotal enabler in fostering integrated, efficient, and sustainable systems. These platforms are not just technological innovations but also foundational elements that can shape the nation's future in multiple domains, ranging from governance and healthcare to agriculture and education. By bridging information silos, enabling cloud computing, and fostering data-sharing ecosystems, digital platforms hold the potential to streamline national efforts towards achieving inclusive growth while addressing the challenges of disparities, inefficiency, and resource mismanagement. As India aims to become a developed nation, these platforms can be crucial in transforming its digital landscape into a dynamic, interconnected, and sustainable ecosystem.

Bridging Silos through Digital Platforms: Historically, India's public and private sectors have been characterized by silos of information. Different government departments, industries, and sectors operated with fragmented data and isolated systems, resulting in inefficiency, duplication of efforts, and slow decision-making. For instance, the healthcare system, education sector, and agricultural policies often lacked coordination, leading to suboptimal outcomes despite significant investments. In this fragmented ecosystem, digital platforms can connect disparate systems, fostering collaboration and streamlining workflows across sectors.

One of the key challenges in India's development is the lack of interconnectivity between various departments and agencies. For example, education policy may not fully align with the employment sector, or healthcare policies may be disconnected from public health infrastructure. Digital platforms can bridge these gaps by acting as a medium through which different sectors can share data, collaborate on initiatives, and ensure that resources are allocated optimally based on real-time information. By allowing centralized access to relevant data, platforms can help policymakers and administrators make more informed, data-driven decisions, leading to a more efficient and responsive governance system.

One example of such a platform is the Digital India initiative, which connects various government departments, businesses, and citizens through a digital infrastructure. Platforms like the National Digital Health Mission (NDHM), Aadhaar, and DigiLocker allow individuals, businesses, and government agencies to access and exchange information seamlessly. These initiatives ensure that data is available in real-time, reducing delays in service delivery, enabling quick responses to emerging issues, and ensuring that resources are deployed effectively.

Digital platforms can also integrate rural and urban areas, ensuring that remote areas benefit from the same services and opportunities as urban centers. In agriculture, digital platforms that aggregate data from farmers, markets, and government schemes allow rural stakeholders to access timely information on crop prices, weather conditions, and market trends. This connectivity fosters a more inclusive economy, where rural areas are not isolated from the national growth trajectory, and rural farmers are empowered with tools to increase their productivity and income.

Cloud Computing – Pillar of Digital Transformation in Viksit Bharat: As India aspires to become a global leader in technology and development, cloud computing stands out as a critical infrastructure for the nation's digital transformation. Cloud platforms provide businesses, government agencies, and individuals access to computing power and storage on-demand without investing in costly hardware or maintaining complex IT infrastructures. This opens up numerous possibilities for cost-effective scalability, resource optimization, and innovation across various sectors, from healthcare and education to smart cities and environmental monitoring.

Cloud computing offers several compelling advantages, particularly regarding data-sharing, storage, and collaborative innovation. By leveraging the cloud, India can create a centralized data repository, where different sectorssuch as healthcare, agriculture, and educationcan share relevant information for better policy-making, forecasting, and planning. This data centralisation promotes collaboration across departments, eliminating the inefficiencies and gaps caused by siloed data systems.

For instance, India's Smart Cities Mission leverages cloud computing to integrate various urban systems such as transportation, waste management, and energy distribution. By hosting smart city data on cloud platforms, the government can monitor real-time data and optimize operations based on predictive analytics, making cities more efficient, livable, and sustainable. Cloud computing allows city planners to design data-driven solutions that can adapt to the needs of a rapidly urbanizing population, ensuring that resources are used effectively and reducing energy consumption and emissions.

Data-Sharing Platforms: Enabling Collaborative Governance and Resource Optimization

In a country as large and diverse as India, data-sharing platforms can accelerate the country's development by ensuring that data is accessible across various levels of governance, from local bodies to central agencies. Data-sharing platforms can help address key issues such as poverty, health disparities, and education gaps while facilitating inter-agency collaboration. By enabling public and private sector organizations to share critical information seamlessly, these platforms create a framework for collaborative governance that benefits all stakeholdersgovernment, citizens, businesses, and non-governmental organizations.

Open data platforms like the India Data Portal and Open Government Data (OGD) Platform allow citizens, businesses, and researchers to access valuable datasets on various topics, including health, education, agriculture, and environmental sustainability. This promotes transparency, encourages innovation, and enables evidence-based policymaking. For example, open data on crop yields, climate patterns, and market demand can be accessed by farmers, enabling them to make informed decisions on when to plant crops, which crops to grow, and when to sell their produce. This can lead to increased agricultural productivity, better financial outcomes for farmers, and improved food security for the nation. Similarly, data-sharing platforms in the healthcare sector, such as the National Health Stack and Ayushman Bharat Digital Mission, facilitate the integration of various health services and insurance systems, improving access to healthcare and reducing inefficiencies in patient care. These platforms enable health providers to securely share patient data across hospitals, clinics, and diagnostic centers, ensuring that individuals receive timely and appropriate care, regardless of location.

The India Stack, a suite of APIs (application programming interfaces) that allow governments and private entities to share data and services securely, also plays a pivotal role in this transformation. With initiatives like Aadhaar (the world's largest biometric ID system) and Digital India infrastructure, data-sharing platforms provide a framework for financial inclusion, e-governance, and service delivery, improving transparency and citizen engagement. By providing citizens access to their digital identities, financial services, and government schemes, these platforms empower individuals to participate more fully in the economy and governance processes.

Role of Data Interoperability in Bridging Silos: To fully harness the potential of cloud computing and data-sharing platforms, it is essential to establish interoperability standards across various systems and sectors. The current data landscape in India is often characterized by fragmented systems, with each government department or sector operating in isolation. This lack of interoperability between systems hinders the flow of information and limits the effectiveness of digital platforms. To overcome this, India must prioritize the development of common data standards that allow data to flow seamlessly across departments and platforms.

Efforts to enhance data interoperability are already underway. For instance, the National Digital Health Mission (NDHM) aims to integrate healthcare services nationwide by ensuring that health records are shared across different healthcare providers and platforms. This standardization ensures that citizens receive the right care at the right time, regardless of which part of the country they are in. Similarly, the Goods and Services Tax (GST) Network has enabled the integration of tax data across states, facilitating seamless tax compliance and reducing administrative burdens. In smart agriculture, data interoperability from various sources such as weather stations, satellite imaging, soil sensors, and market demand platforms can help farmers make more informed decisions and better plan their operations. Agri-tech platforms leveraging cloud computing can also provide personalized advisory services to farmers, making agriculture more productive, sustainable, and profitable.

Unified Digital Infrastructure for Viksit Bharat: As India strives towards Viksit Bharat, digital platformsfrom cloud computing to data-sharing ecosystemsare crucial enablers of holistic development. By bridging silos and fostering collaboration between sectors, these platforms can optimize resource use, increase productivity, and drive sustainable growth. India's digital transformation will depend on integrating these platforms into every facet of governance and public service delivery, ensuring that data-driven decision-making leads to more efficient, transparent, and inclusive systems. To ensure the success of these platforms, India must prioritize the development of robust cybersecurity measures, data governance frameworks, and digital literacy programs that enable citizens and government agencies to leverage these platforms fully. Creating a supportive policy environment that promotes innovation, interoperability, and inclusive access will be critical to unlocking the full potential of these digital platforms.

IT-Enabled Collaborative Ecosystems

In the evolving digital age, cloud computing and artificial intelligence (AI) fusion have become a cornerstone for developing IT-enabled collaborative ecosystems that foster innovation. As nations and businesses strive for growth, technological advancements are essential to unlocking new opportunities, enabling seamless collaboration, and facilitating access to cutting-edge tools and resources. These technologies drive the next wave of business transformations and empower cross-sectoral collaborations that were previously unimaginable. This section delves into the role of **cloud computing** and **AI platforms** in creating dynamic, collaborative ecosystems for innovation, particularly in digital transformation journey.

Rise of IT-Enabled Collaborative Ecosystems: An IT enabled collaborative ecosystem is a networked environment where multiple stakeholderssuch as businesses, governments, institutions, and individualswork together, leveraging technology to achieve shared goals. These ecosystems facilitate data sharing, collective decision-making, and innovative problem-solving through collaboration. Cloud computing and AI technologies are key enablers of such ecosystems, providing the infrastructure, tools, and platforms necessary for real-time interaction, resource optimization, and innovation-driven outcomes.

Cloud Computing – Platform for Collaboration: Cloud computing is pivotal in fostering collaboration across different sectors and regions by offering scalable, on-demand computing resources and data storage. The cloud eliminates the need for local infrastructure, lowering entry barriers for businesses, governments, and individuals to access advanced technologies and tools. It enables distributed teams to work together seamlessly, regardless of their geographic location and facilitates the sharing and processing of vast datasets in real-time, which is particularly crucial for sectors like healthcare, education, agriculture, and urban planning.

The central tenet of cloud computing is its ability to provide shared resources over the internet, significantly enhancing collaborative work. Organizations can access virtualized computing power, software platforms, and storage capacity through cloud-based applications without substantial upfront investment. This democratizes access to high-performance computing resources and fosters innovation by enabling collaboration across different sectors and stakeholders.

For instance, in the healthcare sector, cloud computing has transformed how medical research is conducted, and healthcare services are delivered. Cloud-based medical data platforms allow doctors, researchers, and medical institutions to share patient data, research findings, and treatment protocols, ensuring that knowledge is shared and applied efficiently across the healthcare ecosystem. This collaboration leads to faster innovations in treatments, improved healthcare outcomes, and more efficient use of resources. Similarly, in agriculture, cloud computing supports farmers by providing access to real-time weather data, market prices, crop health diagnostics, and best practices, all of which can be accessed through mobile applications on the cloud. These platforms enable greater collaboration among farmers, agricultural experts, and government agencies, allowing for better planning and decision-making that boosts crop productivity and sustainability.

In urban planning, cloud platforms allow city planners to collect, store, and analyze massive amounts of data from various sources, such as traffic systems, environmental sensors, and utilities. These data-driven platforms enable collaborative governance, where citizens, local governments, and private entities can collaborate in creating smarter, more sustainable cities. Real-time data shared via the cloud allows for the optimization of public services, better resource management, and improved urban infrastructure development.

AI Platforms – Driving Collaborative Innovation: While cloud computing provides the infrastructure, AI platforms enable intelligent data processing and analysis, making them an essential tool for collaborative innovation. AI-driven systems use machine learning algorithms and advanced analytics to extract insights from vast amounts of data, enabling stakeholders to make better decisions, predict future trends, and automate processes. When combined with cloud computing, AI platforms foster collaboration by enabling the real-time analysis of shared data and providing intelligent tools that facilitate problem-solving and innovation.

One of the most significant impacts of AI platforms is their ability to enable predictive analytics, which uses historical data to forecast future events or trends. This predictive capability has profound implications for various sectors. For example, AI algorithms can analyze energy consumption patterns in the energy sector in real-time, optimising energy grids and efficiently distributing resources. Similarly, AI can analyze city traffic patterns to optimize traffic flow, reduce congestion, and improve overall urban mobility.

AI is also revolutionizing the manufacturing sector by enabling the creation of smart factories. In these factories, AI-powered robots, machine learning models, and automation systems work together to streamline production processes, improve supply chain efficiency, and reduce waste. AI can analyze production data in real-time to detect faults, predict machine failures, and recommend maintenance schedules, thereby reducing downtime and improving the overall efficiency of manufacturing operations. The collaborative power of AI is also seen in research and development (R&D). AI platforms enable researchers to accelerate the innovation cycle by analyzing large datasets, identifying patterns, and providing manual insights that would have taken years to discover. By integrating AI into cloud-based ecosystems, researchers worldwide can collaborate in real-time, pooling resources and expertise to tackle global challenges such as climate change, disease outbreaks, and resource management.

Combining Cloud Computing and AI – Synergistic Relationship: The synergy between cloud computing and AI forms the backbone of many successful IT-enabled collaborative ecosystems. Cloud platforms provide the necessary infrastructure to store and share vast datasets, while AI platforms process these datasets, extracting meaningful insights that drive innovation. Combining both technologies allows organizations to create more dynamic, responsive, and scalable solutions that multiple stakeholders can access and refine in real-time.

For example, cloud computing provides the infrastructure for data storage and sharing, while AI platforms enable the analysis and interpretation of the data. By collecting and processing data from various urban sourcessuch as traffic sensors, environmental monitoring systems, and public service databasesAI models can predict traffic congestion, air pollution levels, and utility demand. Cloud platforms allow city planners to access this information in real-time, collaborate with other departments, and implement data-driven policies to optimize city operations and improve citizens' quality of life. Similarly, in the agriculture sector, AI platforms analyze weather patterns, soil health, crop performance, and market demand to provide farmers with actionable insights. These insights are delivered via cloud-based platforms that can be accessed through smartphones and other digital devices, enabling farmers to make informed decisions and improve productivity. Integrating AI-driven advisory systems into cloud platforms helps bridge the digital divide in rural areas by providing farmers with personalized recommendations based on data, even without physical expertise on-site.

Cloud computing and AI also have transformative potential in the education sector. AI-powered learning platforms hosted on the cloud offer personalized learning experiences for students, adapting to individual learning speeds and needs. Cloud computing enables educators to access resources and collaborate on curriculum development in real-time. Additionally, AI's ability to analyze student performance data and predict learning outcomes helps educators intervene early and customize teaching strategies for maximum student success.

Role of Collaborative Ecosystems in Solving Global Challenges: Integrating cloud computing and AI into collaborative ecosystems also holds significant promise for addressing global challenges. For example, in the climate change arena, AI can analyze environmental data to predict climate patterns and assess the impact of various mitigation strategies. Cloud platforms provide the infrastructure for storing and sharing this data across multiple countries and stakeholders, facilitating international collaboration. Global initiatives to mitigate climate change can leverage AI-driven solutions hosted on the cloud, enabling policymakers, researchers, and environmental organizations to work together towards shared sustainability goals.

In healthcare, cloud computing and AI have been pivotal in enabling collaborative efforts to tackle global health challenges, such as pandemics. During the COVID-19 pandemic, AI-driven data analysis and predictive modeling, supported by cloud platforms, helped governments, researchers, and healthcare institutions monitor the spread of the virus, predict hotspots, and allocate resources more effectively. The ability to collaborate in real-time and share data across countries accelerated the development of vaccines, diagnostics, and treatment protocols, highlighting the importance of cloud-based AI ecosystems in solving global health crises.

Case Study

The advent of the Unified Payments Interface (UPI) has marked a transformative milestone in India's financial ecosystem, revolutionizing the way financial transactions are conducted and expanding the reach of financial services across the country. Launched in 2016 by the National Payments Corporation of India (NPCI), UPI is a real-time payment system that allows users to transfer money between bank accounts using smartphones and the internet. The system has had profound implications for the financial inclusion landscape, particularly in a country like India, where the traditional banking infrastructure has historically struggled to reach remote and underserved populations. This case study examines the role of UPI in catalyzing India's FinTech revolution, how it has transformed financial transactions, and its impact on financial inclusion, economic growth, and digital literacy.

Role of UPI in the Financial Ecosystem: Before the launch of UPI, the Indian financial system was heavily reliant on physical banking branches, ATMs, and outdated payment systems that were not only inefficient but also difficult to access for large portions of the population. Digital payments in India had traditionally been limited to debit and credit card transactions, net banking, and mobile wallets, but these methods still required infrastructure and user knowledge that were unavailable to many, especially in rural and remote areas. UPI's unique offering lies in its simplicity, ease of use, and accessibility. The system allows users to link multiple bank accounts to a single mobile application and make instantaneous peer-to-peer (P2P) transactions using a mobile number or UPI ID, eliminating the need for bank account numbers or IFSC codes. This streamlined process has made digital payments more accessible, especially for the unbanked and underbanked populations, offering a solution to India's longstanding challenges of financial exclusion.

UPI's accessibility is further enhanced by its integration into a vast ecosystem of third-party applications that allow for seamless transactions not only in personal payments but also in bill payments, merchant transactions, government services, and remittances. Major FinTech players such as PhonePe, Google Pay, Paytm, and Amazon Pay have all integrated UPI into their platforms, allowing users to make payments directly from their bank accounts without relying on intermediaries or traditional credit systems. UPI has transformed the smartphone into a gateway for financial services, enabling users to make transactions easily and confidently.

UPI and Financial Inclusion: One of the primary goals of UPI's design was to improve financial inclusion in India, which has long been hindered by several factors: a fragmented banking infrastructure, geographical barriers, lack of awareness about banking products, and a low level of digital literacy, especially in rural areas. According to data from the Reserve Bank of India (RBI), more than 80% of India's population now has access to banking services through UPI, compared to just 35% in 2011, when India's financial inclusion index was first measured.

UPI has particularly benefited rural and underserved communities with sparse physical banking infrastructure. By enabling digital transactions through basic smartphones, UPI has enabled people in rural areas to access financial services and participate in the formal economy, without needing to travel long distances to the nearest bank branch. The widespread adoption of UPI has also facilitated direct benefit transfers (DBT) from government programs like PMGDISHA (Pradhan Mantri Gramin Digital Saksharta Abhiyan) and PM-KISAN, where subsidies, pensions, and other welfare benefits are directly transferred to citizens' bank accounts via UPI. This system has minimized leakages, reduced fraud, and made the welfare process more efficient, transparent, and accountable. Another significant impact of UPI has been on remittance flows. India is the world's largest recipient of international remittances, and UPI has streamlined the process of sending money back home from abroad. By integrating UPI into remittance services, the cost of cross-border remittances has been significantly reduced, and the time taken for funds to reach beneficiaries has been shortened, thus improving the quality of life for millions of migrant workers and their families.

FinTech Revolution in India: The FinTech revolution in India, driven by platforms like UPI, has a ripple effect on the country's broader economic landscape, fueling innovation, entrepreneurship, and job creation. India is currently one of the largest FinTech hubs in the world, with thousands of startups and entrepreneurs developing innovative solutions to address the country's complex financial needs. The explosion of UPI, along with its integration into digital wallets, mobile payment platforms, and innovative financial products, has spurred a shift from traditional banking models to more dynamic, tech-driven financial solutions.

A key aspect of India's FinTech transformation has been the surge in digital payments, which have been integrated into a wide range of consumer activities, from paying for groceries to booking ride-sharing services. For instance, UPI-based apps such as PhonePe, Google Pay, and Paytm have become part of everyday life for millions of users, enabling them to easily transfer money to friends, pay bills, and shop online. According to the NPCI, UPI recorded over 9 billion transactions in a single month in late 2023, with a cumulative value of INR 15 trillion (approximately $180 billion), a clear indication of UPI-driven services' rapid adoption and scalability. UPI has helped increase the financial literacy of millions of Indians, particularly in the rural and semi-urban areas. The simplicity of the UPI interface, combined with government initiatives such as Digital India, has played a crucial role in educating the population about the benefits of digital payments, personal finance management, and online financial services. The digital divide is being bridged through these initiatives, creating new economic growth and development opportunities, especially in agriculture, education, and small business enterprises.

Limitations: While the impact of UPI and the FinTech revolution has been overwhelmingly positive, several challenges and limitations still need to be addressed to realize its full potential.

- **Security and Privacy Concerns:** Despite UPI's success, security remains a primary concern. With the rise in digital payments, there has been an increase in cybercrime, identity theft, and fraud. Phishing attacks, unauthorized transactions, and digital wallet hacks have become more common. To address this, the RBI and the NPCI have implemented several measures, including multi-factor authentication, secure socket layer (SSL) encryption, and periodic security updates to UPI apps. However, a stronger emphasis on educating users about online safety and better enforcing cybersecurity protocols is essential to build trust in digital payments.
- **Technological Barriers:** While UPI has revolutionized financial inclusion, it is essential to note that digital literacy remains a barrier in some population segments. While smartphones are becoming increasingly ubiquitous, many of India's population still lacks the basic digital skills to navigate UPI apps effectively. In rural areas, poor internet connectivity and lack of digital infrastructure can impede the smooth functioning of UPI and other FinTech services.
- **Financial Exclusion of the Last Mile:** While UPI has done a tremendous job of addressing the needs of urban and semi-urban users, the last-mile delivery of digital financial services remains a challenge. The unbanked population, particularly in the most remote corners of India, still faces difficulties accessing banking services due to a combination of factors such as lack of identification documents, unreliable internet connectivity, and absence of banking agents.
- **Regulatory Hurdles:** As FinTech grows, the regulatory environment must evolve to keep pace with new technologies, products, and services. In particular, issues like data privacy, consumer protection, and cross-border transactions need to be addressed by regulatory bodies like the RBI and the Ministry of Finance to ensure the sustainability and growth of the sector.

Future Scenario

As the world moves towards achieving sustainable development goals (SDGs) outlined by the United Nations, the need for more advanced and integrated approaches to global sustainability is paramount. The future scenario of sustainability hinges on the effective integration of artificial intelligence (AI), data interoperability, and collaborative platforms to enhance decision-making, improve policy implementation, and ensure that sustainability efforts are scalable, equitable, and impactful. AI-driven collaborative platforms and interoperable data platforms will play a pivotal role in streamlining global sustainability initiatives by connecting various ecosystems, driving smarter decisions, and enabling systemic change that benefits people, the planet, and prosperity.

Role of AI in Sustainability: AI is poised to transform how we approach sustainability by leveraging machine learning algorithms, predictive analytics, and real-time data processing to address complex global challenges. These technologies can help optimize resource use, mitigate the impacts of climate change, and drive innovation in sectors such as energy, agriculture, water management, and urban development. By processing vast amounts of data from diverse sourcessuch as satellite imagery, IoT sensors, and social media platformsAI can help monitor environmental conditions, forecast natural disasters, and guide decision-making on the most effective interventions.

One of the most promising applications of AI in sustainability is the development of AI-powered collaborative platforms that enable different stakeholders from governments and businesses to NGOs and local communities to work together towards shared sustainability goals. These platforms leverage AI's ability to analyze data and provide actionable insights, creating a data-driven ecosystem for collaborative problem-solving. Such platforms can foster cross-sector partnerships, improve communication, and accelerate progress towards global sustainability targets by ensuring all actors are aligned and working towards a common purpose.

AI-driven Collaborative Platforms – Enabling Global Sustainability Initiatives: The future of global sustainability initiatives will increasingly rely on AI-driven collaborative platforms that enable real-time decision-making and integration across sectors. These platforms are built on the foundation of AI, big data, and cloud computing, allowing global networks to share information, tools, and resources. By democratizing access to data, these platforms will empower diverse stakeholders to collaborate more effectively, ultimately driving more inclusive and equitable sustainability outcomes.

AI can enhance the effectiveness of sustainability initiatives by enabling predictive modelling and scenario analysis. For example, AI algorithms can be employed to forecast climate change impacts, identify trends in resource consumption, or evaluate the environmental footprint of various activities across regions. By integrating AI models into collaborative platforms, stakeholders can simulate different policy responses or interventions, identifying those with the greatest positive impact in the shortest time. AI can also help create personalized solutions for stakeholders based on their needs, resources, and objectives. By processing data from various sources, AI-powered platforms can offer customized recommendations for businesses, governments, and individuals, enabling them to adopt the most effective strategies for improving sustainability. This tailored approach fosters engagement and encourages active participation in sustainability efforts, driving greater accountability and transparency.

Interoperable Data Platforms – Connecting Ecosystems for Smarter Decision-making: The interoperability of data is a crucial enabler of sustainability in the future. For AI-driven collaborative platforms to function effectively, they must be supported by interoperable data systems that connect various stakeholders and ecosystems seamlessly. These platforms enable different sectors environmental organizations, research institutions, policy-makers, industries, and local communities to share data, insights, and resources without encountering compatibility issues or silos. Interoperable data platforms allow standardised data exchange across borders, sectors, and technologies, ensuring all parties can access the same information.

A key element of this vision is the creation of open data platforms that allow for data integration across various domains. To comprehensively understand sustainability issues, these platforms can aggregate data from diverse sources, such as sensor networks, geospatial data, weather models, and economic indicators. For instance, AI could use data from smart grids, weather stations, and renewable energy systems to optimize energy consumption, ensuring a more sustainable energy future. Similarly, big data analytics can assess agricultural practices' environmental impact, while machine learning models can help improve supply chain efficiency and reduce emissions. Interoperability in data systems will also facilitate the development of smart cities. Smart city platforms rely on a network of IoT devices, AI algorithms, and cloud computing to optimize urban management and improve resource efficiency. For example, data from smart meters can be integrated with AI-driven decision-making tools to monitor and manage water usage, waste management, and energy consumption in real-time. By connecting various urban systems, interoperable data platforms allow cities to become more efficient, sustainable, and resilient to climate change impacts.

Smarter Decision-making – Data-Driven Governance for Sustainability: As the complexity of global sustainability challenges increases, the need for data-driven decision-making becomes even more critical. Interoperable data platforms enable smarter governance, where data from multiple sources is integrated to inform policy, strategy, and action. Governments can use these platforms to track the progress of sustainability goals, identify gaps, and develop policies that align with climate goals and economic growth.

AI can also enhance the capacity of governments and organizations to respond to environmental challenges. With access to vast amounts of data, decision-makers can make evidence-based decisions that are more likely to yield positive outcomes. For example, AI-driven platforms can help governments predict the effects of climate change on vulnerable communities, allowing them to implement preventive measures and adaptation strategies in advance. Additionally, AI models can optimize resource allocation, ensuring that financial and human resources are directed where they are needed most.

One of the most promising areas of AI-driven decision-making is managing natural resources. By integrating data from satellite imaging, sensor networks, and IoT devices, AI can monitor forests, water bodies, and ecosystems in real-time, detecting patterns of deforestation, pollution, or overuse of resources. This enables policymakers to proactively mitigate damage, ensure sustainable resource use, and protect biodiversity. In agricultural practices, AI-driven decision-making tools can optimize irrigation schedules, pesticide use, and crop selection, reducing waste and increasing productivity while maintaining environmental sustainability. Also, AI's ability to analyze large-scale data sets enables more accurate risk assessments of environmental hazards, such as floods, droughts, and hurricanes. Governments and relief agencies can leverage this capability to improve disaster response and resilience strategies, ensuring communities are better prepared for climate-related emergencies.

Bridging the Gap: Implementing AI-driven collaborative platforms and interoperable data systems can bridge the gap between the Global North and Global South for sustainability. One of the key challenges in global sustainability efforts is the uneven distribution of resources, technological access, and expertise between developed and developing countries. However, AI and interoperable data platforms can help address these disparities by democratizing access to data and technological tools. For instance, countries in the Global South often face limited access to cutting-edge technologies, hindering their ability to implement sustainable practices in agriculture, water management, and energy production. AI-driven platforms can provide these countries access to best practices, data, and insights from more developed regions, enabling them to leapfrog traditional development models and adopt more sustainable approaches. Similarly, global partnerships between governments, NGOs, international organizations, and the private sector can help transfer knowledge, resources, and technology to needy countries, fostering greater collaboration in achieving the SDGs.

Future of Smarter Sustainability: Global sustainability initiatives will be shaped by the seamless integration of AI-driven collaborative platforms and interoperable data systems that enable smarter decision-making, greater resource efficiency, and more effective governance. By harnessing the power of AI to analyze complex datasets, these platforms will provide actionable insights that can guide policy-making and accelerate progress towards the SDGs. Interoperable data systems will ensure that stakeholders across borders, sectors, and regions can collaborate more effectively, sharing resources, tools, and knowledge to address sustainability challenges on a global scale. As we look towards a future of sustainable development, integrating AI and data interoperability will be crucial in driving innovative solutions, bridging gaps between developed and developing countries, and ensuring that global sustainability efforts are inclusive, efficient, and impactful. Through AI-driven collaborative platforms, the world will be better equipped to navigate the complex challenges of sustainability and create a more equitable, resilient, and prosperous future for all.

8

Climate Change Mitigation and IT

Introduction

Climate change mitigation refers to the efforts and strategies to reduce or prevent the release of greenhouse gases (GHGs) into the atmosphere to limit global warming and its subsequent impacts on ecosystems, human health, and economies. This process is critical to achieving the global climate targets set under frameworks like the Paris Agreement. Technology, particularly Information Technology (IT), has emerged as a vital tool in the fight against climate change, providing a range of solutions to monitor, manage, and mitigate the environmental impacts caused by human activities. IT-driven strategies enable real-time monitoring of environmental variables, optimising resource use, and developing innovative solutions across various energy, transportation, agriculture, and urban planning sectors.

Role of IT in Climate Change Mitigation: The role of IT in climate change mitigation is multifaceted and spans several domains, ranging from data analytics and artificial intelligence (AI) to cloud computing and internet of things (IoT). At the core of these IT innovations is the ability to process vast amounts of environmental data, analyze patterns, and provide actionable insights that inform local, national, and global decision-making. Big data analytics and AI algorithms can be used to predict environmental changes, optimize resource usage, and reduce emissions, thereby playing an essential role in the design of climate-resilient systems.

Monitoring and Predicting Climate Change through IT: The first step in addressing climate change is understanding its current and projected impact. Satellite imaging, IoT sensors, and remote sensing technologies enable the continuous collection of data related to temperature variations, air quality, ocean levels, deforestation, and other critical environmental indicators. These technologies not only monitor real-time environmental changes but also facilitate the forecasting of long-term climate trends, enabling policymakers to develop informed strategies for mitigation. For instance, climate models powered by AI can simulate future climate scenarios, helping scientists predict the potential impacts of rising global temperatures on different regions and sectors, thereby informing global strategies for climate mitigation. One example of IT's contribution to climate change monitoring is NASA's Earth Observing System Data and Information System (EOSDIS), which uses satellite data to track global environmental changes. The system collects and analyzes climate data daily, helping monitor key indicators like global temperatures, sea ice levels, and forest coverage. These insights are critical for policymakers and international organizations working to reduce GHG emissions and adapt to the changing climate.

IT for Energy Efficiency and Renewable Energy Integration: A key area where IT can make a significant contribution to climate change mitigation is through the optimization of energy usage and the integration of renewable energy sources. The energy sector is a major contributor to GHG emissions, with traditional fossil fuels accounting for a significant proportion of global emissions. However, IT-driven solutions such as smart grids, energy management systems, and distributed energy resources (DERs) are helping to transform the energy landscape by improving energy efficiency, reducing wastage, and enabling the integration of solar, wind, and other renewable energy sources into national and global grids.

Smart grids are one of the most impactful innovations in the energy sector. They use IT infrastructure, such as IoT sensors and advanced metering systems, to provide real-time data on energy consumption and supply, enabling more efficient distribution and usage. Smart grids allow for the dynamic integration of renewable energy, such as solar or wind, into the grid by adjusting energy flows in real-time, responding to fluctuations in energy production and demand. The use of AI in managing these systems further enhances their ability to predict energy needs and optimize power distribution. Additionally, cloud-based energy management platforms can track energy consumption patterns across industries, homes, and commercial spaces, recommending strategies for reducing energy waste. For example, smart home systems can automatically adjust heating, cooling, and lighting based on occupancy, reducing unnecessary energy use. Large-scale cloud computing resources allow for processing and storing vast amounts of energy consumption data, helping governments, corporations, and individuals make more informed decisions about reducing emissions and transitioning to renewable energy sources.

IT in Transportation and Mobility Solutions: The transportation sector, responsible for a significant share of global carbon emissions, can also benefit from IT-driven innovations to reduce carbon footprints. One of the most effective strategies for mitigating climate change is the transition to electric vehicles (EVs) and shared mobility platforms that minimize the number of vehicles on the road. IT plays a central role in supporting the development and adoption of these technologies.

Electric vehicle (EV) networks and charging infrastructure are becoming increasingly integrated with smart cities, leveraging cloud computing, IoT, and AI to optimize the placement of EV chargers and ensure efficient charging cycles. AI-based route optimization can significantly reduce transportation fuel consumption and carbon emissions. For instance, ride-sharing platforms such as Uber and Lyft are using IT to optimize the fleet management of electric vehicles, enabling consumers to access shared mobility options while reducing emissions per passenger mile. The advent of autonomous vehicles also holds the potential for reducing fuel consumption and improving the overall efficiency of transportation systems. AI algorithms can predict traffic patterns, optimize routes, and minimize delays, reducing fuel consumption and decreasing emissions by lowering the number of vehicles idling on the road. Additionally, vehicle-to-vehicle (V2V) and vehicle-to-infrastructure (V2I) communication, facilitated by 5G networks and IoT technologies, can contribute to a more seamless and energy-efficient transportation system.

IT in Agriculture – Precision Farming and Carbon Sequestration: Agriculture, a critical sector for food security and climate change mitigation, benefits significantly from IT-driven innovations. Agricultural practices, including crop production and livestock management, contribute substantially to global GHG emissions. However, through precision farming, drone technologies, and soil monitoring sensors, IT can help farmers reduce emissions, conserve water, and optimize fertilizer usage, all of which can help mitigate climate change.

Precision agriculture relies on data analytics to make better-informed decisions about irrigation, pesticide application, and crop rotation, reducing resource waste and GHG emissions. IT platforms that combine remote sensing technologies, IoT sensors, and machine learning algorithms enable farmers to monitor their fields in real-time, helping them make data-driven decisions that improve productivity while minimizing environmental impacts. IT can also support carbon sequestration efforts by monitoring soil health and facilitating carbon farming initiatives. By leveraging IT to collect soil data and track carbon storage in agricultural lands, farmers can be incentivized to adopt practices that improve the ability of soils to store carbon, ultimately reducing the amount of CO2 in the atmosphere.

Opportunities for IT in Climate Change Mitigation: While IT offers significant opportunities for climate change mitigation, several challenges remain in its full-scale implementation. One of the most important barriers is the digital divide, where access to advanced technologies such as AI, IoT, and cloud computing is limited in developing regions. Bridging this gap and ensuring equitable access to IT tools will be essential for achieving global climate mitigation goals. Additionally, the high energy consumption associated with large-scale IT infrastructureespecially in data centers and blockchain systemsposes a concern in terms of sustainability. Nonetheless, IT also offers opportunities to address these challenges. For example, by cloud computing solutions that improve efficiency in resource management, and by encouraging the use of renewable energy in powering data centers, the environmental impact of IT itself can be reduced. Governments and industries must collaborate to ensure that IT solutions are not only accessible but also designed with sustainability in mind.

Current Scenario in IT Tools for Environmental Monitoring and Emission Reduction

Environmental monitoring has become one of the critical components in the fight against climate change, offering essential insights into the health of ecosystems, air quality, water resources, and the effects of human activities on the planet. Information Technology (IT) has played an increasingly prominent role in advancing environmental monitoring systems, contributing significantly to climate change mitigation in recent years. IT tools such as satellite-based remote sensing, Internet of Things (IoT) sensors, artificial intelligence (AI), and big data analytics have transformed how environmental data is collected, analyzed, and acted upon. These tools enable real-time monitoring of greenhouse gas (GHG) emissions, air pollution, deforestation, and other climate-related metrics, providing governments, businesses, and researchers with the data necessary to mitigate and adapt to the impacts of climate change.

One of the key advancements in environmental monitoring has been using satellite data and remote sensing technologies. Satellites, such as those operated by NASA, the European Space Agency, and other space agencies, have provided the world with a wealth of global-scale environmental data that covers atmospheric conditions, ocean temperatures, ice coverage, and land use patterns. These satellites capture high-resolution imagery and sensor data, enabling scientists to track the effects of climate change across vast geographical regions. Through these platforms, global environmental changessuch as rising temperatures, changing weather patterns, deforestation, and the spread of wildfirescan be tracked in real-time. In addition to satellite-based monitoring, IoT sensors have made significant strides in environmental monitoring. IoT sensors are now widely deployed across urban and rural areas to track air and water quality, soil health, and various environmental indicators directly affecting human and ecological health. These sensors are particularly valuable in measuring air pollutants such as CO2, NOx, methane, and particulate matter (PM2.5), which are directly linked to climate change and human health risks. IoT devices can also be integrated into smart city frameworks to help manage resources more efficiently, reduce energy consumption, and minimize carbon emissions in real-time.

AI and machine learning (ML) have emerged as powerful tools for improving the accuracy and efficiency of environmental monitoring systems. AI algorithms can process massive amounts of environmental data collected by satellites, IoT sensors, and ground-based stations, identifying trends and making predictions supporting proactive measures to reduce emissions and mitigate environmental degradation. For instance, AI can analyze data related to urban traffic patterns to recommend adjustments in transportation systems that reduce congestion and, consequently, emissions. AI-powered predictive models can also forecast weather events, such as hurricanes or heatwaves, helping to prepare for and minimize their impact on vulnerable regions. Additionally, machine learning models detect deforestation, biodiversity loss, and illegal fishing activities, which are key contributors to carbon emissions and environmental degradation. Another critical application of IT in environmental monitoring is in emission reduction strategies. The growing interest in carbon markets and emission trading schemes has made it crucial to track global, national, and local emissions accurately. IT tools are now central to implementing these systems, providing real-time data on emissions from industrial activities, power plants, and transportation systems. For instance, advanced sensors in factories can continuously monitor GHG emissions, allowing organizations to take corrective action before emissions exceed set thresholds. Similarly, cloud-based platforms enable governments and businesses to aggregate emission data from various sources and track progress toward emission reduction targets.

In the industrial sector, energy management systems (EMS) powered by IT are designed to optimize energy use, reduce waste, and minimize emissions. These systems can be integrated with smart grids, enabling dynamic energy distribution, optimizing renewable sources such as solar and wind power. By utilizing IT in this way, organizations can better manage their energy consumption and align their operations with national or international carbon reduction goals.

Current Limitations in Real–Time Climate Analytics: While IT has advanced environmental monitoring and emission reduction capabilities significantly, several limitations still exist in the current landscape that hinders the real-time analysis and actionable outcomes of climate-related data. One of the primary challenges is data quality and availability. Despite the widespread use of sensors and satellites, the accuracy and coverage of environmental data remain inconsistent. For example, limited infrastructure and a lack of standardized measurement techniques in remote or developing regions can lead to incomplete or imprecise data. In these regions, real-time data may be sparse, so decision-makers may lack the timely information to implement effective climate change mitigation strategies. Data interoperability is another challenge in the real-time analysis of environmental data. Organizations, governments, and industries use various formats, protocols, and systems for collecting and analyzing environmental data, creating barriers to seamless data integration. This fragmentation of data creates silos that can impede efforts to develop comprehensive, holistic views of the climate situation. As a result, it can be difficult to correlate data from multiple sources and generate actionable insights quickly.

Even when real-time data is available, its interpretation and actionability can be a significant barrier. The volume of environmental data generated daily can be overwhelming, making it difficult to extract meaningful insights promptly. Many decision-makers still struggle to process and interpret the data effectively despite using advanced AI and machine learning models. In some cases, predictive models might lack the accuracy to forecast future climate events reliably. Additionally, the complexity of climate systems means that many predictions are based on assumptions that may not fully account for unforeseen variables or non-linear processes.

Another key limitation lies in the cost and scalability of IT tools for environmental monitoring. While satellite data and IoT sensors are instrumental in collecting environmental data, deploying these technologies can be expensive, especially in low-income or developing regions. Governments and businesses in these areas may face financial constraints in adopting advanced technologies or scaling up their monitoring systems. Similarly, while cloud-based data platforms and AI tools can potentially improve environmental monitoring, the high computational power required for processing large datasets can be cost-prohibitive for many organizations, especially in the Global South. Integrating real-time environmental data into policy and decision-making processes remains a challenge. Even when high-quality data is available, there are often gaps in how policymakers and industries use it. Many government systems lack the necessary infrastructure to integrate climate data into policy formulation, and industries may not be motivated to act on the data unless incentivized.

In some cases, bureaucratic hurdles, such as regulatory delays or lack of coordination between departments, can prevent the rapid deployment of solutions based on real-time data. Additionally, the fragmented regulatory environment across countries makes implementing universal data collection and reporting standards difficult, which hinders global cooperation in addressing climate change. Finally, the technological barriers in real-time climate analytics must be considered. While AI, machine learning, and IoT advances have revolutionized environmental monitoring, these technologies still face certain technical limitations. IoT sensors, for example, can be subject to data inaccuracies due to sensor malfunctions or poor calibration, which may lead to unreliable results. IoT sensors deployed in remote or harsh environments may face difficulties related to battery life, network connectivity, and data transmission, which can delay or compromise the timeliness of climate data.

IT for Monitoring and Forecasting

Information Technology (IT) integration in environmental monitoring and climate forecasting has significantly transformed over the past few decades. Leveraging tools like satellite-based climate data, predictive models, geospatial analytics, and Internet of Things (IoT) technologies, researchers, policymakers, and businesses now have unprecedented access to real-time and predictive information that aids in addressing the global challenges of climate change. By incorporating various IT tools, we can monitor climate patterns more effectively, predict future climate-related events, and generate actionable insights to reduce the environmental impact of human activities.

Satellite–based Climate Data: Satellites have long been essential tools for environmental monitoring. Since the launch of the first Earth-observing satellites in the 1960s, they have provided invaluable insights into the Earth's atmosphere, land surfaces, oceans, and ice caps. Using satellite-based climate data has become integral to monitoring various environmental conditions, such as global warming, sea-level rise, deforestation, and extreme weather patterns. These satellites, equipped with sensors and imaging tools, collect high-resolution data on key climate variables, such as temperature, humidity, cloud cover, air quality, and greenhouse gas (GHG) concentrations.

Using satellite data for climate forecasting is vital because it offers a global perspective that is difficult to achieve through traditional ground-based monitoring. For instance, NASA's Earth Observing System (EOS) satellites, which include the Terra, Aqua, and Suomi NPP satellites, provide data on key climate indicators like atmospheric composition, radiation balance, and ocean currents. These datasets have been crucial in modeling the Earth's climate system, identifying long-term trends in temperature and precipitation, and observing the behavior of climate anomalies such as El Niño and La Niña. Similarly, the European Space Agency's Copernicus program offers satellite data related to land surface temperature, sea surface height, and vegetation cover, providing essential inputs for understanding the impacts of climate change.

The role of satellite-based data in climate monitoring extends beyond just data collection. With the advent of advanced data assimilation techniques and machine learning algorithms, scientists can now integrate satellite observations with predictive models to generate more accurate forecasts of future climate conditions. For example, satellite data is used with numerical weather prediction models to improve the accuracy of long-range forecasts of temperature, precipitation, and extreme weather events, such as hurricanes or droughts. These forecasts are crucial for early warning systems, disaster preparedness, and mitigating the impacts of climate-related events on vulnerable populations and ecosystems.

Predictive Models for Climate Forecasting: Predictive models are at the forefront of climate science and climate change mitigation. These models use historical climate data derived from satellites, weather stations, and other observational toolsto simulate how the climate system behaves under different scenarios. By incorporating factors such as greenhouse gas emissions, solar radiation, ocean currents, and atmospheric dynamics, predictive models help us understand the potential future effects of climate change.

The climate models used today are often highly complex and computationally intensive. For example, the Coupled Model Intercomparison Project (CMIP) provides a framework for comparing global climate models, simulating interactions between the atmosphere, oceans, and land. These models predict how global temperatures will change over the next century based on various emissions scenarios, such as those outlined in the Intergovernmental Panel on Climate Change (IPCC) reports.

AI and machine learning techniques are increasingly important in enhancing the accuracy and efficiency of climate forecasting. By applying AI algorithms to vast datasets, researchers can identify patterns and correlations that may not be apparent through traditional modeling approaches. For instance, AI can assist in downscaling global climate models to predict regional climate trends with higher resolution and accuracy. Similarly, ensemble modeling techniqueswhere multiple models are combined to account for different uncertaintiesare enhanced by machine learning, enabling more robust climate predictions. However, despite the advancements in predictive models, uncertainties remain, primarily due to the complexity of the climate system and the difficulty in accurately predicting future human activities, such as emissions and land-use changes. While these models provide crucial insights into future scenarios, they are not always precise enough to inform specific local-level policy decisions. This uncertainty highlights the need for continuous refinement of models and more comprehensive datasets to support better climate predictions and strategies.

Geospatial Analytics in Climate Monitoring: Geospatial analytics has revolutionized our ability to monitor and understand the complexities of climate change. Geospatial analytics involves using geographic information systems (GIS) and spatial data to analyze geographic space patterns, trends, and relationships. These tools allow for the mapping of climate-related phenomena, such as deforestation, land-use change, urbanization, and water availability, all of which are essential in assessing the impacts of climate change. For example, geospatial analytics can be used to track changes in forest cover, monitor water resources, and assess the vulnerability of ecosystems to climate-related stresses such as rising temperatures or altered precipitation patterns. By overlaying climate projections with current environmental conditions, geospatial tools provide insights into potential risks and help plan adaptation and mitigation strategies. For example, researchers can model the impacts of different climate scenarios on coastal infrastructure or agriculture to inform decision-making on where to implement flood defences or irrigation systems.

Integrating satellite data with spatial modelling is one of the most significant innovations in geospatial analytics. This integration allows for continuous, real-time monitoring of environmental changes at a global scale. For instance, the

Global Forest Watch initiative uses geospatial analytics to track deforestation globally, enabling stakeholders to monitor illegal logging activities and detect patterns of environmental degradation. Similarly, climate vulnerability maps, incorporating geospatial data with climate forecasts, help identify regions most at risk of experiencing extreme weather events or resource scarcity, allowing policymakers to prioritize actions and allocate resources more effectively.

Geospatial tools are also critical in evaluating the effectiveness of climate adaptation and mitigation strategies. For example, cities can use GIS-based climate modeling to evaluate the impact of urban green spaces on urban heat island effects, or countries can assess the feasibility of large-scale renewable energy projects by mapping wind and solar potential in specific locations. This spatial dimension is critical for making informed decisions about allocating resources, mitigating climate impacts, and planning for a sustainable future.

Real–Time Climate Data Through IoT: Integrating IoT (Internet of Things) technology into environmental monitoring systems has further enhanced the ability to track real-time climate data. IoT devices, which include environmental sensors placed in urban areas, forests, oceans, and agricultural fields, can continuously measure parameters like temperature, humidity, air quality, soil moisture, and greenhouse gas emissions. These sensors provide data that can be transmitted in real-time to centralized cloud platforms, where advanced analytics can process and interpret the data.

For instance, IoT sensors deployed in agricultural fields can monitor soil conditions, optimize irrigation schedules, and minimize water waste. Similarly, urban air quality sensors can detect pollution levels in real time, providing cities with the data to reduce pollution immediately. Smart grids, which use IoT technology to manage energy distribution, enable more efficient use of renewable energy sources and help reduce overall carbon emissions in cities.

One of the biggest advantages of IoT-based monitoring is its scalability. IoT sensors are inexpensive and can be deployed in large numbers across various environments, from urban centers to remote locations. This makes them a powerful tool for real-time climate monitoring across the globe. IoT systems can integrate data from diverse sources, including satellites, weather stations, and local sensors, providing a more comprehensive picture of climate dynamics. However, while IoT offers significant advantages, it also faces challenges such as data security and the need for interoperability between diverse sensor networks and platforms. Additionally, the sheer volume of real-time data generated by IoT sensors can overwhelm existing data processing and storage systems, necessitating advanced big data technologies and cloud platforms to manage and analyze the information.

Energy Transition through IT

The global transition from fossil fuels to renewable energy sources is one of the most critical challenges and opportunities of the 21st century. This transition, called the energy transition, is essential for mitigating climate change, reducing dependency on finite resources, and ensuring a sustainable energy future. Information technology (IT) is increasingly pivotal in integrating and optimising renewable energy sources into the existing energy grid. IT's capacity to enhance the efficiency, reliability, and scalability of renewable energy systems, while addressing challenges related to intermittency, storage, and grid stability positions it as a key enabler in the ongoing energy transition.

Role of IT in Renewable Energy Integration: At the core of the energy transition is the shift towards low-carbon, sustainable energy sources, including solar power, wind energy, hydropower, geothermal, and biomass. However, integrating these renewable sources into traditional energy grids presents numerous challenges. Unlike conventional fossil fuel-based energy generation, which is steady and controllable, renewable energy sources are often intermittent and variable. Solar power, for instance, depends on sunlight, and wind energy is contingent on wind speed and direction. This variability makes it difficult to ensure a constant, reliable supply of electricity, which is crucial for meeting the demand in real-time.

IT systems can help address these challenges by enabling smart grid technologies, energy management systems, and advanced forecasting tools. These technologies leverage data analytics, machine learning (ML), and artificial intelligence (AI) to optimize the operation and performance of renewable energy systems, making them more efficient and responsive to fluctuations in demand and supply. One of the most impactful IT solutions is developing and deploying smart grids, which use advanced sensors, communication networks, and data analytics to monitor and manage electricity distribution in real-time. Smart grids enable utilities to balance supply and demand more effectively by incorporating renewable energy sources into the grid, improving grid stability and reducing the risk of blackouts. Through smart grid technologies, utilities can dynamically adjust to changes in renewable energy generation (e.g., during cloud cover or shifts in wind speed) and use predictive data analytics to forecast energy production and consumption patterns.

Advanced forecasting tools powered by IT are crucial for integrating renewable energy into the grid. By leveraging big data, weather modeling, and AI, these tools enable more accurate energy production predictions from renewable sources. This allows grid operators to anticipate fluctuations in renewable generation, which is key to maintaining grid stability. For instance, machine learning algorithms can process vast amounts of historical weather data and real-time sensor information to predict solar radiation and wind speed patterns with high accuracy, allowing operators to anticipate better and manage fluctuations in energy supply. AI-driven forecasting can assist in the optimization of demand-response strategies, adjusting consumer energy consumption patterns to better match the availability of renewable energy.

IT in Energy Storage Solutions: A key challenge in integrating renewable energy into the grid is energy storage. Because renewable energy sources like wind and solar are intermittent, they must be stored for use during periods of low generation or high demand. Energy storage systems (ESS) such as batteries, pumped hydro storage, and compressed air energy storage are critical components of the energy transition, ensuring that electricity generated from renewable sources can be stored and dispatched when needed.

IT systems play a crucial role in optimising and managing these storage systems. Energy management software and IoT sensors are used to monitor the performance of storage systems, track energy levels, and optimize charge/discharge cycles based on real-time energy demand and supply conditions. Artificial intelligence is increasingly applied to predict energy storage needs by analyzing historical data on energy consumption, grid load, and weather patterns. This allows for smarter, more efficient storage management, reducing the reliance on fossil fuels during periods of peak demand and optimizing the use of renewable energy. For example, smart batteries can integrate with the grid to store excess solar or wind energy during the day and release it during periods of high demand or at night when renewable generation is low. IoT sensors monitor the state of charge of these batteries and provide real-time data on their health, ensuring that energy is efficiently stored and released. Additionally, AI algorithms can predict energy demand spikes and optimize the release of stored energy to prevent outages or grid instability.

Blockchain technology is also emerging as a potential solution for managing decentralized energy storage systems. Through blockchain, individual consumers, businesses, and utilities can securely trade stored energy, creating a peer-to-peer market. This decentralized approach to energy storage enhances grid stability, encourages renewable energy use, and empowers consumers to participate actively in the energy transition.

IT for Decentralized and Distributed Energy Systems: Decentralized energy systems are at the heart of the energy transition, with local energy generation and storage becoming more prevalent. These systems reduce reliance on centralized, fossil fuel-based power plants and increase the resilience and efficiency of energy grids. IT is central in enabling these systems by facilitating distributed energy resources (DERs), such as rooftop solar panels, small-scale wind turbines, and home battery storage. These resources are often coupled with smart meters and advanced control systems that enable consumers to monitor and manage their energy usage more effectively.

Through digital platforms and cloud computing, IT enables aggregating distributed energy resources into virtual power plants (VPPs). A VPP is a network of decentralized energy resources that can be operated as a single, cohesive entity. VPPs use advanced IT systems to optimize renewable energy generation, storage, and consumption across multiple distributed locations. These systems allow for the real-time monitoring of energy production and consumption across a wide geographic area, enabling efficient dispatch and integration of renewable energy into the grid.

Integrating smart meters and IoT devices further enhances the capabilities of decentralized systems. Smart meters enable consumers to track their energy consumption in real-time, while IoT-enabled devices can communicate with the grid to manage energy demand and supply dynamically. For example, when the supply of renewable energy exceeds demand, excess energy can be directed to local storage or sold back to the grid, ensuring that renewable resources are utilized efficiently. Blockchain technology also plays a critical role in enabling decentralized energy systems. By providing a secure, transparent, and tamper-proof ledger for energy transactions, blockchain allows for the seamless energy exchange between prosumers (producers and consumers) in a decentralized energy market. This creates an efficient and transparent energy trading platform, incentivizing the use of renewable energy and facilitating the transition to a more sustainable and resilient energy system.

IT in Renewable Energy Optimization: Once renewable energy sources are integrated into the grid, IT plays a key role in optimizing their performance to ensure that they generate and deliver energy as efficiently as possible. Big data analytics, machine learning, and artificial intelligence are all used to optimize renewable energy generation, transmission, and consumption. For example, machine learning algorithms are used to optimize the operation of wind turbines and solar panels by analyzing data on weather patterns, energy output, and turbine performance. This allows operators to predict the most efficient operating conditions and proactively maintain equipment, reducing downtime and maximizing energy output. Similarly, AI-driven predictive maintenance tools monitor the condition of renewable energy infrastructure and anticipate failures before they occur, further optimizing the performance of renewable energy systems.

In the case of solar energy, IT is also being used to enhance solar panel efficiency. Advanced solar management software can monitor weather conditions, temperature, and sunlight intensity to adjust the tilt and orientation of solar panels in real-time, ensuring that they capture the maximum amount of solar radiation. These systems also optimize energy output by automatically adjusting the panels to avoid shading and reduce energy losses. Additionally, demand-side management systems, powered by IT, can optimize energy consumption patterns to match renewable energy generation better. These systems use real-time data on energy production and consumption to adjust the operation of appliances, industrial equipment, and HVAC systems, reducing energy use during periods of low renewable generation and shifting consumption to times when renewable energy is plentiful.

Future Directions: While IT is critical in integrating and optimising renewable energy, several challenges must be overcome. One major challenge is the interoperability of various IT systems across different regions, grid infrastructures, and energy markets. Ensuring that renewable energy technologies, smart grids, and decentralized systems can work together seamlessly is critical for realizing the full potential of the energy transition. As energy systems become more digitalized and interconnected, the potential for cyberattacks increases. Protecting sensitive energy data and ensuring IT infrastructure security is essential to maintaining grid stability and consumer trust. Finally, the scalability of IT solutions for renewable energy is a critical issue. Many renewable energy solutions face financial and infrastructural barriers to large-scale adoption, particularly in emerging markets. Ensuring that IT-driven solutions can be scaled up to meet global energy needs while remaining cost-effective and accessible is key to achieving a sustainable energy future.

The future of IT in renewable energy looks promising, with continued advancements in artificial intelligence, blockchain, 5G networks, and edge computing enabling even greater optimization, efficiency, and integration. As IT continues to evolve, it will play an increasingly central role in realizing the full potential of renewable energy, driving the global energy transition towards a more sustainable, resilient, and equitable future. IT is at the forefront of the energy transition, enabling the integration and optimization of renewable energy sources and overcoming intermittency, storage, and grid stability challenges. Through smart grids, energy management systems, predictive models, blockchain, and IoT, IT helps ensure that renewable energy is efficiently produced, distributed, and consumed, creating a more sustainable energy future. As the energy transition continues, the role of IT will only become more critical in achieving the global sustainability goals of reducing carbon emissions and mitigating climate change.

Case Study: As the world grapples with the urgent need to mitigate climate change, many nations have turned to information technology (IT) as a powerful tool to address the growing environmental challenges. Among these efforts, India stands out for its innovative, large-scale initiatives to reduce carbon footprints while fostering economic growth and sustainable development. By integrating IT-enabled solutions into various sectors, such as energy, transportation, agriculture, and urban development, India is positioning itself as a key player in the global fight against climate change. This case study explores the national efforts undertaken by India to reduce carbon emissions, focusing on the role of IT-driven solutions in achieving sustainability goals.

IT in Renewable Energy and Carbon Reduction: India's commitment to expanding its renewable energy capacity has been a cornerstone of its efforts to reduce carbon emissions. With one of the largest solar power programs in the world, India has set ambitious targets to achieve 500 GW of non-fossil fuel capacity by 2030, with a specific focus on solar, wind, and other renewable energy sources. The role of IT in this transition is indispensable, as it provides the tools for integrating renewable energy into the national grid, optimizing energy storage, and improving energy efficiency.

One of the primary IT-enabled solutions driving this transformation is using smart grids. A smart grid is an advanced electrical grid that uses information and communication technologies (ICT) to monitor and manage electricity distribution efficiently. In India, the Smart Grid Mission launched by the Ministry of Power is an example of a national initiative that leverages IT to enable more efficient energy distribution, integration of renewable energy sources, and demand-side management. Smart meters and real-time monitoring systems, which collect and analyze data on energy consumption, are being installed across the country to enhance the management of renewable energy. This data-driven approach allows grid operators to better balance supply and demand, avoid wastage, and seamlessly integrate intermittent energy sources like solar and wind into the grid. IT plays a key role in optimizing energy storage systems, which are vital for addressing the intermittent nature of renewable energy. Using AI and machine learning algorithms, energy storage systems can predict demand patterns and optimize the charge-discharge cycles of batteries, ensuring that renewable energy is stored efficiently and used when needed. These technologies, which are supported by cloud computing platforms, enable remote monitoring, control, and diagnostics, ensuring that the stored energy is utilized effectively while minimizing carbon emissions.

IT in Smart Transportation Systems: The transportation sector is one of India's largest contributors to carbon emissions, accounting for nearly 14%. India's transition to a more sustainable transportation system is crucial to achieving its carbon reduction targets. IT-enabled solutions are transforming mobility systems, shifting towards electric vehicles (EVs) and intelligent transportation systems (ITS).

The government of India has launched several initiatives to promote the use of EVs, including the Faster Adoption and Manufacturing of Hybrid and Electric Vehicles (FAME) scheme and various state-level subsidies. The use of IT is central to the success of these initiatives. AI-based routing and vehicle management systems are helping optimize the performance of EVs, increasing their efficiency and range. Cloud-based platforms are also being used to track the availability of EV charging stations, manage the operation of charging networks, and monitor the performance of EVs in real-time. These solutions enable seamless integration of EVs into urban mobility systems, reducing reliance on fossil-fuel-powered vehicles and lowering overall emissions. Also, smart transportation systems (ITS) use a combination of IT tools such as GPS, IoT sensors, big data analytics, and cloud computing to optimize traffic flow, reduce congestion, and promote sustainable mobility solutions. ITS applications include real-time traffic monitoring, congestion prediction, and route optimization, all of which help minimize fuel consumption and reduce emissions. For instance, AI-powered traffic signal optimisation can reduce idling time and promote smoother traffic flow, contributing to lower carbon emissions. Similarly, smart public transport systems that rely on data analytics help optimize the scheduling and routing of buses and trains, further reducing fuel consumption and emissions.

IT in Agriculture: Agriculture remains a key sector in India in terms of employment and environmental impact. Conventional farming practices are resource-intensive, leading to high carbon footprints, water scarcity, and soil degradation. In response to these challenges, IT-enabled solutions are being increasingly adopted to promote sustainable agriculture practices, optimize resource usage, and reduce carbon emissions.

Precision farming, which uses data-driven approaches to optimize water, fertilizers, pesticides, and seeds, is one of the most promising IT solutions for reducing agriculture's carbon footprint. By leveraging IoT sensors, satellite imagery, and data analytics, precision farming enables farmers to make informed irrigation, crop management, and input usage decisions. For instance, smart irrigation systems powered by IoT sensors and AI can determine the precise amount of water crops need, thereby minimizing water wastage and reducing energy consumption. Additionally, drones and satellite imagery are used to monitor crop health, detect pests, and optimize pesticide application, reducing the need for chemical inputs and lowering emissions associated with agriculture. IT also enables better supply chain management in agriculture, reducing food waste and carbon emissions. By leveraging blockchain and AI, farmers can track the movement of goods from farm to market, ensuring that produce is transported efficiently and minimizing spoilage. This reduces the overall carbon footprint of food production, storage, and distribution.

IT in Urban Sustainability: As India's urban population grows, it becomes even more urgent to manage urbanisation and promote sustainable urban development. The concept of smart cities has emerged as a key solution to address the environmental and social challenges associated with urbanization. In India, the Smart Cities Mission launched by the Ministry of Housing and Urban Affairs aims to create 100 smart cities that utilize IT solutions to improve infrastructure, reduce emissions, and enhance the quality of life for citizens.

IT-enabled solutions such as smart waste management, energy-efficient buildings, and green infrastructure are central to India's smart city initiatives. For example, smart waste management systems use IoT sensors to monitor the level of waste in bins and optimize the collection schedule, reducing fuel consumption and carbon emissions from garbage trucks. Building energy management systems (BEMS), which integrate IoT and AI technologies, help monitor and optimize energy consumption in commercial and residential buildings, reducing the overall carbon footprint of urban infrastructure. Smart lighting systems, powered by IoT and AI, adjust lighting levels based on real-time data, reducing electricity consumption and emissions. These systems are designed to minimize energy waste, ensuring that lights are only on when needed and adjusting brightness based on weather conditions and time of day. Additionally, green spaces and urban forestry are being integrated into urban planning to enhance carbon sequestration and improve the overall environmental quality of cities.

IT in Policy and Governance for Carbon Reduction: Effective policy and governance are critical to reducing carbon emissions at the national level. IT solutions are essential in enhancing policy formulation, monitoring, and implementation. By leveraging big data analytics, AI, and IoT, policymakers can make data-driven decisions that align with environmental goals and track progress toward achieving sustainability targets.

For instance, the National Action Plan on Climate Change (NAPCC), which outlines India's strategy for mitigating climate change, relies heavily on IT-enabled monitoring systems to track the implementation of its eight national missions. IT tools help gather and analyze data related to emissions, energy consumption, and climate impacts, allowing policymakers to identify trends, gaps, and opportunities for improvement. Cloud-based platforms and dashboards provide real-time insights into the performance of various climate mitigation initiatives, enabling swift corrective actions. In addition, blockchain technology is being explored for carbon trading and creating carbon credits, incentivising companies and individuals to reduce carbon emissions. Blockchain ensures transparency and traceability in carbon credit transactions, making it easier for stakeholders to participate in carbon markets and contribute to national emissions reduction goals.

Future Scenario

As global awareness of climate change intensifies and nations commit to more ambitious sustainability goals, artificial intelligence (AI) and the Internet of Things (IoT) are poised to advance climate action. With the year 2047 serving as a key milestone for India's carbon neutrality goals, integrating these advanced technologies into climate models and adaptive response systems holds immense potential for predicting long-term impacts, optimizing resource management, and fostering climate resilience. AI and IoT are transforming how we understand climate dynamics and offering innovative pathways for achieving sustainability at the national and global levels.

AI-Enabled Climate Models: The ability to accurately predict and mitigate the long-term impacts of climate change is essential for crafting effective policies and interventions. AI-driven climate models are set to revolutionize how we approach climate prediction, offering a level of precision and adaptability previously unattainable through traditional methods. AI, particularly machine learning (ML) and deep learning algorithms, has already shown promise in enhancing the accuracy of climate simulations, enabling the integration of complex datasets, and providing more detailed predictions of weather patterns, sea-level rise, temperature shifts, and extreme weather events.

One of the most significant advantages of AI-enabled climate models is their ability to process vast amounts of data from multiple sources, including satellite imagery, atmospheric measurements, oceanic data, and climate simulations. These models can use historical data to identify patterns and predict future climate conditions. For example, AI algorithms can forecast the likelihood of extreme weather events, such as hurricanes, floods, or droughts, by recognizing patterns in historical climate data and adjusting for current trends. This predictive capability is crucial for policymakers to anticipate and prepare for future climate events, ensuring timely interventions and reducing vulnerability to climate-related disasters. AI models can simulate the interactions between various environmental variables, such as temperature, precipitation, and vegetation, to create more nuanced and accurate long-term climate change projections. These models can assess the impact of different emission scenarios, guiding mitigation strategies. By incorporating AI into climate science, governments and researchers can gain a deeper understanding of the long-term effects of human activity on global ecosystems, allowing for more informed decision-making and targeted interventions.

For India, AI-powered climate models can be tailored to local contexts, considering geographic diversity, socioeconomic conditions, and regional environmental challenges. These models can provide insights into how climate change will affect specific regions, enabling the country to design region-specific climate adaptation strategies that protect vulnerable communities, ecosystems, and infrastructure. As India seeks to achieve carbon neutrality by 2047, AI-driven models will be indispensable for tracking progress, refining strategies, and ensuring that emissions reduction targets are met in line with national and global climate agreements.

IoT Networks for Adaptive Climate Response Systems: The integration of IoT into climate response systems is another critical component of India's journey toward carbon neutrality. IoT refers to the interconnected network of devices that collect and exchange data, facilitating real-time monitoring and control of physical systems. IoT networks have the potential to significantly enhance adaptive climate responses by providing the real-time data necessary for informed decision-making and rapid action.

One of the key applications of IoT in climate response is environmental monitoring. IoT-enabled sensors can monitor various environmental parameters, such as air quality, temperature, humidity, soil moisture, and water quality. These sensors can be deployed across vast areas, from urban centers to remote rural locations, providing granular data on climate conditions and enabling more accurate tracking of environmental changes. For instance, IoT-based air pollution monitoring systems can detect and analyze particulate matter, greenhouse gas emissions, and pollutants in real-time. This data can then be used to implement corrective measures, such as adjusting industrial emissions, regulating traffic, or enhancing public health responses to air pollution.

IoT technologies empower farmers to adapt to climate change through precision farming techniques in the agricultural sector. Using soil moisture sensors, weather prediction tools, and remote sensing data, IoT systems help optimize water usage, improve crop management, and reduce the carbon footprint of farming operations. These systems enable farmers to make real-time irrigation, fertilization, and pest control decisions, ensuring more sustainable agricultural practices resilient to the changing climate. IoT networks can play a significant role in disaster management and early warning systems. By integrating IoT with AI-driven predictive models, governments can create adaptive climate response systems that provide early warnings for extreme weather events, such as floods, hurricanes, or heat waves. These systems can use data from IoT sensors to monitor environmental conditions in real-time and trigger alerts when conditions indicate a potential disaster. Such proactive measures can help save lives, protect infrastructure, and minimize the economic losses associated with climate-related disasters.

Smart cities with IoT networks are emerging as key components of adaptive climate responses in urban areas. These cities use IoT-based solutions to optimize energy consumption, manage waste, and monitor air and water quality. Smart grids, powered by IoT and AI, enable the efficient distribution and consumption of energy, integrating renewable energy sources and reducing dependency on fossil fuels. Additionally, smart transportation systems that rely on IoT sensors can reduce congestion, lower emissions, and promote the use of electric vehicles, all of which contribute to reducing the carbon footprint of urban centers.

Carbon Neutrality Goals for 2047: India's ambitious target of achieving carbon neutrality by 2047 requires a comprehensive and integrated approach that spans multiple sectors, including energy, transportation, industry, agriculture, and urban development. IT-enabled solutions, including AI-driven climate models and IoT networks, are central to the country's strategy for achieving this goal. These technologies will enable India to track and reduce emissions and ensure the transition to a low-carbon economy is inclusive, sustainable, and resilient. The first step toward achieving carbon neutrality is decarbonising the energy sector. As one of the world's largest energy consumers, India's shift to renewable energy is critical in meeting its carbon neutrality targets. AI-enabled climate models will help optimize the integration of renewable energy sources into the national grid, manage intermittency, and predict future energy demand. IoT networks will play a key role in smart grids, enabling efficient energy distribution and reducing transmission losses. By leveraging these technologies, India can achieve a cleaner, more reliable energy system that reduces its carbon footprint and supports the growth of renewable energy.

The transportation sector, a significant contributor to India's emissions, will benefit from AI and IoT technologies. Adopting electric vehicles (EVs), smart transportation systems, and intelligent traffic management will reduce emissions from conventional fossil fuel-powered vehicles. AI-based models will enable more efficient routing of public transport systems, while IoT networks will optimize traffic flow and reduce congestion. Integrating EVs into the national grid, supported by IoT networks and AI, will further reduce the sector's carbon emissions and promote sustainable urban mobility.

In agriculture, precision farming techniques powered by IoT and AI will help optimize resource usage, reduce waste, and minimize the carbon footprint of food production. The shift toward sustainable agricultural practices, including organic fertilizers, water-efficient irrigation, and carbon sequestration techniques, will play a crucial role in reducing emissions from this sector. For India to meet its carbon neutrality goals, it must also promote circular economy principles, emphasising resource efficiency, waste reduction, and product lifecycle management. AI-driven models will support circularity by optimizing resource use, identifying waste reduction opportunities, and enabling better recycling systems. IoT networks will ensure data flows seamlessly across sectors, connecting ecosystems and creating a more integrated, sustainable economy. Finally, achieving carbon neutrality will require substantial investment in climate resilience. As India faces the dual challenges of mitigating climate change and adapting to its unavoidable impacts, IoT-enabled adaptive climate response systems will help manage and reduce risks associated with extreme weather events. AI-driven predictive models will enable better disaster preparedness, ensuring India can effectively respond to climate-related crises while minimizing their socioeconomic and environmental impacts.

9

IT and Energy Efficiency

Introduction

Information Technology (IT) has become a cornerstone of modern society, driving innovations across industries and redefining operational efficiencies. However, energy consumption has increased exponentially as the IT sector expands, making energy efficiency a critical focus area. The interplay between IT and energy efficiency is multifaceted, encompassing hardware optimization, sustainable software development, advanced cooling technologies, and integrating renewable energy sources. This relationship is pivotal in mitigating the environmental impact of IT while sustaining its growth trajectory.

Hardware Optimization and Energy Efficiency: Modern IT infrastructures are built on high-performance hardware components, often consuming significant energy. Data centers, the backbone of IT services, exemplify this challenge. These facilities house thousands of servers, networking devices, and storage units, all requiring substantial electrical power. Innovations in hardware design have aimed to address this issue. For instance, energy-efficient processors, such as those designed with reduced instruction set computing (RISC) architecture, minimize power consumption without compromising computational capabilities. Similarly, the advent of solid-state drives (SSDs) has reduced energy usage compared to traditional hard disk drives (HDDs), as SSDs have no moving parts and require less power for data retrieval. Advancements in server virtualization have significantly enhanced energy efficiency in data centers. By enabling multiple virtual machines to run on a single physical server, virtualization reduces the number of servers needed, leading to lower energy consumption. This optimization reduces the energy demand and decreases the associated cooling requirements, enhancing overall efficiency.

Software Development and Energy Optimization: Software plays a crucial role in determining the energy efficiency of IT systems. Poorly optimized software can lead to excessive computational cycles and higher energy consumption. Developers now prioritise energy-aware programming, employing algorithms and coding practices that minimize processing demands. Techniques such as efficient memory management, adaptive algorithms, and load balancing ensure that IT systems perform tasks with minimal energy expenditure. Cloud computing has emerged as a transformative force. Cloud service providers can optimise energy use across shared infrastructures by centralizing computing resources and employing advanced resource allocation algorithms. Platforms like Microsoft Azure and Amazon Web Services (AWS) utilize predictive analytics to adjust resource allocation dynamically, ensuring energy is used efficiently.

Cooling Technologies and Energy Management: Cooling systems are essential for maintaining the operational integrity of IT hardware, yet they are among the largest contributors to energy consumption in data centers. Traditional air-cooling systems often consume nearly as much energy as the IT equipment they support. To address this, innovative cooling solutions have been developed. Liquid cooling, for instance, uses specialized coolants to dissipate heat more efficiently than air-based systems. Immersion cooling furthers this concept by submerging hardware components in non-conductive cooling fluids, achieving superior thermal management. In addition, free cooling techniques, which leverage natural environmental conditions such as cold air or water, have gained traction. Free cooling significantly lowers energy usage by reducing reliance on mechanical cooling systems. Companies like Google and Facebook have implemented such systems in their data centers, achieving remarkable energy savings.

Renewable Energy Integration in IT: Integrating renewable energy sources into IT operations represents a sustainable pathway toward energy efficiency. Solar, wind, and hydroelectric power are increasingly being used to meet the energy demands of IT facilities. For example, Apple has committed to powering its global operations entirely with renewable energy, including its data centers. Similarly, Google has invested heavily in wind and solar farms to offset the carbon footprint of its IT infrastructure. Renewable energy is often coupled with energy storage solutions, such as lithium-ion batteries and grid-scale energy storage systems, to ensure a stable power supply. These technologies allow IT facilities to operate on renewable energy even during periods of low generation, such as cloudy or windless days.

Energy-Efficient Networking and IoT: Networking infrastructure, including routers, switches, and transmission lines, is another significant energy consumer within the IT ecosystem. Energy-efficient networking protocols, such as Energy-Efficient Ethernet (EEE), have been developed to reduce power consumption during periods of low network activity. These protocols adjust the power usage of network devices based on real-time data transmission requirements, ensuring that energy is not wasted. The Internet of Things (IoT) further exemplifies the importance of energy efficiency in IT. IoT devices often operate on limited power sources, such as batteries or energy-harvesting systems. To extend their operational lifespans, these devices rely on energy-efficient sensors, low-power communication protocols, and edge computing, which processes data locally to reduce energy-intensive transmissions to centralized servers.

Current Scenario in Smart Grids and Energy Monitoring Systems: The transition toward sustainable energy systems has underscored the importance of smart grids and advanced energy monitoring systems. These technologies have emerged as pivotal components in modern energy management, facilitating the integration of renewable energy sources (RES), optimizing energy distribution, and improving grid reliability. Despite significant advancements, integrating RES into smart grids remains challenging, characterized by technical, economic, regulatory, and social barriers. This chapter explores the current landscape of smart grids and energy monitoring systems, emphasizing the gaps that hinder the seamless incorporation of RES.

Smart Grids and Their Role in Modern Energy Systems: Smart grids represent a transformative evolution of traditional electrical grids, incorporating digital communication, advanced sensors, and automation to improve energy efficiency and resilience. Unlike conventional grids, which rely on centralized power generation and unidirectional energy flow, smart grids enable bidirectional energy and information flow, accommodating distributed energy resources (DERs) such as solar photovoltaic (PV) systems, wind turbines, and energy storage solutions. At the core of smart grids lie energy monitoring systems, which provide real-time data on energy production, consumption, and grid performance. These systems enable dynamic load balancing, fault detection, and demand response, ensuring optimal grid operation. For instance, advanced metering infrastructure (AMI) allows utilities to monitor energy usage at granular levels, offering insights into consumption patterns and enabling time-of-use pricing strategies.

Integration of Renewable Energy Sources: Renewable energy sources, such as solar, wind, hydro, and biomass, are crucial for reducing greenhouse gas emissions and combating climate change. Their integration into smart grids offers significant opportunities to decarbonise the energy sector and enhance energy security. However, RES's intermittent and variable nature substantially challenges grid stability and reliability.

- **Intermittency and Variability:** The output of RES is inherently variable, influenced by weather conditions and diurnal cycles. Solar energy generation, for instance, fluctuates with cloud cover and daylight hours, while wind energy depends on wind speed and direction. This variability complicates grid management, requiring advanced forecasting tools and flexible grid infrastructure to balance supply and demand.
- **Grid Stability and Reliability:** High penetration of RES can lead to issues such as voltage fluctuations, frequency deviations, and power quality disturbances. Traditional grids are designed for predictable power flows from centralized generators, making them ill-suited for RES's decentralized and variable nature. Smart grids address these challenges through real-time

monitoring, automated controls, and grid-forming inverters that stabilize voltage and frequency.

- **Energy Storage and Demand Response:** Energy storage systems (ESS), such as lithium-ion batteries and pumped hydro storage, play a critical role in mitigating the intermittency of RES. ESS enhances grid flexibility by storing excess energy during periods of high generation and releasing it during low generation. Additionally, demand response programs incentivize consumers to adjust their energy usage in response to grid conditions, aligning demand with RES availability.

Gaps in Integrating Renewable Energy Sources: Despite the potential of smart grids and energy monitoring systems, several gaps hinder the effective integration of RES:

- **Technological Gaps: Grid Infrastructure**: Many existing grids lack the infrastructure required to support high levels of RES penetration. Upgrading grids with advanced sensors, communication networks, and automation technologies is costly and time-intensive.
- **Interoperability**: The lack of standardized communication protocols and data formats limits the interoperability of diverse energy assets and systems, impeding seamless integration.
- **Energy Storage Limitations**: Current energy storage technologies face cost, scalability, and efficiency challenges. The limited lifespan and environmental impact of lithium-ion batteries, for instance, highlight the need for alternative storage solutions.

- **Data and Communication Gaps**
  - **Real-Time Data Integration**: Efficient RES integration requires real-time data from generation, transmission, and consumption points. However, many grids lack comprehensive data integration capabilities, leading to suboptimal decision-making.
  - **Cybersecurity**: As smart grids become more reliant on digital communication, they are increasingly vulnerable to cyberattacks. Ensuring robust cybersecurity measures is critical for safeguarding grid operations.

- **Regulatory and Policy Gaps**
  - **Inconsistent Policies**: The absence of harmonized policies and incentives across regions creates uncertainty for renewable energy projects and grid modernization efforts.

- **Grid Access and Fair Compensation**: Small-scale RES producers often face challenges in gaining access to the grid and receiving fair compensation for their contributions, discouraging broader participation.

- **Economic and Financial Gaps**
    - **High Upfront Costs**: Deploying smart grid technologies and energy storage systems involves significant upfront investments, which can deter utilities and consumers.
    - **Market Structures**: Traditional energy markets are often designed around centralized power generation, making accommodating distributed and variable RES difficult.

- **Social and Behavioral Gaps**
    - **Public Awareness**: Limited awareness of smart grid benefits and demand response programs among consumers can hinder adoption.
    - **Resistance to Change**: Stakeholders, including utilities and policymakers, may resist transitioning from conventional systems due to perceived risks and complexities.

Pathways to Bridging the Gaps: To address these gaps, a multifaceted approach involving technological innovation, policy reforms, and stakeholder engagement is essential:

- **Advancing Technology**
    - **Smart Inverters and Controllers**: Developing advanced inverters capable of managing voltage and frequency deviations can enhance grid stability.
    - **Grid-Edge Technologies**: Distributed energy resource management systems (DERMS) and microgrids enable localized energy optimization, reducing the burden on central grids.
    - **Emerging Storage Solutions**: Research into alternative storage technologies, such as flow batteries, solid-state batteries, and hydrogen storage, can address the limitations of current systems.

- **Enhancing Data Integration**
    - **IoT and Big Data**: Leveraging IoT devices and big data analytics can provide actionable insights for grid operators, enabling proactive decision-making.
    - **Cybersecurity Frameworks**: Implementing robust cybersecurity frameworks ensures the resilience of smart grid systems against cyber threats.

- **Policy and Regulatory Innovations**
  - **Standardization**: Developing standardized protocols and regulations for grid interoperability fosters seamless integration of diverse energy resources.
  - **Incentives and Subsidies**: Policymakers can encourage RES adoption and smart grid deployment through financial incentives, tax credits, and grants.
- **Economic Models**
  - **Innovative Financing**: Public-private partnerships, green bonds, and performance-based funding models can address financial barriers.
  - **Dynamic Pricing**: Time-of-use and real-time pricing models incentivize consumers to align energy usage with RES availability.
- **Engaging Stakeholders**
  - **Consumer Education**: Awareness campaigns and educational programs can highlight the benefits of smart grids and RES integration.
  - **Collaborative Governance**: Involving diverse stakeholders, including utilities, consumers, and policymakers, ensures a holistic approach to energy transition.

Energy Infrastructure Challenges: The global energy landscape is transforming profoundly, driven by the need to decarbonize power systems and transition to renewable energy sources (RES) like solar, wind, and hydropower. While this shift promises significant environmental and economic benefits, it also brings forth critical challenges in the energy infrastructure, particularly concerning transmission inefficiencies and the seamless integration of RES. These challenges are rooted in the aging grid infrastructure, geographical disparities between energy generation and consumption, and the technical complexities of incorporating variable renewable energy sources into existing grids. This section provides an in-depth exploration of these issues, highlighting their implications and potential pathways to overcome them.

Transmission Inefficiencies in Existing Energy Infrastructure: The efficiency of energy transmission systems is a cornerstone of any reliable power network. However, traditional energy grids face inherent inefficiencies that result in significant energy losses during transmission and distribution (T&D). According to estimates, global electricity transmission losses average around 8-15%, varying based on the infrastructure quality and geographical conditions.

1. **Aging Transmission Infrastructure:** Many countries rely on decades-old transmission networks designed for centralized power generation systems.

These networks are ill-suited for modern energy demands, particularly in accommodating distributed and variable renewable energy. Aging transformers, conductors, and substations often operate below optimal efficiency, leading to energy dissipation as heat, increased maintenance costs, and heightened vulnerability to outages. Modernizing these components is essential but requires substantial financial investment and logistical coordination.

2. **Long-Distance Transmission Losses:** A key challenge arises from the geographical separation of renewable energy generation sites and consumption hubs. For instance, wind farms are often located in remote, wind-rich regions, while solar farms thrive in deserts far from urban centers. Transmitting electricity over long distances leads to resistive losses in conventional alternating current (AC) lines. High-voltage direct current (HVDC) systems offer a solution by reducing resistive losses and enabling more efficient long-distance energy transfer, but their implementation is capital-intensive and technologically complex.
3. **Bottlenecks in Transmission Capacity:** Integrating large-scale RES into grids often encounters bottlenecks in transmission capacity. Existing infrastructure may lack the capacity to handle the sudden influx of electricity during peak generation periods, such as midday for solar farms or windy nights for wind farms. This mismatch leads to curtailment, where surplus renewable energy is wasted due to insufficient grid capacity.

Defies in Renewable Energy Integration: Incorporating renewable energy sources into traditional grids is a multifaceted challenge, influenced by the intermittent nature of RES, their spatial dispersion, and the need for robust grid flexibility.

1. **Intermittency and Variability:** Unlike fossil fuel-based power plants, Renewable energy sources depend on inherently unpredictable natural conditions. Solar power generation fluctuates with cloud cover and night cycles, while wind energy depends on wind speed and direction. This variability necessitates real-time adjustments in grid operations to balance supply and demand. Conventional grids, designed for stable and controllable power flows, struggle to accommodate these dynamic changes.
2. **Decentralization of Energy Generation:** Unlike centralized power plants, RES installations are often distributed across multiple small-scale sites, such as residential solar panels and community wind turbines. This decentralization challenges traditional grid designs, which are optimized for a unidirectional flow of electricity from centralized plants to consumers.

Integrating decentralized RES requires advanced grid management technologies and bi-directional power flow capabilities.

3. **Grid Stability and Reliability Issues:** The integration of RES can lead to power quality issues, including voltage fluctuations, harmonic distortions, and frequency deviations. High penetration of RES may also reduce the availability of inertia in the grid, which is traditionally provided by the rotational mass of conventional generators. Inertia is critical for stabilizing grid frequency during sudden disturbances, and its absence necessitates alternative solutions such as synthetic inertia from advanced inverters.

4. **Curtailment of Renewable Energy:** A growing challenge in renewable energy integration is the curtailment of RES due to grid constraints. During periods of high renewable generation and low demand, the inability of the grid to absorb surplus energy leads to wastage. This undermines the environmental benefits of RES and reduces the economic returns for renewable energy producers.

Technical and Economic Barriers: The technical and economic barriers to addressing transmission inefficiencies and renewable energy integration are significant:

1. **Insufficient Energy Storage Solutions:** Energy storage systems (ESS) are critical for mitigating the intermittency of RES by storing surplus energy during peak generation periods and releasing it during low generation periods. However, current ESS technologies face cost, scalability, and efficiency challenges. Lithium-ion batteries, the most widely used storage technology, have limitations such as high initial costs, limited lifespan, and environmental concerns.

2. **Inadequate Grid Flexibility:** Grid flexibility is essential for integrating RES, allowing grids to adapt to rapid changes in generation and consumption. However, most traditional grids lack the flexibility needed to manage variable RES. Implementing demand-side management (DSM) strategies and enhancing grid responsiveness requires substantial investment in advanced technologies like smart grids and real-time monitoring systems.

3. **Economic Constraints:** Modernizing energy infrastructure, including upgrading transmission lines, deploying HVDC systems, and integrating advanced control technologies, requires significant financial resources. Developing countries, in particular, face challenges in mobilizing the necessary funding, delaying the transition to a more resilient and efficient grid.

4. **Regulatory and Policy Barriers:** Regulatory frameworks often lag behind technological advancements, creating barriers to renewable energy integration. Inconsistent policies, lack of incentives for grid modernization, and outdated market structures designed for centralized power generation hinder progress.

Pathways to Overcome Infrastructure Defies: Addressing the challenges of transmission inefficiencies and renewable energy integration requires a combination of technological innovation, policy reform, and stakeholder collaboration.

Technological Solutions: High-Voltage Direct Current (HVDC) Lines: HVDC systems are highly efficient for long-distance electricity transmission, reducing energy losses and enabling the integration of remote RES. Expanding HVDC networks can bridge the geographical gap between renewable generation sites and consumption hubs.

- **Smart Grids and IoT**: Smart grids equipped with IoT devices and advanced sensors enable real-time monitoring and dynamic grid management, improving efficiency and flexibility.
- **Advanced Inverters**: Grid-forming inverters with synthetic inertia capabilities enhance grid stability, compensating for the reduced inertia in RES-dominated grids.
- **Energy Storage Systems**: Investing in diverse ESS technologies, such as flow batteries, compressed air storage, and hydrogen storage, can complement lithium-ion batteries and address their limitations.

- **Policy and Regulatory Reforms**
  - **Incentives for Grid Modernization**: Governments can provide subsidies, tax credits, and low-interest loans to encourage investments in grid upgrades and renewable energy projects.
  - **Harmonized Standards and Protocols**: Developing standardized protocols for grid interoperability ensures seamless integration of diverse energy assets.
  - **Dynamic Pricing Models**: Time-of-use pricing and real-time energy markets incentivize consumers to align their energy consumption with renewable generation patterns.
- **Economic Models and Financing**
  - **Public-Private Partnerships**: Collaboration between public and private entities can mobilize resources for large-scale infrastructure projects.

- **Green Bonds**: Issuing green bonds provides an effective mechanism for financing renewable energy and grid modernization projects.

- **Research and Development**
  - Continued R&D in energy storage technologies, grid management systems, and renewable energy forecasting tools is essential for overcoming existing barriers and enhancing grid resilience.

Smart Grids and Decentralized Energy Systems: The convergence of artificial intelligence (AI) with smart grids and decentralized energy systems is reshaping the energy landscape, unlocking unprecedented opportunities for optimizing energy management and integrating renewable energy sources (RES). Traditional centralized grids, which rely on unidirectional energy flow from large power plants to consumers, are evolving into smart, decentralized systems that leverage AI-driven technologies to enable dynamic, adaptive, and efficient energy distribution. The interplay between AI-optimized smart grids and decentralized energy trading is a critical area of innovation, addressing challenges in load management, renewable integration, and grid resilience. This chapter delves into the transformative potential of AI in this domain, exploring its applications, benefits, and challenges.

Emergence of AI-Optimized Smart Grids: Smart grids are advanced energy networks equipped with sensors, communication systems, and automation technologies to monitor and control energy flow in real-time. Incorporating AI enhances these capabilities by enabling predictive analytics, adaptive control, and decision-making. AI-optimized smart grids analyze vast amounts of data from diverse sources, including energy meters, weather forecasts, and consumer behavior patterns, to dynamically balance energy supply and demand.

1. **Dynamic Load Forecasting:** AI algorithms, such as machine learning (ML) and deep learning (DL), excel at processing historical and real-time data to forecast energy demand accurately. By predicting load variations, AI enables utilities to optimize energy generation, minimize waste, and avoid grid overloads. For instance, recurrent neural networks (RNNs) and long short-term memory (LSTM) models are widely used for time-series load forecasting, allowing grid operators to anticipate peaks and troughs in energy demand.
2. **Real-Time Grid Optimization:** AI enhances grid efficiency through real-time energy flow optimisation. For example, reinforcement learning (RL) algorithms enable smart grids to autonomously adjust energy distribution based on dynamic conditions, such as changes in renewable generation or demand fluctuations. These algorithms continuously learn from their interactions with the grid environment, improving decision-making.

3. **Fault Detection and Self-Healing:** AI-powered anomaly detection systems monitor grid performance to identify faults, predict failures, and initiate corrective actions. Machine learning models can detect patterns indicative of equipment wear, voltage instability, or cyberattacks, enabling proactive maintenance and reducing downtime. Additionally, self-healing grids, driven by AI, can isolate faulted segments and reroute power automatically, enhancing grid resilience.

Decentralized Energy Systems and AI-Driven Energy Trading: Decentralized energy systems, characterized by distributed energy resources (DERs) such as rooftop solar panels, wind turbines, and energy storage units, transform traditional power grids. These systems empower consumers to become "prosumers" who both produce and consume energy. AI facilitates decentralized energy trading by optimizing energy exchange among prosumers, utilities, and markets.

1. **Peer-to-Peer (P2P) Energy Trading:** P2P energy trading platforms, enabled by blockchain and AI, allow prosumers to trade surplus energy with their neighbors or the grid. AI algorithms optimize trading decisions by analyzing market conditions, energy availability, and pricing signals. For instance, predictive analytics can determine the optimal timing and volume of energy trades, maximizing economic benefits for prosumers while stabilizing the grid.
2. **Dynamic Pricing and Market Optimization:** AI supports dynamic pricing models that reflect real-time energy supply and demand. By analyzing market trends and grid conditions, AI ensures pricing mechanisms incentivize energy conservation during peak periods and encourage renewable energy utilization. Reinforcement learning techniques further optimize market operations, balancing economic efficiency with grid stability.
3. **Microgrid Management:** AI plays a pivotal role in managing microgrids, which are localized energy systems that operate independently or in conjunction with the main grid. AI algorithms optimize microgrid operations by coordinating energy generation, storage, and consumption. For example, hybrid systems combining solar panels and battery storage can use AI to maximize self-consumption of renewable energy and minimize reliance on external power sources.

Role of AI in Load Management: Load management is critical to grid operations, ensuring that energy supply meets demand while maintaining grid stability. AI-driven load management systems enhance efficiency and flexibility, particularly of integrating RES.

1. **Demand Response Optimization:** Demand response (DR) programs incentivize consumers to adjust their energy usage based on grid conditions. AI improves DR by predicting consumer behavior, segmenting user groups, and personalizing incentives. Advanced analytics identify high-potential participants, optimizing DR event outcomes. For instance, AI can automate load-shifting strategies, such as delaying non-critical appliances during peak hours and reducing strain on the grid.

2. **Predictive Maintenance:** AI enhances load management by proactively predicting equipment failures and scheduling maintenance. Machine learning models identify potential issues before they escalate by analyzing sensor data from transformers, substations, and other grid components, ensuring uninterrupted power delivery and efficient load handling.

3. **Load Balancing in Hybrid Grids:** Hybrid grids, which integrate conventional and renewable energy sources, benefit from AI-driven load balancing systems. AI algorithms dynamically allocate loads between fossil fuel plants and RES based on availability, cost, and environmental impact. This approach minimizes reliance on fossil fuels during periods of high renewable generation, reducing carbon emissions.

Renewable Energy Integration and AI's Contribution: The integration of RES into smart grids poses challenges due to their intermittent and variable nature. AI addresses these challenges through advanced forecasting, grid flexibility enhancement, and coordination of energy assets.

1. **Renewable Generation Forecasting:** Accurate forecasting of renewable energy generation is essential for grid planning. AI-powered models analyze meteorological data, historical generation patterns, and satellite imagery to predict solar and wind energy output. Convolutional neural networks (CNNs) and ensemble learning methods improve forecast accuracy, enabling grid operators to make informed decisions about energy dispatch.

2. **Energy Storage Optimization:** AI optimizes the operation of energy storage systems by predicting energy demand, renewable generation, and market prices. Battery management systems use AI to determine optimal charging and discharging schedules, extending battery life and maximizing economic returns. For instance, AI can prioritize energy storage during low demand and high renewable generation periods, ensuring availability during peak periods.

3. **Grid Flexibility and Stability:** AI enhances grid flexibility by coordinating DERs, energy storage, and demand response programs. Multi-agent systems, where AI agents represent different grid components, enable

collaborative decision-making to maintain grid stability. For example, AI can dynamically adjust the output of virtual power plants (VPPs) to respond to frequency deviations, ensuring grid reliability.

Case Study

India's Rooftop Solar Initiative represents a significant milestone in pursuing a sustainable energy future. As one of the fastest-growing economies with an ever-increasing energy demand, India has strategically focused on renewable energy to address its energy security and environmental challenges. The rooftop solar initiative, a critical component of India's broader renewable energy strategy, exemplifies the transition toward decentralized and sustainable energy systems. This case study explores rooftop solar's evolution, impact, challenges, and future potential in India, emphasizing its role in transforming the country's energy landscape.

Background and Policy Framework: India launched its National Solar Mission (NSM) in 2010 as part of the National Action Plan on Climate Change (NAPCC). The mission aimed to promote the development and deployment of solar energy technologies across the country, aiming to achieve 20 GW of installed solar capacity by 2022. In 2015, this target was revised to an ambitious 100 GW, including 40 GW allocated for rooftop solar installations. The rooftop solar initiative aligns with India's commitments under the Paris Agreement to reduce carbon emissions intensity and increase the share of non-fossil fuel energy in its total capacity. It also supports providing affordable and reliable electricity to all, as outlined in the Saubhagya Scheme and UDAY (Ujwal DISCOM Assurance Yojana). To facilitate rooftop solar adoption, the government introduced several policy measures and financial incentives, including:

1. **Subsidies and Incentives:** The Ministry of New and Renewable Energy (MNRE) provides up to 40% capital subsidies for residential rooftop systems and various industrial and commercial installations incentives.

2. **Net Metering Regulations:** Net metering policies allow consumers to feed excess electricity from rooftop solar systems back into the grid, receiving credits or payments in return.

3. **Financing Mechanisms:** Public and private sector banks, supported by initiatives like the World Bank's Solar Rooftop Financing Program, offer low-interest loans for solar installations.

4. **State-Level Policies:** States like Gujarat, Rajasthan, and Karnataka have implemented rooftop solar schemes to complement national efforts.

Current Status and Achievements: India's rooftop solar capacity has grown significantly over the past decade, driven by declining technology costs, supportive policies, and increased consumer awareness. As of 2023, the country has achieved over 10 GW of installed rooftop solar capacity, with a significant share contributed by commercial and industrial (C&I) consumers, followed by residential and institutional segments.

- **Adoption Trends**
  - **Commercial and Industrial (C&I) Sector:** The C&I sector accounts for most rooftop solar installations, as these consumers benefit from high electricity tariffs and substantial energy cost savings.
  - **Residential Sector:** Though adoption has been slower in the residential segment, recent policy changes, such as higher subsidies and streamlined net metering processes, have spurred growth.
  - **Institutional Installations:** Schools, hospitals, and government buildings have also embraced rooftop solar as a cost-effective and sustainable energy solution.

Geographical Distribution: Rooftop solar adoption varies across states, with Gujarat, Maharashtra, and Karnataka leading the way. These states have favorable policies, higher solar irradiance, and proactive distribution companies (DISCOMs) facilitating installations.

- **Impact on Energy Security and Environment**
  - **Energy Security:** Rooftop solar reduces dependence on centralized power plants and fossil fuels, enhancing energy security and resilience.
  - **Environmental Benefits:** The initiative has significantly reduced carbon emissions, contributing to India's climate goals. Each megawatt (MW) of rooftop solar capacity offsets approximately 1,500 tons of CO_2 annually.

Innovative Solutions and Best Practices: To address these challenges, various stakeholders have introduced innovative solutions and practices to accelerate rooftop solar adoption.

- **Technology Innovations**
  - **Advanced Inverters and Energy Storage:** Smart inverters and battery storage systems enhance the reliability of rooftop solar by addressing intermittency and grid integration issues.
  - **Digital Platforms:** Solar Marketplace and Solar Rooftop Calculator simplify installation by providing cost estimates, vendor comparisons, and financing options.

Community Solar Projects: Community solar initiatives allow multiple consumers to share the benefits of a single solar installation, overcoming space and financial constraints. These projects are particularly effective in urban and rural settings with limited rooftop availability.

- **Innovative Financing Models**
  - **Pay-As-You-Go (PAYG) Systems:** PAYG models allow consumers to pay for solar installations in affordable instalments, reducing the burden of upfront costs.
  - **Solar Leasing and Power Purchase Agreements (PPAs):** These models enable consumers to adopt solar systems without owning them, paying only for the energy consumed.

Public-Private Partnerships: Partnerships between government agencies, private companies, and non-governmental organizations (NGOs) have driven awareness campaigns, capacity building, and technology deployment.

Future Outlook and Opportunities: The rooftop solar initiative is pivotal in India's transition to a decentralized and sustainable energy system. The government's ambitious goal of achieving 40 GW of rooftop solar capacity by 2030 underscores the potential of this segment. Key drivers of future growth include:

1. **Integration with Smart Grids:** Integrating rooftop solar systems with smart grids and Internet of Things (IoT) technologies will enhance monitoring, optimization, and load management capabilities.
2. **Expansion of Energy Storage:** Coupling rooftop solar with energy storage solutions will enable round-the-clock power availability, reducing reliance on the grid and enhancing energy independence.
3. **Emerging Business Models:** Innovative business models like solar cooperatives and decentralized energy trading platforms will empower consumers and promote equitable energy access.
4. **Government Initiatives:** Continued government support through subsidies, regulatory reforms, and capacity-building programs will drive adoption, particularly in underserved segments.

India's Rooftop Solar Initiative is a testament to the country's commitment to clean energy and sustainable development. While significant progress has been made, addressing the challenges of grid integration, financing, and policy inconsistency will be crucial to realizing its full potential. India can transform its rooftop solar vision into a reality by fostering innovation, enhancing stakeholder collaboration, and prioritizing consumer awareness. The initiative contributes to national energy goals and is a model for other countries striving to embrace decentralized and renewable energy systems.

10

Water Resource Management with IT

Introduction

Given increasing demands, dwindling resources, and climate-induced uncertainties, water resource management is a critical challenge in the 21st century. Integrating Information Technology (IT) into water resource management offers transformative solutions to optimize water usage, monitor resources, predict shortages, and enhance distribution efficiency. IT tools such as remote sensing, geographic information systems (GIS), artificial intelligence (AI), Internet of Things (IoT), and big data analytics are at the forefront of this evolution, enabling more effective planning, decision-making, and resource sustainability.

Role of IT in Water Resource Management

- **Monitoring and Data Collection:** IT provides advanced methods for real-time monitoring of water resources. Remote sensing and IoT-enabled sensors collect data on groundwater levels, reservoir capacities, and water quality. These systems provide granular, real-time insights into water systems, enabling immediate detection of anomalies such as contamination or overuse. For example, IoT sensors deployed in rivers and reservoirs can monitor parameters like pH, temperature, and dissolved oxygen levels, transmitting data to centralized systems for analysis. This facilitates timely interventions to prevent waterborne diseases or ecological imbalances.
- **Data Integration and GIS Mapping:** Geographic Information Systems (GIS) are pivotal in visualizing and analyzing spatial data related to water resources. GIS creates comprehensive maps of water availability, usage patterns, and potential risks by integrating data from satellite imagery, hydrological models, and local monitoring systems. In India, GIS-based systems have been utilized for watershed management, helping policymakers identify areas requiring soil and water conservation measures. Similarly, in urban settings, GIS aids in mapping water distribution networks, identifying leakage hotspots, and planning pipeline repairs.
- **Predictive Analytics and AI:** Artificial intelligence and machine learning algorithms analyze large datasets to predict water-related trends, such as

droughts, floods, and demand surges. Predictive analytics enables water managers to make proactive decisions, such as optimizing reservoir releases during dry periods or preemptively enhancing flood defences. AI models trained on historical weather and hydrological data can accurately predict river flows and rainfall patterns. These insights are invaluable for irrigation planning, hydroelectric power generation, and disaster preparedness.

- **Optimizing Water Distribution:** IT solutions optimize water distribution by minimizing waste and ensuring equitable allocation. Smart water management systems equipped with AI optimize pump operations, valve settings, and flow rates to reduce losses in supply networks. Advanced algorithms dynamically adjust water delivery based on demand forecasts and real-time consumption patterns. Such systems are particularly effective in urban areas where aging infrastructure contributes to significant water loss.
- **Water Quality Management:** Water quality is vital for public health and ecosystem sustainability. IT tools enhance water quality management by enabling continuous monitoring and rapid response mechanisms. For instance, machine learning algorithms can analyze sensor data to detect contamination sources, while blockchain technology ensures transparency and traceability in water quality records. Cloud-based platforms consolidate data from multiple sources, providing stakeholders a unified view of water quality metrics. These platforms facilitate collaboration among municipal authorities, industries, and communities to address pollution effectively.

Applications in Key Sectors

- **Agriculture:** Agriculture accounts for most global water use, making efficient water management crucial. IT-based irrigation systems, such as drip and sprinkler irrigation integrated with IoT sensors, ensure precise water delivery based on soil moisture levels and crop requirements. Decision-support systems (DSS) powered by AI analyze weather forecasts, crop types, and soil conditions to recommend optimal irrigation schedules. These technologies reduce water consumption while maintaining or enhancing crop yields.
- **Urban Water Management:** Urban areas face unique water challenges, including high demand, aging infrastructure, and pollution. IT systems enable smart water grids that monitor usage, detect leaks, and optimize distribution. For example, smart meters provide consumers real-time feedback on their water usage, encouraging conservation. Integrated urban water management platforms use data analytics to balance supply and

demand across residential, industrial, and commercial users. Such platforms are essential for ensuring water sustainability in rapidly growing cities.

- **Disaster Management:** IT is critical in managing water-related disasters like floods and droughts. Flood forecasting models combine meteorological data, hydrological simulations, and GIS to predict flood extents and issue early warnings. Similarly, drought monitoring systems analyze rainfall, vegetation health, and groundwater levels to guide rationing and relief measures. Mobile applications and social media platforms disseminate critical information to affected populations, enhancing community resilience and reducing casualties.

Challenges in IT-Driven Water Resource Management

1. **Data Reliability and Integration:** The effectiveness of IT systems depends on the availability of accurate and comprehensive data. Disparate data sources and inconsistent data quality can hinder analysis and decision-making. Developing standardized protocols for data collection and integration is a key challenge.

2. **Infrastructure and Cost:** Implementing IT systems requires substantial investments in infrastructure, such as sensors, communication networks, and data centers. Developing countries may face financial and logistical barriers to adopting these technologies.

3. **Capacity Building:** Effective use of IT in water management requires skilled personnel to operate, maintain, and interpret technological tools. Training programs and capacity-building initiatives are essential to bridge this gap.

4. **Cybersecurity Risks:** Increasing reliance on IT systems exposes water infrastructure to cybersecurity threats like hacking and data breaches. Robust security measures, including encryption, firewalls, and regular audits, are necessary to safeguard critical systems.

Case Studies

5. **Israel's Water Management System:** Israel has emerged as a global leader in IT-driven water management, employing advanced technologies for desalination, wastewater recycling, and precision irrigation. Its National Water Carrier integrates IoT sensors and GIS to monitor water flows and optimize resource allocation.

6. **India's Jal Jeevan Mission:** India's Jal Jeevan Mission leverages IT tools to provide safe drinking water to rural households. The mission uses GIS to

plan infrastructure projects and mobile applications for real-time monitoring and grievance redressal.

7. **Smart Water Networks in Singapore:** Singapore's Public Utilities Board (PUB) uses smart water sensors, predictive analytics, and centralized control systems to manage its water supply efficiently. These technologies ensure minimal losses and high-quality water for residents.

Future Outlook: Integrating emerging technologies, such as AI, blockchain, and edge computing, will further revolutionize water resource management. AI-driven systems will enable hyper-localized water management, tailoring solutions to specific communities or ecosystems. Blockchain will enhance transparency in water transactions, ensuring accountability in water trading and conservation initiatives. As climate change intensifies water-related challenges, IT will be indispensable in building resilient and adaptive water management systems. Governments, industries, and communities must collaborate to harness the full potential of these technologies, ensuring sustainable water use for generations to come. Water resource management with IT represents a paradigm shift in addressing global water challenges. By leveraging advanced technologies for monitoring, optimization, and decision-making, IT empowers stakeholders to manage water resources sustainably and equitably. While challenges persist, continued innovation and collaboration will pave the way for a future where water is managed as a vital and sustainable resource.

Current Scenario

Water is essential for life and economic development, but its management remains a worldwide challenge. Integrating advanced technologies like sensors and Geographic Information Systems (GIS) has revolutionized water quality monitoring and distribution management, addressing issues such as resource scarcity, pollution, and inefficiencies in water networks. The current scenario highlights the role of these technologies in enhancing water governance, ensuring sustainability, and mitigating the impacts of climate change and urbanization.

Role of Sensors in Water Quality Monitoring

1. **Real-Time Monitoring:** Sensors are at the forefront of modern water quality management, enabling real-time monitoring of critical parameters such as pH, turbidity, dissolved oxygen, conductivity, and temperature. These devices provide continuous data streams, allowing authorities to detect contamination events, monitor pollution levels, and ensure compliance with regulatory standards.

2. For instance, smart water quality sensors deployed in rivers, reservoirs, and industrial effluent discharge points enable instant identification of

pollutants, such as heavy metals, nitrates, or organic contaminants. This capability helps prevent ecological damage and safeguards human health.

3. **Advanced Detection Capabilities:** Recent advancements in sensor technology include the development of biosensors, which use biological recognition elements to detect specific pathogens, toxins, or chemicals in water. These sensors are crucial in addressing microbial contamination, a significant cause of waterborne diseases in developing countries.

4. Similarly, optical sensors using spectroscopic techniques provide high-precision water quality measurements, detecting even trace amounts of harmful substances. These technologies enhance the capacity for early warning systems in water resource management.

5. **Networked Systems for Comprehensive Coverage:** Sensors are often deployed as part of larger networked systems that provide comprehensive coverage of water resources. IoT (Internet of Things) technology connects these sensors to centralized platforms, enabling remote monitoring and management. Cloud-based systems store and analyze data from multiple sources, providing actionable insights to stakeholders. For example, Singapore's Smart Water Monitoring System integrates IoT sensors across its water supply network to ensure high-quality drinking water and prevent distribution losses.

GIS in Water Quality and Distribution

6. **Mapping and Visualization:** GIS is vital in visualizing spatial variations in water quality and identifying pollution hotspots. By integrating data from sensors, satellite imagery, and field surveys, GIS provides detailed maps that illustrate the spatial distribution of water quality indicators, aiding in resource planning and decision-making. For instance, GIS-based maps showing nitrate contamination levels in groundwater can guide agricultural policy reforms to reduce fertilizer usage in vulnerable areas. Similarly, pollution dispersion models generated through GIS help assess the impact of industrial effluents on downstream water bodies.

7. **Integration with Hydrological Models:** GIS integrates seamlessly with hydrological and hydrogeological models, enabling simulations of water flow, contaminant transport, and recharge dynamics. These models inform strategies for groundwater management, watershed protection, and pollution control. For example, the U.S. Geological Survey (USGS) uses GIS-based models to study the impact of land-use changes on water quality and quantity, providing critical insights for sustainable development.

8. **Decision Support Systems (DSS):** GIS is a foundation for Decision Support Systems (DSS) in water management. DSS platforms combine GIS maps, sensor data, and predictive models to assist stakeholders in making informed decisions. These systems are particularly valuable in urban settings, where competing demands for water require efficient allocation and distribution. For instance, GIS-enabled DSS helps municipal authorities plan pipeline networks, prioritize repairs, and reduce non-revenue water (NRW) losses due to leaks or theft.

Integration of Sensors and GIS in Water Distribution Management

- **Smart Water Grids:** Integrating sensors and GIS has given rise to smart water grids, which optimize water distribution by monitoring flow rates, pressure levels, and leakage points in real-time. These grids combine sensor data with GIS-based spatial analysis to provide a comprehensive view of the distribution network, enabling proactive maintenance and efficient resource allocation. For example, in Tokyo, smart water grids equipped with pressure sensors and GIS mapping have reduced leakage rates to less than 3%, one of the lowest levels globally.

- **Leak Detection and Loss Prevention:** Non-revenue water (NRW), primarily caused by leaks and unauthorized consumption, is a significant challenge in water distribution systems. Sensors detect anomalies in flow rates or pressure that indicate leaks, while GIS maps pinpoint the exact location for targeted repairs. Advanced acoustic sensors combined with GIS analysis identify underground leaks in pipelines, minimizing water loss and repair costs. These technologies are particularly valuable in arid regions where every drop of water counts.

- **Equitable Water Distribution:** GIS aids in planning equitable water distribution by analyzing demographic data, infrastructure capacity, and resource availability. By integrating these insights with sensor data on consumption patterns, authorities can design distribution strategies that meet the needs of all stakeholders, including marginalized communities. For instance, GIS-based planning in Bangalore, India, ensures that underserved neighbourhoods receive adequate water supply by optimizing pipeline routes and storage capacities.

Applications in Specific Contexts

1. **Urban Water Management:** Cities face complex challenges in managing water quality and distribution due to high population densities, aging infrastructure, and pollution. Sensors and GIS address these issues by

enhancing monitoring and enabling data-driven decision-making. Smart meters, equipped with sensors, track household water usage, providing consumers with feedback on their consumption and promoting conservation. GIS-based models guide the expansion of urban water networks, ensuring efficient service delivery to new developments.

2. **Rural Water Resource Management:** In rural areas, where water resources are often limited and infrastructure is sparse, sensors and GIS are invaluable for ensuring sustainability. Sensors monitor groundwater levels and quality, while GIS maps identify suitable locations for boreholes, rainwater harvesting systems, and small-scale irrigation projects. For example, in Sub-Saharan Africa, GIS-based groundwater mapping combined with sensor data has improved remote villages' access to safe drinking water.
3. **Disaster Management:** Water-related disasters, such as floods and droughts, significantly impact water quality and distribution. Sensors and GIS enhance disaster preparedness and response by providing real-time data and predictive insights. During floods, water quality sensors detect contamination from overflowed sewage systems, while GIS maps visualize affected areas for targeted relief efforts. Similarly, in drought conditions, GIS identifies vulnerable regions requiring immediate water delivery or rationing.

Future Prospects: The convergence of emerging technologies such as artificial intelligence (AI), blockchain, and edge computing with sensors and GIS holds great promise for water quality and distribution management. AI will enhance the predictive capabilities of sensors, enabling more accurate forecasting of contamination events or infrastructure failures. Blockchain will ensure transparency and traceability in water transactions and quality reporting. Edge computing will process sensor data locally, reducing latency and improving response times. Advances in satellite remote sensing and unmanned aerial vehicles (UAVs) will complement ground-based sensors and GIS, providing high-resolution data for inaccessible or large-scale regions. These innovations will drive the evolution of integrated water management systems, ensuring sustainability and resilience in the face of growing challenges.

Using sensors and GIS in water quality and distribution has transformed how water resources are managed, providing real-time insights, enhancing decision-making, and promoting efficiency. While challenges remain, the continued integration of these technologies with emerging innovations offers a path toward sustainable and equitable water governance. By harnessing the power of sensors and GIS, societies can build resilient water systems that meet the demands of a growing population while safeguarding this vital resource for future generations.

Water Resource Challenges in India

India, home to over 1.4 billion people, faces a profound water resource crisis that threatens its agricultural, industrial, and domestic sectors. The country's water resources are being stretched to unsustainable levels due to over-extraction, inefficiencies in water use, and unequal distribution across regions. While India is endowed with substantial water resources, the challenge lies in its management. Over-extraction of surface and groundwater, coupled with highly uneven spatial and temporal distribution, has resulted in severe water stress across many parts of the country. The issues of over-extraction and inequitable distribution are central to India's water crisis, leading to water scarcity, environmental degradation, and inequitable access to clean water. These challenges are compounded by rapid urbanization, climate change, and population growth, making it imperative to develop effective water resource management strategies that address regional disparities and sustainable water use.

Over-extraction of Water Resources: India's dependence on water for agriculture, industry, and domestic consumption has led to the over-extraction of its available water resources, particularly groundwater. Groundwater is a critical resource in India, as it accounts for approximately 60% of the country's irrigation needs and is a major source of drinking water, especially in rural areas. However, the extraction rate of groundwater is unsustainable in many regions. According to the Central Ground Water Board (CGWB), approximately 60% of India's aquifers are over-exploited, meaning the water extraction rate exceeds the natural recharge rate. This phenomenon is particularly prevalent in agriculturally intensive regions like Punjab, Haryana, Uttar Pradesh, and Rajasthan, where extensive irrigation practices rely heavily on groundwater. In these regions, groundwater depletion has reached alarming levels, with water tables falling by several meters yearly. In Punjab, for example, the water table is depleted by around 30 cm per year.

The root cause of groundwater over-extraction is the extensive use of water for irrigation. India has long relied on flood irrigation techniques that are highly inefficient and waste large amounts of water. Despite the introduction of water-saving technologies like drip and sprinkler irrigation, the adoption rate of these methods remains low, primarily due to the lack of financial incentives, insufficient awareness, and policy support. This results in the continuation of water-intensive farming practices, which, coupled with the absence of effective groundwater management systems, leads to the progressive depletion of aquifers. The over-extraction of surface water also contributes to the overall water stress. India's major rivers, including the Ganges, Yamuna, and Godavari, have been increasingly exploited for agricultural, industrial, and domestic purposes. Over the years, many of these rivers have been dammed, diverted, and drained to meet the water needs of growing populations, contributing to reduced river flows and significant ecological damage. While beneficial in controlling floods and providing water for irrigation and power generation, dam construction has resulted in altering natural flow regimes, impacting aquatic ecosystems and the communities dependent on them. The mismanagement of surface water and the over-extraction of groundwater are accelerating India's water scarcity crisis.

Inequitable Distribution of Water Resources: Apart from over-extraction, the unequal distribution across India is a critical issue. India's water availability is highly uneven, both geographically and temporally. While some regions, such as the northeastern states, the Ganges-Brahmaputra Basin, and the Western Ghats, enjoy relatively abundant rainfall and water resources, other regions, especially in the northwest and south, face persistent water scarcity. The northern and northeastern parts of India, characterized by heavy monsoon rainfall, have a higher density of rivers and groundwater resources. States like Uttar Pradesh, West Bengal, and Bihar benefit from significant annual rainfall and a dense river network.

In contrast, states like Rajasthan, Gujarat, Tamil Nadu, Andhra Pradesh, and parts of Maharashtra and Karnataka face chronic water scarcity. These regions receive significantly less rainfall and have fewer perennial rivers. Consequently, these states rely highly on groundwater, which is rapidly depleting due to over-extraction. The challenge of ensuring water security for these water-scarce regions has been exacerbated by the increasing frequency and intensity of droughts, which are attributed to natural variability and human-induced climate change. In addition to spatial inequalities, the temporal distribution of water is highly variable, with most rainfall occurring during the monsoon season, from June to September. This concentration of rainfall during just a few months results in severe water shortages during the rest of the year, particularly in regions with poor water storage infrastructure. While reservoirs and dams have been constructed to store water for dry seasons, these infrastructures cannot often meet growing demands, especially in urban areas. The seasonal nature of water availability impacts agriculture, which remains highly dependent on monsoon rainfall, leading to irrigation shortfalls during dry spells.

Socioeconomic factors further complicate the inequitable distribution of water. In urban areas, wealthier households and industrial sectors tend to secure priority access to water, while poorer communities, particularly those in informal settlements, often face inadequate access to clean water and sanitation. In rural areas, especially in water-scarce regions, farmers are often the most affected by the over-extraction of groundwater, as they lack the financial means to adopt water-efficient technologies or access alternative water sources. Women and children in rural areas are disproportionately impacted by water scarcity, as they are responsible for collecting water, often from distant sources, which affects their health and educational opportunities. The inequitable access to water is a significant barrier to addressing social and economic inequalities in India.

Impact of Over-extraction and Inequitable Distribution

The consequences of over-extraction and inequitable distribution of water are wide-ranging and severe. Groundwater depletion is a major concern, as it affects agricultural productivity and threatens the availability of drinking water for millions of people. In areas like Punjab and Haryana, the water table has dropped to alarming depths, requiring deeper and more energy-intensive pumping to extract water. This, in turn, raises the cost of irrigation and further burdens farmers, many of whom are already facing economic stress due to low crop yields and poor prices for their produce. The growing cost of water extraction, coupled with the declining quality of groundwater due to contamination from fertilizers, pesticides, and industrial pollutants, has led to a decline in rural livelihoods and increased poverty in water-scarce regions. In addition, water scarcity has profound implications for public health. As groundwater levels fall and surface water bodies dry up or become contaminated, people increasingly rely on untreated water sources, leading to outbreaks of waterborne diseases. Poor sanitation infrastructure, especially in rural areas, exacerbates this problem. Contaminated water sources are a significant cause of diseases like diarrhoea, cholera, and dysentery, which affect vulnerable populations, particularly children.

Ecological damage is another critical consequence of over-extraction and inequitable distribution. Over-extraction of water from rivers, lakes, and aquifers disrupts aquatic ecosystems, affecting biodiversity and the livelihoods of communities dependent on these ecosystems. For instance, excessive water extraction from the Yamuna River has led to a dramatic decline in its flow, affecting the aquatic species that depend on the river and reducing the water available for the millions of people living along its banks.

Addressing the Challenges: Addressing the water resource challenges in India requires comprehensive reforms in water management policies, technological innovation, and sustainable practices. First and foremost, India needs to prioritize water conservation and the sustainable management of its groundwater resources. This can be achieved by implementing efficient irrigation systems, such as drip and sprinkler irrigation, and promoting rainwater harvesting and groundwater recharge practices. Policies promoting water use efficiency in agriculture, industry, and households are essential to reduce the pressure on surface and groundwater resources.

In terms of equitable distribution, it is crucial to improve water governance at the local and regional levels. Decentralized water management, including community-based water systems and better allocation of resources based on need, can help ensure that water is distributed more equitably. Improved infrastructure for water storage, such as larger reservoirs, interlinking of rivers, and better groundwater recharge systems, is also essential to mitigate the temporal variations in water availability. Investments in wastewater treatment and reuse can reduce the demand for freshwater, particularly in urban areas. Technologies like desalination and wastewater recycling for non-potable uses can help ease water stress in coastal and urban regions. Finally, public awareness campaigns and capacity-building initiatives are critical to changing attitudes towards water use and conservation, particularly in rural areas where water inefficiency and waste are rampant.

IoT and AI for Water Conservation

Water scarcity is one of the most pressing challenges of the 21st century, and its impacts are felt across various sectors, particularly agriculture, industry, and urban water management. As global populations increase and climate change exacerbates extreme weather events, efficient water use and conservation have become essential for sustainable development. Integrating Internet of Things (IoT) and Artificial Intelligence (AI) technologies presents a promising solution for water conservation, particularly in smart irrigation systems, water quality monitoring, and drought prediction. By enabling real-time data collection, analysis, and automation, IoT and AI offer significant potential to improve water management practices, enhance agricultural productivity, and mitigate the impacts of water scarcity.

Smart Irrigation Systems: Enhancing Water Use Efficiency: One of the primary sectors where water conservation is crucial is agriculture, which accounts for around 70% of global water consumption. Traditional irrigation practices, especially flood irrigation, are often inefficient, leading to significant water wastage. Smart irrigation, powered by IoT sensors and AI-based analytics, offers a highly efficient and sustainable solution to this problem. Smart irrigation systems use IoT sensors to monitor soil moisture levels, weather conditions, and other environmental factors, enabling precise control over irrigation schedules and water distribution.

IoT-Enabled Smart Irrigation Systems: IoT devices play a critical role in modern irrigation systems by providing real-time data on various parameters such as soil moisture, temperature, and atmospheric humidity. These sensors are placed in the field or on irrigation equipment and are connected to a central management system via wireless networks. The data collected by the sensors is transmitted to cloud-based platforms where it is processed and analyzed to determine the optimal irrigation requirements for the crops.

For instance, soil moisture sensors can inform whether the soil is sufficiently moist or needs additional water. By monitoring the moisture levels in real-time, farmers can avoid over- or under-irrigating, which can lead to inefficiencies, either by wasting water or stressing the plants. Weather data, such as temperature and rainfall forecasts, can also be integrated into the system, allowing for adjustments in irrigation schedules to avoid irrigation during periods of expected rainfall. IoT-enabled irrigation systems can also automate irrigation processes by connecting with automated valves or pumps. This automation allows for water delivery to be optimized based on real-time conditions, ensuring that crops receive the right amount of water at the right time. These systems also allow for remote monitoring and control, enabling farmers to adjust irrigation settings from anywhere, reducing the need for manual labor and improving water management efficiency.

AI-Based Optimization in Smart Irrigation: While IoT devices collect critical data for irrigation, AI algorithms enhance the decision-making process by analyzing the collected data and predicting optimal irrigation schedules. Machine learning (ML) algorithms can be trained to recognize patterns in soil moisture, weather data, and plant requirements to make more accurate and context-aware decisions about irrigation.

For example, AI can forecast crop water requirements based on historical data, weather forecasts, and real-time sensor readings. Deep learning models can be developed to understand the water consumption patterns of different crop types under various climatic conditions. The AI system can then generate predictive models suggesting the most efficient irrigation timings, water quantities, and methods for each field or crop. By analysing sensor data trends, AI can help detect problems in irrigation systems, such as leaks or equipment malfunctions. If the AI detects abnormal readings that suggest inefficiencies, it can send alerts to the farmer, enabling timely intervention and maintenance. By optimizing irrigation schedules, AI-driven systems can significantly reduce water consumption, improve crop yields, and lower energy costs associated with irrigation, contributing to overall water conservation.

Water Quality Monitoring: Water quality is critical for water supply and irrigation. Contaminated water poses significant risks to human health and agricultural productivity, affecting food security and public health. Traditional water quality monitoring methods, such as periodic manual sampling and laboratory testing, are time-consuming, expensive, and unable to provide real-time data. In contrast, IoT-based water quality monitoring systems, powered by AI analytics, enable continuous, real-time monitoring of water sources, offering timely insights into water quality parameters.

IoT-Enabled Water Quality Sensors: IoT-enabled water quality sensors are capable of measuring key parameters such as pH levels, turbidity, dissolved oxygen, temperature, conductivity, and the presence of harmful chemicals or pollutants (e.g., heavy metals, pesticides, nitrates, and bacteria). These sensors are often deployed in rivers, lakes, reservoirs, groundwater, pipelines, and storage tanks, enabling continuous, real-time water quality monitoring across different sources. For instance, turbidity sensors can detect the level of suspended particles in water, indicating pollution or contamination, while dissolved oxygen sensors monitor the water's capacity to support aquatic life. Chemical sensors can detect contaminants, such as harmful heavy metals or toxic agricultural runoff, that may affect both human health and crop productivity. IoT sensors can help track water quality throughout treatment facilities in industrial and municipal settings, ensuring drinking water meets safety standards and remains free from contaminants.

Once the data is collected from IoT sensors, it is transmitted to a cloud-based platform for storage and analysis. Real-time data analytics allow water quality issues to be identified quickly, enabling prompt action to address contamination before it poses serious risks. For instance, in case of a sudden increase in turbidity or the presence of hazardous substances, the system can send immediate alerts to the relevant authorities, triggering a response such as water treatment, filtering, or the closure of affected areas.

AI in Water Quality Monitoring and Analysis: AI plays a crucial role in analysing water quality data collected through IoT sensors. AI-powered algorithms can process vast amounts of real-time sensor data, identify trends, and predict future water quality based on historical patterns and external factors, such as rainfall or industrial activities.

Machine learning models can be trained to detect anomalies in water quality data and predict potential contamination events. For example, AI can recognize patterns in data that signal the early stages of contamination, such as a sudden drop in dissolved oxygen levels or an increase in nitrate concentration, which could indicate agricultural runoff. By detecting these patterns early, AI systems can help authorities mitigate the impacts of contamination and take preventive measures before they affect public health or the environment. AI models can assist in the optimization of water treatment processes. By analyzing real-time data from water treatment plants, AI can suggest adjustments to chemical dosing, filtration methods, or other treatment parameters to ensure the highest water quality with minimal resource consumption. AI algorithms can also predict future water quality trends based on environmental factors, allowing for better long-term planning and risk management in water supply systems.

IoT and AI for Drought Prediction and Management: Droughts are among the most devastating natural disasters, often leading to water shortages, crop failures, and environmental degradation. Predicting drought conditions and managing water resources in anticipation of such events is critical for minimizing their impacts. Traditional methods of drought prediction, based on historical data and meteorological forecasts, often lack the precision and timeliness needed to prepare for the consequences of droughts. IoT and AI offer powerful tools for enhancing drought prediction and management, enabling proactive measures to conserve water and mitigate the effects of drought.

IoT for Drought Monitoring: IoT sensors can be crucial in monitoring environmental variables indicative of drought conditions. These sensors can measure soil moisture, temperature, precipitation, and atmospheric pressure in real-time, providing valuable data that can help detect early warning signs of drought. For example, soil moisture sensors can be used to monitor the drying of soil in agricultural regions, which is one of the first signs of water scarcity during a drought. These sensors provide continuous data on the soil's water content, which can help farmers and water managers track the severity of the drought and adjust their irrigation practices accordingly. In addition to soil moisture, weather stations equipped with IoT sensors can provide real-time data on rainfall, temperature, and humidity, offering further insights into the evolving weather conditions contributing to droughts. This real-time data can be integrated into a centralized monitoring system, enabling water authorities and farmers to make informed water allocation and drought management decisions.

AI for Drought Prediction and Risk Assessment: AI can significantly enhance drought prediction and risk assessment by analyzing large amounts of data from IoT sensors, historical weather data, and other sources. Machine learning models can be trained to recognize patterns in environmental data that precede drought conditions, such as extended periods of low rainfall or higher-than-normal temperatures. These models can predict the likelihood and severity of future drought events based on real-time and historical data, allowing for early warning and preparation.

AI-based models can also help optimize water allocation during droughts. By analyzing the data on soil moisture, crop water requirements, and available water resources, AI can recommend the most efficient ways to allocate limited water supplies. This could involve prioritizing water for essential uses, such as drinking water and critical crops while minimizing non-essential water use. AI can predict droughts' economic and environmental impacts, enabling stakeholders to plan for mitigation strategies such as water conservation campaigns, irrigation scheduling, and water-saving technologies. In drought-prone areas, AI models can also simulate various water management scenarios to identify the most effective strategies for maintaining water availability. These models can assess the impact of different water conservation measures, such as rainwater harvesting, wastewater recycling, or the adoption of drought-resistant crop varieties, allowing policymakers to make data-driven decisions that help mitigate droughts' effects on agriculture and urban water supplies.

Case Study

India's vast population faces significant water resource challenges, particularly in over-extraction, pollution, and the erratic water distribution across regions. The country's increasing demand for waterdue to population growth, urbanization, and agricultural needscombined with the effects of climate change has led to a growing water scarcity crisis. Against this backdrop, the Indian government launched the Jal Shakti Abhiyan (JSA) in 2019, an ambitious water conservation initiative to improve water management nationwide. The JSA's primary goal is to ensure the sustainability of water resources by enhancing water conservation, rainwater harvesting, and reducing water wastage. A key feature of this initiative is the integration of modern technologies, including Geographic Information System (GIS), Internet of Things (IoT), Artificial Intelligence (AI), and remote sensing, to optimize water use and improve resource management across urban and rural areas. By focusing on these technologies, the Jal Shakti Abhiyan seeks to address the water scarcity issue in a more data-driven, precise, and efficient manner.

Background and Objectives of Jal Shakti Abhiyan: Launched by the Ministry of Jal Shakti, the Jal Shakti Abhiyan is a nationwide mission aimed at improving the management and conservation of water resources, particularly focusing on rainwater harvesting, groundwater recharge, and reducing water wastage. The initiative encompasses multiple facets of water resource management, including increasing the efficiency of water distribution systems, promoting community engagement, improving the infrastructure for water conservation, and reducing the dependency on groundwater. The campaign was designed to operate in two phasesone focusing on water conservation in rural areas and the second targeting urban areas and water-intensive sectors. The core objectives of the Jal Shakti Abhiyan include:

1. **Increasing Water Use Efficiency**: This is achieved through modern irrigation techniques, rainwater harvesting, and water-saving technologies.
2. **Groundwater Recharge**: The initiative replenishes groundwater resources by recharging depleted aquifers and optimizing water table levels.
3. **Community Participation**: The initiative stresses community involvement in water management, including local stakeholders, farmers, and rural communities.
4. **Infrastructure Improvements**: Developing infrastructure to improve water storage, distribution, and conservation.
5. **Technology Integration**: Integrating modern technological tools to monitor, manage, and optimize water use.

Technology Integration in Jal Shakti Abhiyan: Technology integration is central to the success of the Jal Shakti Abhiyan. Advanced technologies such as GIS, IoT, AI, and remote sensing enable real-time data collection, monitoring, and decision-making, which is crucial for optimizing water resources. These technologies help identify water-scarce areas, track water levels, plan water distribution systems, and detect inefficiencies in water usage.

Geographic Information System (GIS): GIS is integral to the Jal Shakti Abhiyan. This technology allows for the mapping and analysing water resources across the country. GIS tools help identify critical areas facing water shortages, poor water quality, and regions where water conservation efforts are most needed. Using GIS mapping, planners can track the spatial distribution of water resources and optimize the locations for intervention projects such as rainwater harvesting, groundwater recharge, and water conservation activities.

GIS also allows for detailed analysis of water availability in different regions, such as identifying areas with inadequate storage capacity or regions where water table levels have fallen significantly. GIS-based applications can analyze past and present rainfall patterns, predict future water availability, and highlight the most vulnerable to water scarcity. For instance, in Rajasthan, GIS mapping has been used to pinpoint areas where groundwater recharge structures, such as check dams and ponds, can be constructed to improve water availability in the region. GIS has been instrumental in monitoring and managing existing water infrastructure, helping the authorities keep track of the condition of canals, pipes, and water storage tanks, which are crucial for maintaining water supply systems. The technology supports the efficient allocation of resources, ensuring that limited water supplies are distributed equitably and sustainably.

Internet of Things (IoT) for Real-Time Monitoring: IoT technology has also been employed in the Jal Shakti Abhiyan to optimize water management through real-time monitoring and data collection. IoT-enabled devices such as sensors are deployed across water bodies, wells, and reservoirs to collect continuous data on ality parameters such as pH, turbid This data is sent to central servers or cloud platforms, where it is analyzed to provide actionable insights into the state of water resources in real-time.

IoT devices can help farmers monitor soil moisture levels in irrigation, enabling them to irrigate only when necessary, thereby preventing water wastage. These smart irrigation systems, often paired with weather forecast data, ensure water is used efficiently, reducing unnecessary water consumption. In some areas, IoT sensors monitor and optimize water use in canals and pipelines, detecting leaks and optimizing water distribution across the region. IoT technology plays a critical role in monitoring groundwater levels. In areas where groundwater extraction has caused significant depletion, IoT sensors in wells help track the water table's health. This real-time data is crucial for managing the sustainable extraction of groundwater and ensuring that over-extraction is avoided. Integrating IoT with GIS allows a more comprehensive view of surface water and groundwater resources, enabling better decision-making and proactive measures.

Artificial Intelligence (AI) for Predictive Analysis and Optimization: Artificial Intelligence (AI) and Machine Learning (ML) technologies enhance the efficiency of water conservation efforts by providing predictive analytics and optimizing resource management. AI algorithms can process vast amounts of data from various sources, including IoT sensors, satellite imagery, and weather forecasts, to identify patterns and predict water availability, usage, and demand. These predictive models can forecast potential drought conditions, allowing for early warnings and preparedness measures. AI can also optimize water distribution systems in real-time. AI algorithms can use data on water usage, weather forecasts, and current water levels to predict water demand and supply for different regions. This ensures that water is allocated efficiently, prioritizing critical areas while preventing overuse in other areas. For instance, AI models have been used to predict water demand in urban areas, ensuring that water is delivered optimally based on real-time needs, thereby avoiding wastage and enhancing the system's sustainability.

In agriculture, AI-driven systems can be used to predict crop water requirements based on factors such as weather conditions, soil moisture, and crop type. This can guide farmers in making informed irrigation decisions, reducing water wastage while ensuring optimal crop yields. Additionally, AI-powered satellite imagery and remote sensing data analysis can identify water-stressed areas and suggest targeted interventions, such as constructing rainwater harvesting structures or promoting water-efficient farming practices.

Remote Sensing for Monitoring Water Resources: Remote sensing technologies, particularly satellites, have become indispensable for monitoring water resources. Through satellite-based imagery and sensors, it is possible to track changes in water bodies, monitor the health of aquifers, and assess the impact of climate change on water availability. Remote sensing data is used to measure water levels in rivers, lakes, and reservoirs, providing valuable real-time information about the overall state of water resources.

In the Jal Shakti Abhiyan context, remote sensing data is used to monitor the effectiveness of water conservation measures such as check dams, rainwater harvesting structures, and groundwater recharge projects. By comparing pre- and post-intervention satellite images, authorities can assess the impact of these interventions on water availability and quality. Remote sensing also allows identifying areas vulnerable to flooding, providing critical data for flood control and water management planning. Additionally, remote sensing helps in mapping catchment areas for rainwater harvesting. Using satellite topographical data, engineers can identify suitable sites for constructing rainwater harvesting structures. This data-driven approach ensures that rainwater is captured most efficiently and sustainably, enhancing the initiative's overall water conservation efforts.

Community Engagement and Capacity Building: While technology plays a crucial role in the Jal Shakti Abhiyan, community participation is equally important for the initiative's success. Technology alone cannot solve the problem of water scarcity; local communities need to be actively involved in water management. Mobile apps and online platforms developed under the JSA help engage communities by providing real-time updates on water conservation activities, enabling citizens to track water levels, and encouraging them to adopt water-saving practices. Workshops and training programs are also organized to raise awareness about water conservation techniques, such as rainwater harvesting, the efficient use of water in agriculture, and wastewater management. Technology enables the widespread dissemination of knowledge, ensuring that communities understand the benefits of adopting sustainable water practices and how technology can aid in these efforts.

Future Scenario

The future of water management is increasingly defined by advanced technologies that can offer precision, efficiency, and sustainability in handling water scarcity and surpluses. Predictive algorithms, driven by artificial intelligence (AI) and machine learning (ML), are revolutionizing how we anticipate and manage water-related disasters such as droughts and floods. These technologies enable better forecasting, risk mitigation, and proactive responses, thus reducing the adverse impacts of water variability on agriculture, infrastructure, and communities. Simultaneously, integrating blockchain technology into water governance offers new solutions for managing water distribution and ensuring fairness, transparency, and equitable access. As water scarcity intensifies and climate change leads to increasingly erratic weather patterns, the convergence of predictive algorithms and blockchain will define the next generation of water resource management.

Predictive Algorithms for Drought and Flood Management: Water management is at the forefront of discussions around climate adaptation, especially in light of the increasing unpredictability of rainfall and extreme weather events. Droughts and floodsexpected to become more frequent and severe due to climate changepose critical challenges to water resource management, agriculture, and urban infrastructure. Traditional methods of managing these challenges, relying on historical data and reactive approaches, are no longer sufficient to deal with the increasing complexity and scale of water-related disasters. Predictive algorithms, powered by AI and ML, are helping mitigate these issues by enabling proactive management strategies, offering earlier warning systems, and optimizing responses to water-related crises.

Drought Prediction and Early Warning Systems: Drought is one of the most insidious water-related disasters, often emerging slowly and affecting water supplies, agriculture, and the economy over extended periods. Traditional methods of drought forecasting typically rely on meteorological data, historical rainfall patterns, and manual observations. While these methods provide some insight, they often lack the precision and real-time responsiveness needed to address droughts' dynamic and complex nature. Predictive algorithms, however, are transforming the landscape by utilizing big data, machine learning models, and real-time environmental sensing to predict drought conditions with much greater accuracy.

Predictive models can integrate various data sources, including weather forecasts, soil moisture levels, remote sensing data, and historical trends, to develop a more comprehensive understanding of potential drought risks. Machine learning algorithms are particularly effective in identifying complex patterns in this multi-dimensional data, allowing for better early warning and preparedness. For example, AI models can analyze satellite data to detect changes in vegetation health, soil moisture, and water availability, providing valuable insights into areas at drought risk. This predictive capability enables governments, farmers, and water managers to take early actions, such as implementing water conservation measures, adjusting irrigation schedules, or enhancing water storage before the drought reaches critical levels. In agriculture, AI-driven predictive systems can estimate crop water requirements based on weather patterns, evapotranspiration, and soil moisture data, helping farmers optimize irrigation practices and avoid overuse of water resources. By incorporating AI into water management systems, stakeholders can make more informed decisions, reduce water waste, and increase crop resilience to dry conditions. Predictive algorithms can be used to develop more robust and accurate drought models, which can factor in local environmental variables, geographic features, and socioeconomic conditions, allowing for tailored interventions that improve long-term drought management.

Flood Prediction and Management: Flooding, particularly in riverine areas and urban environments, is another major challenge climate change exacerbates. Like droughts, flood events are often sudden, catastrophic, and difficult to predict with traditional methods. Conventional flood forecasting relies on river discharge data, rainfall amounts, and floodplain maps, but it tends to be reactive and cannot provide timely warnings for rapidly developing flood scenarios. Predictive algorithms can vastly improve flood forecasting by combining real-time data streams from IoT sensors, weather stations, satellite images, and hydrological models to forecast floods more accurately and with greater lead time.

Machine learning models can analyze historical flood data, topographical information, and real-time data to predict flood occurrences and intensity. For example, AI algorithms can process rainfall data and river flow rates to predict when a river may exceed its banks, or identify the likelihood of flash floods in urban areas. To provide a more nuanced and accurate prediction, these models can also factor in other environmental variables, such as soil saturation levels, snowmelt, and human activities like urban development or deforestation. With this information, authorities can issue flood warnings to affected areas well in advance, reducing loss of life, damage to infrastructure, and economic costs. Additionally,

AI-powered flood prediction models can optimize the management of flood control infrastructure, such as dams, levees, and reservoirs. By predicting the volume and timing of floodwater, these models can help water authorities regulate water flow through such structures, ensuring that they operate efficiently to prevent overflow and minimize downstream flooding. In urban areas, where impervious surfaces like roads and buildings exacerbate flood risks, predictive models can inform the design of stormwater drainage systems and floodplain zoning, ensuring that cities are better equipped to handle extreme rainfall events.

Blockchain for Equitable Water Sharing: Water scarcity is an issue that is both geographically and socially inequitable. In many regions, water resources are unevenly distributed, leading to conflicts over access, especially in areas where water is a shared resource. Additionally, the lack of transparency in water distribution systems, inefficient infrastructure, and governance challenges often exacerbate these inequalities. Blockchain technology, a decentralized, secure, and transparent digital ledger system, offers a promising solution to these challenges by enabling equitable water sharing and governance.

Blockchain for Water Rights Management: In many parts of the world, water rights are poorly managed, often leading to disputes and unsustainable water extraction. Blockchain technology has the potential to address these issues by providing a transparent, tamper-proof system for tracking water ownership, usage rights, and transactions. Through blockchain-based smart contracts, water rights can be digitized, enabling real-time, transparent, and automated water allocations based on predefined rules.

For example, in regions with scarce water resources, blockchain could facilitate the fair distribution of water by tracking who is using the water, how much is being consumed, and ensuring that water is allocated according to existing rights or needs. This would help prevent over-extraction by users who may otherwise take more than their share, promoting more equitable and sustainable water usage practices. Smart contracts could automatically enforce water-sharing agreements, ensuring that water is distributed in compliance with legal frameworks and that transactions are recorded immutable, auditable. Blockchain can create a digital marketplace for water rights, where buyers and sellers can trade water based on supply and demand. Such a marketplace would allow water rights to be transferred between users, ensuring that water is allocated to those who need it most, such as farmers in drought-affected areas or urban populations facing water shortages. By allowing for the transparent transfer of water rights, blockchain can reduce the occurrence of water conflicts and promote more efficient use of limited water resources.

Blockchain for Water Quality and Usage Tracking: Another promising application of blockchain in water management is monitoring water quality and usage in real-time. Blockchain can provide a secure, transparent, and immutable record of water quality data by integrating IoT sensors that track water quality parameters such as pH levels, turbidity, temperature, and contaminants. This data can be used by water authorities, communities, and industries to ensure that water quality standards are maintained and that water distribution systems are operating efficiently.

Blockchain technology can also track water usage across industries and communities, ensuring water is not being misused or wasted. For example, businesses or households that use large amounts of water can be incentivized to reduce consumption by providing proof of their reduced water usage via blockchain. Blockchain could enable the creation of digital water credits or certificates that reward individuals or organizations that conserve water, thus promoting sustainable water use. Additionally, blockchain can help verify compliance with water conservation policies or regulations. For example, in regions where wastewater recycling or rainwater harvesting is encouraged, blockchain can record the amount of water harvested or recycled by individual users or organizations, ensuring that these efforts are tracked and rewarded appropriately.

Transparency in Water Distribution and Governance: Governance in water management systems is often opaque, leading to inefficiencies, corruption, and inequitable distribution of resources. Blockchain can introduce transparency that ensures all stakeholders government agencies, private entities, or communitiescan access accurate, real-time data about water usage, distribution, and quality. Using a decentralized ledger system, blockchain eliminates the need for intermediaries and central authorities, which often create bottlenecks and reduce the efficiency of water distribution systems.

For example, by using blockchain to track water allocations in real-time, authorities can monitor compliance with water-sharing agreements, prevent unauthorized usage, and ensure that water distribution systems operate optimally. Citizens can also access this data, which can help them hold local authorities accountable and demand better water management practices. In addition, blockchain can play a critical role in building trust among stakeholders by offering transparent mechanisms for water management decisions. Whether in setting water tariffs, implementing water-saving policies, or regulating water usage, blockchain's transparency ensures that all parties have equal access to information and can make decisions based on verified, tamper-proof data.

11

Sustainable Agriculture through IT

Introduction

Sustainable agriculture refers to farming practices that meet current food production needs without compromising the ability of future generations to meet their own needs. It aims to balance environmental health, economic profitability, and social equity, ensuring that agricultural systems can be productive, resource-efficient, and resilient over time. The role of Information Technology (IT) in achieving these objectives has become increasingly crucial as it provides innovative solutions to optimize agricultural practices, minimize resource consumption, and improve overall productivity while safeguarding the environment. Through the integration of IT tools such as Geographic Information Systems (GIS), Internet of Things (IoT), Big Data analytics, Artificial Intelligence (AI), and cloud computing, sustainable agriculture can be advanced in ways that were previously unimaginable. These technologies facilitate precision farming, enhance resource management, monitor crop health, predict weather patterns, and provide farmers with the information to make data-driven decisions that foster sustainability.

Precision Agriculture: One of the most significant ways IT supports sustainable agriculture is through precision farming. Precision agriculture utilizes IT to optimize water, fertilizers, pesticides, and labor resources. This approach uses sensors, GPS technology, IoT devices, and remote sensing tools to gather real-time data on various environmental and soil conditions. For instance, soil moisture sensors can provide insights into when and how much irrigation is needed, reducing water waste and ensuring that crops receive the right amount of water at the right time. Similarly, GPS-guided tractors and harvesters enable farmers to plant seeds and harvest crops with high precision, minimizing fuel consumption and improving crop yield.

IT-based solutions in precision agriculture also help judiciously apply fertilizers and pesticides. By utilizing remote sensing technologies, satellite imagery, and drones, farmers can monitor crop health on a granular level, identifying areas of the field that are underperforming or affected by pests. This allows for targeted interventions, such as spot application of pesticides or fertilizers, thereby reducing the overall use of chemicals and minimizing their negative environmental impact. This targeted approach reduces costs and minimizes the risk of chemical runoff into surrounding ecosystems, contributing to environmental sustainability.

Big Data and AI in Predictive Analytics: Another transformative application of IT in sustainable agriculture is using Big Data analytics and AI for predictive modeling. By collecting and analyzing vast amounts of data from diverse sources, such as weather forecasts, soil conditions, market trends, and historical crop performance, AI-driven algorithms can predict optimal planting and harvesting times, anticipate weather-related risks, and improve crop yield forecasts. This predictive capability allows farmers to plan more effectively, reduce risks, and make informed decisions contributing to higher productivity and resource conservation.

For example, predictive models can forecast drought conditions, enabling farmers to adjust irrigation schedules or adopt water-saving techniques before a potential water shortage occurs. Similarly, AI algorithms can help farmers predict pest infestations or disease outbreaks based on patterns observed in environmental data, allowing for early intervention and reducing the need for widespread pesticide application. AI-powered systems can integrate data from multiple sensors, drones, and satellite imagery to assess crop health, enabling farmers to make real-time adjustments and ensure that crops receive the necessary inputs at the right time. Big Data also aids in improving agricultural supply chain efficiency. By analyzing historical data, including market demands, price fluctuations, and transportation patterns, farmers can better predict demand for specific crops and adjust their planting strategies accordingly. This minimizes food waste by ensuring that crops are grown aligned with market needs, reducing the environmental footprint of unnecessary production and transportation.

IoT for Real–Time Monitoring and Management: The Internet of Things (IoT) has become a cornerstone of modern agriculture, providing farmers with real-time data from various sensors deployed across the farm. These sensors monitor environmental variables such as soil moisture, temperature, humidity, and nutrient levels, offering actionable insights that help optimize resource use. For instance, IoT-enabled weather stations can provide hyper-localized weather data, allowing farmers to make real-time decisions about irrigation, pest control, and other essential tasks. In addition to soil and environmental monitoring, IoT can be integrated with smart irrigation systems to ensure that water is used efficiently. IoT-based irrigation controllers can adjust watering schedules based on real-time data from soil moisture sensors and weather forecasts, preventing both over-irrigation and water stress. This conserves water and reduces energy consumption, as irrigation systems are optimized to run only when necessary.

IoT technology is also instrumental in monitoring livestock health and productivity. Sensors can track vital parameters such as body temperature, heart rate, and movement patterns of animals, providing farmers with early warnings of potential health issues. This enables timely intervention and minimizes antibiotics and other medications, contributing to healthier livestock and reducing the environmental impact of intensive animal farming.

Cloud Computing for Collaboration and Data Sharing: Cloud computing has revolutionized the agricultural sector by providing a real-time platform for storing, analyzing, and sharing data. Farmers, agronomists, and researchers can access critical information from any location by moving data storage and computational resources to the cloud, facilitating collaboration and knowledge exchange. Cloud-based platforms can host agricultural management systems that integrate data from IoT sensors, weather stations, satellite imagery, and other sources, enabling farmers to monitor and manage their farms remotely.

The cloud also allows farmers to store historical data, which can be analyzed to identify trends and patterns over time. This long-term data storage enables farmers to compare different growing seasons, track the effects of various agricultural practices, and make evidence-based decisions that optimize resource use and yield over the long term. Additionally, cloud platforms can host AI models and analytics tools that help farmers make data-driven decisions about crop rotation, fertilizer use, pest management, and farm management. Collaboration between farmers, agricultural extension services, researchers, and policymakers is also facilitated through cloud-based platforms. This fosters knowledge-sharing and disseminating best practices, empowering farmers to adopt more sustainable practices and improve their productivity while reducing environmental impact.

Sustainability through Resource Conservation: A key principle of sustainable agriculture is the efficient use of natural resources, particularly water, soil, and energy. IT tools play a significant role in conserving these resources by providing farmers with the information to reduce waste and optimize input use. For example, precision irrigation systems powered by IoT and AI technologies help ensure that water is applied only where and when needed, minimizing water waste and preventing the depletion of local water supplies. These systems are critical in regions where water scarcity is a pressing concern, such as arid and semi-arid areas. Similarly, farmers can monitor soil health and fertility in real-time by using soil sensors and remote sensing technology. This allows them to apply fertilizers and soil amendments more accurately, reducing the risk of nutrient runoff and soil degradation. Precision agriculture practices also help maintain soil biodiversity and reduce the need for chemical inputs, which helps preserve the soil's long-term health and ability to support healthy crops.

Energy conservation is another aspect of sustainable agriculture that IT can address. By optimizing the use of irrigation systems, mechanized equipment, and farm operations, farmers can reduce their energy consumption and associated carbon footprint. Smart grid systems and renewable energy sources such as solar-powered irrigation pumps are increasingly integrated into farming operations, reducing reliance on fossil fuels and contributing to the transition toward more sustainable agricultural practices.

Digital Platforms for Knowledge and Capacity Building: In addition to its technological applications, IT also plays a critical role in knowledge dissemination and capacity building for farmers. Digital platforms and mobile applications provide farmers expert advice, market information, and educational resources to help them adopt sustainable farming practices. For example, farmers can use mobile apps to get real-time weather updates, pest and disease forecasts, and tips on crop management, all tailored to their specific region and needs.

E-learning platforms and online agricultural communities allow farmers to share experiences, exchange ideas, and learn from experts. These platforms empower smallholder farmers, who often lack access to formal agricultural education and extension services, to improve their practices and sustainably enhance their productivity. Farmers can access microloans, insurance, and other financial tools that support their transition to more sustainable practices by integrating financial services into these platforms.

Defies and the Future of Sustainable Agriculture through IT: Despite the transformative potential of IT in sustainable agriculture, several challenges need to be addressed. The digital divide between urban and rural areas, particularly in developing countries, limits access to the technologies and infrastructure required to implement IT-based solutions. High initial hardware, sensors, and software costs also pose barriers for smallholder farmers. There is a need for proper training and capacity building to ensure that farmers can effectively use these technologies. As the adoption of IT in agriculture continues to grow, solutions must be developed to overcome these barriers and make these technologies accessible and affordable for all farmers, particularly those in underserved regions. In addition, greater collaboration between governments, private companies, research institutions, and farmers is needed to create policies and incentives that promote the widespread use of IT in agriculture.

In the future, as IT continues to advance, we can expect even greater innovations in sustainable agriculture. Integrating blockchain for transparent supply chains, expanding 5G networks for better connectivity, and developing more affordable and efficient agricultural technologies will further accelerate the shift toward sustainable, data-driven farming practices. By continuing to leverage IT, sustainable agriculture can play a pivotal role in feeding the growing global population while preserving the environment and ensuring the long-term health of the planet's ecosystems.

Current Scenario

The integration of Information Technology (IT) into agriculture, particularly through precision farming and crop monitoring, has ushered in a new era of agricultural efficiency, sustainability, and productivity. By leveraging technologies such as the Internet of Things (IoT), Geographic Information Systems (GIS), drones, and data analytics, precision farming allows farmers to optimize resource usage, enhance yields, and reduce environmental impact. Farmers can make data-driven decisions on irrigation, fertilization, pest control, and harvesting through real-time monitoring of soil health, crop conditions, and environmental factors, promoting more sustainable and efficient agricultural practices. However, the potential benefits of these technologies are hindered by a range of challenges, particularly in rural and underserved regions. Issues such as inadequate infrastructure, high costs, and limited access to technical knowledge prevent many farmers from fully embracing and scaling agri-tech solutions. In parallel, broader systemic issues such as food security concerns, supply chain inefficiencies, resource overuse, and productivity gaps further complicate the efforts to integrate IT into farming practices on a large scale.

Precision Farming and Crop Monitoring: Precision farming represents a paradigm shift from traditional agricultural methods, focusing on optimizing every aspect of farming to improve yield and resource efficiency. By utilizing technologies like IoT, Big Data, and machine learning, precision farming enables farmers to monitor soil moisture, temperature, humidity, crop growth stages, and pest infestations, among other factors, in real-time. This allows for more precise application of resources like water, fertilizers, and pesticides, reducing waste and improving efficiency. IoT devices play a crucial role in collecting real-time data from the field. Sensors embedded in the soil monitor moisture content, temperature, pH levels, and nutrient availability, providing insights into soil health and enabling farmers to decide about irrigation and fertilization. Remote sensing technologies, including satellite imagery and drone-based monitoring, allow crop surveillance from a distance, capturing detailed data on crop health, stress factors, and pest damage. These technologies enable farmers to intervene at the right time and with the appropriate resources, enhancing crop quality and yield.

Machine learning and artificial intelligence (AI) further enhance precision farming by processing large volumes of data to identify patterns and predict outcomes. For example, AI-driven models can forecast weather patterns, predict pest outbreaks, and recommend optimal planting and harvesting times. Additionally, AI-powered crop health monitoring systems use images from drones or satellites to detect early signs of disease or stress, allowing farmers to address issues before they escalate into widespread problems. These technologies enable farmers to act with greater foresight and precision, leading to more sustainable and resource-efficient farming practices.

Challenges in Scaling Agri – Tech Solutions in Rural Areas: While the potential benefits of precision farming are immense, scaling agri-tech solutions in rural and remote regions presents significant challenges. A major barrier to the widespread adoption of IT in agriculture is the lack of adequate infrastructure in many rural areas. High-speed internet connectivity, essential for real-time data transmission and cloud-based solutions, remains limited or unavailable in large parts of rural landscapes, particularly in developing countries. Without reliable internet access, farmers cannot fully take advantage of cloud-based platforms that offer crop management tools, weather forecasts, and market information, thus hindering the widespread adoption of digital solutions. The high initial investment costs associated with deploying IoT devices, drones, sensors, and other precision farming technologies are prohibitive for smallholder farmers, who comprise most of the agricultural workforce in rural areas. The cost of purchasing, installing, and maintaining these technologies can be a significant financial burden, particularly for those who lack access to credit or financial services. Even when farmers do manage to acquire these technologies, the ongoing costs for data storage, software updates, and device maintenance can be difficult to bear, particularly for those working with tight margins.

In addition to financial constraints, technical literacy and training are significant hurdles to adopting IT-based solutions. Many rural farmers are unfamiliar with digital tools or data-driven farming methods, and there is often a lack of accessible training programs to teach them how to use these technologies effectively. As a result, even if they have access to advanced agri-tech solutions, farmers may struggle to use them to their full potential, leading to underutilization of resources and suboptimal outcomes. The agricultural sector in many rural areas is still predominantly traditional, with farming practices passed down through generations. This deeply ingrained cultural resistance to new technologies can slow the adoption of precision farming methods, as farmers may hesitate to trust and invest in unfamiliar systems. Overcoming this reluctance requires not only the development of affordable and accessible technologies but also the creation of support systems that encourage farmers to embrace change and adopt sustainable, IT-enabled farming practices.

Food Security, Supply Chain Inefficiencies, and Resource Overuse: While precision farming technologies can improve resource efficiency on individual farms, broader systemic challenges, such as food security and supply chain inefficiencies, complicate ensuring sustainability and productivity at a national or global scale. Food security remains a critical issue, particularly in developing countries where rising populations, climate change, and economic instability have increased the vulnerability of agricultural systems. Even as individual farms become more efficient through IT, inefficiencies within the broader food supply chain can lead to waste, reduced food availability, and market distortions.

One major issue is the inefficiency of food distribution systems. In many developing regions, poor infrastructure, inadequate cold storage facilities, and unreliable transportation networks result in significant food losses between the farm and the market. According to the Food and Agriculture Organization (FAO), nearly one-third of the food produced globally is lost or wasted, much of which occurs post-harvest. IT solutions, such as IoT-enabled cold chains, blockchain for traceability, and AI-driven logistics systems, can address some inefficiencies by improving transparency, reducing waste, and ensuring that food reaches consumers promptly. However, adopting these technologies requires significant investment in infrastructure and the development of partnerships between farmers, supply chain operators, and technology providers.

Resource overuse is another major challenge that precision farming aims to address, but it is still a persistent issue on a global scale. Overusing water, fertilizers, and pesticides has resulted in soil degradation, water scarcity, and environmental pollution. Despite the potential of IT to optimize resource use on individual farms, large-scale overuse of resources, driven by unsustainable farming practices and poorly regulated industries, continues to deplete the world's natural capital. Addressing this problem requires technological solutions, policy interventions, and regulatory frameworks that incentivize sustainable resource use and penalize wasteful practices.

Productivity Gaps and Market Inefficiencies: Another critical issue in sustainable agriculture is the productivity gap between different regions and types of farms. While precision farming has the potential to boost yields and efficiency, productivity gaps persist between smallholder farms and large commercial operations. Smallholder farmers often lack access to the capital, technology, and expertise to fully leverage precision farming techniques. This disparity results in lower productivity, reduced profitability, and a widening income gap between small and large-scale farmers.

IT-based solutions can help reduce these productivity gaps by providing smallholders access to knowledge, training, and market information. Mobile apps and cloud-based platforms can offer farmers advice on crop management, weather forecasting, and pest control, which can help increase yields and improve farm productivity. However, the gap in access to these technologies between smallholder and large-scale farmers must be addressed to create a more equitable agricultural system. Subsidies, microloans, and community-based initiatives can help facilitate the adoption of precision farming technologies among smallholder farmers, levelling the playing field and promoting greater food security and economic stability. Market inefficiencies also contribute to problems with food security and agricultural productivity. In many regions, farmers face difficulties in accessing fair markets, and they are often forced to sell their produce at low prices to intermediaries who control the supply chain. This market power imbalance leads to economic disparities and can reduce farmers' incentives to invest in sustainable agricultural practices. IT-based platforms, such as e-marketplaces and blockchain, can help address these market inefficiencies by enabling farmers to access a wider range of buyers, receive better product prices, and gain more control over their income.

Precision Agriculture and Farmer Empowerment

Precision agriculture has emerged as a transformative force in modern farming practices, revolutionizing how agricultural activities are managed by integrating advanced technologies. Among these technologies, drones, sensors, and Artificial Intelligence (AI) stand out as key enablers of precision farming, offering unprecedented capabilities for crop monitoring, resource management, and farm productivity enhancement. Precision agriculture uses detailed data to optimize inputs, reduce waste, and improve crop yields, all while minimizing environmental impact. At its core, it empowers farmers with the tools and insights needed to make data-driven decisions that ensure the sustainability of agricultural practices. This empowerment is critical, especially as the global population grows and the challenges of climate change, resource scarcity, and food insecurity. Using drones, sensors, and AI in precision farming enhances crop monitoring and plays a vital role in fostering a more equitable and resilient agricultural system, particularly by empowering farmers to manage their resources more efficiently and increase their productivity.

Drones – Aerial Surveillance and High-Resolution Monitoring: Drones, or Unmanned Aerial Vehicles (UAVs), have emerged as powerful tools in precision agriculture, providing farmers with aerial surveillance capabilities that were once only available through satellite imagery or manned aircraft. Drones are equipped with high-resolution cameras, multispectral sensors, and thermal imaging devices that enable them to capture real-time data on crop health, soil moisture, temperature, and plant stress. This ability to monitor crops from the air gives farmers a comprehensive view of their fields, enabling them to identify issues such as disease outbreaks, pest infestations, water stress, and nutrient deficiencies more efficiently than traditional methods.

The high-resolution imagery provided by drones can detect subtle changes in crop health that may not be visible to the naked eye. For example, drones equipped with multispectral sensors can capture images across different wavelengths of light, including infrared, allowing farmers to identify plant stress long before visible symptoms appear. This early detection capability is crucial for implementing targeted interventions, such as applying pesticides, herbicides, or fertilizers to specific field areas, rather than treating the entire crop. This targeted approach helps reduce the use of chemicals, minimizes environmental impact, and ensures that resources are applied more efficiently, leading to cost savings for farmers. In addition to crop health monitoring, drones are valuable for mapping and surveying fields. By capturing detailed, high-resolution images, drones can create precise, geo-referenced maps of farms. These maps can analyze field variability, helping farmers understand the spatial distribution of soil properties, crop performance, and environmental factors. Armed with this knowledge, farmers can implement variable rate technology (VRT), which allows for the tailored application of water, fertilizers, and pesticides based on the specific needs of different areas within a field. This precision reduces waste, conserves resources, and improves crop yields, contributing to the overall sustainability of farming operations.

Sensors – Real-Time Data for Precision Resource Management: Sensors play a critical role in precision agriculture by providing real-time, on-the-ground data essential for effective decision-making. Sensors deployed in the field can monitor various variables influencing crop growth, including soil moisture, temperature, pH, salinity, and nutrient levels. This data is invaluable for managing irrigation, fertilization, and other critical aspects of crop production. By continuously monitoring these variables, sensors allow farmers to respond promptly to changes in field conditions, ensuring that crops receive the right amount of resources at the right time.

One of the most important applications of sensors in precision agriculture is in irrigation management. Traditional irrigation methods often result in water wastage, with fields watered on fixed schedules regardless of soil moisture levels. However, with soil moisture sensors, farmers can monitor the moisture content at various depths in the soil, providing a more accurate measure of when and where irrigation is needed. This enables farmers to implement smart irrigation systems that water crops only when necessary, conserving water and reducing costs. This is especially critical in water-scarce regions, as it helps ensure that available water resources are used as efficiently.

Nutrient management is another area where sensors can significantly improve efficiency. Soil sensors can measure the availability of key nutrients such as nitrogen, phosphorus, and potassium, allowing farmers to apply fertilizers only when and where needed. This precision reduces the over-application of fertilizers, which can lead to nutrient runoff and environmental pollution, particularly in water bodies. By using sensors to track nutrient levels, farmers can optimize fertilization practices, improving both crop yields and soil health while minimizing environmental harm. In addition to soil and environmental sensors, sensors attached to equipment such as tractors, harvesters, and irrigation systems can collect data on the performance and efficiency of machinery. This data can be used to optimize the operation of these machines, improving fuel efficiency and reducing wear and tear, ultimately leading to cost savings and more sustainable farm management practices.

AI in Crop Monitoring – Data Processing and Decision Support: Artificial Intelligence (AI) is the backbone of modern precision agriculture, as it enables the processing of vast amounts of data collected by drones, sensors, and other technologies. AI algorithms can analyze data in real-time to detect patterns, identify trends, and generate actionable insights that can be used to optimize farming practices. In crop monitoring, AI is critical in interpreting data from multiple sources, such as satellite images, drone footage, and sensor readings, to provide a holistic view of farm performance.

One of the key applications of AI in crop monitoring is predictive analytics. AI algorithms can process historical data on weather, crop growth patterns, soil health, and pest cycles to predict future conditions and recommend proactive interventions. For instance, AI models can forecast the likelihood of a pest outbreak based on environmental factors such as temperature and humidity, allowing farmers to take preventative measures before an infestation occurs. Similarly, AI can predict weather patterns, such as rainfall or drought conditions, which enables farmers to adjust irrigation schedules accordingly, preventing water waste and ensuring optimal crop growth.

AI-driven crop monitoring systems can also integrate data from various sources to create real-time crop health maps. By processing images from drones and satellites, AI can identify areas of a field experiencing stress, such as nutrient deficiencies or pest damage. This allows farmers to pinpoint specific areas that require attention, enabling them to implement targeted interventions that are more efficient and effective than traditional methods. AI systems can also track changes in crop growth over time, helping farmers monitor their interventions' effectiveness and adjust their strategies as needed. Machine learning, a subset of AI, improves crop monitoring systems over time. As more data is collected, machine learning models can be trained to identify new patterns and make more accurate predictions, continuously improving the system's ability to monitor and optimize farm operations. This adaptability is one of the key advantages of AI in precision agriculture, as it enables the system to evolve and respond to changing conditions, making it a valuable tool for long-term sustainability.

Farmer Empowerment – Making Data-Driven Decisions: The ultimate goal of using drones, sensors, and AI in precision agriculture is to empower farmers with the tools, information, and insights they need to make data-driven decisions that improve their productivity and sustainability. By providing farmers with real-time data and predictive insights, these technologies enable them to manage their farms more efficiently, reduce waste, and minimize environmental impact. However, the empowerment of farmers goes beyond access to technology; it also involves the development of skills, knowledge, and support systems that enable farmers to leverage the capabilities of these tools fully.

For farmers, adopting precision agriculture technologies can lead to improved decision-making at every stage of crop production. Farmers can optimize their water use, fertilizers, pesticides, and other inputs with better information about soil conditions, crop health, and weather patterns. This leads to cost savings and improves the quality and yield of crops, which translates to higher profits and greater food security. Precision farming also promotes environmental sustainability by reducing the need for excessive chemical applications and water usage, helping farmers protect the health of their land and water resources. In addition to improving farm operations, precision agriculture technologies enable farmers to gain greater control over their livelihoods. By providing access to real-time market data, AI-driven platforms can help farmers make better decisions about when and where to sell their produce, maximizing their income. Technology empowers farmers to monitor their farms remotely, reducing the time and effort spent on manual inspections and allowing them to focus on more strategic decisions.

Barriers to Adoption: Despite the numerous benefits, there are significant challenges to the widespread adoption of precision agriculture technologies, particularly in developing countries and among smallholder farmers. The high initial costs of drones, sensors, and AI systems can be prohibitive for many farmers, especially those with limited access to capital. Additionally, the technical skills required to operate and interpret the data generated by these technologies can be a barrier, particularly for farmers unfamiliar with digital tools. Another challenge is the lack of infrastructure in rural areas, such as reliable internet connectivity and electricity, which are essential for the smooth operation of IoT devices and cloud-based platforms. Without proper infrastructure, farmers cannot take full advantage of the real-time data and remote monitoring capabilities offered by precision farming technologies. To overcome these barriers, governments, NGOs, and private sector players must collaborate to provide affordable financing options, training programs, and infrastructure development to support the adoption of precision agriculture.

Digital Marketplaces for Farmers: E-NAM as a Case Study

The agricultural sector is a cornerstone of the global economy, particularly in countries like India, where a significant portion of the population depends on farming for their livelihoods. However, traditional agricultural marketing systems often face challenges such as a lack of transparency, fragmented supply chains, and exploitation by intermediaries. Digital marketplaces have emerged as transformative solutions to address these issues, empowering farmers with direct access to buyers, fair pricing mechanisms, and a more efficient supply chain. One prominent initiative is the Electronic National Agriculture Market (e-NAM) in India, a flagship nationwide program to integrate agricultural markets. This case study delves into the workings of e-NAM, its impact on farmers, challenges in implementation, and its role as a model for global digital marketplaces.

Introduction to E-NAM – Unified Digital Platform: Launched in 2016 by the Government of India, e-NAM was envisioned as a unified national market for agricultural commodities, leveraging technology to integrate existing physical wholesale markets, known as mandis, across the country. Managed by the Small Farmers' Agribusiness Consortium (SFAC), the platform connects farmers, traders, and buyers through an online marketplace, facilitating seamless agricultural product trading. e-NAM operates on the principle of fair and transparent price discovery by allowing buyers and sellers to interact directly, thereby reducing the dependence on intermediaries who often exploit farmers through low prices and high commission rates. The platform uses robust digital infrastructure to support the trading process. It includes features such as e-bidding, quality testing, and online payment systems, ensuring farmers receive fair prices for their produce promptly. Additionally, e-NAM provides access to real-time market data, such as prevailing prices, demand-supply trends, and buyer preferences, enabling farmers to make informed decisions about where and when to sell their produce.

Key Features and Functionalities of E-NAM

1. **Unified Market Access**: e-NAM connects over 1,000 mandis across 18 states and union territories in India, creating a single platform for trading agricultural commodities. This integration allows farmers to access a larger pool of buyers, expanding their market reach beyond local mandis.
2. **Transparent Price Discovery**: Through e-bidding, e-NAM ensures that the price of agricultural produce is determined transparently. Farmers can compare offers from multiple buyers and choose the best deal, eliminating the need for intermediaries who often manipulate prices.
3. **Quality Assurance**: The platform incorporates quality testing facilities at mandis, ensuring buyers receive produce that meets their standards. This focus on quality incentivizes farmers to produce higher-quality crops, which can command better prices.
4. **Efficient Payments**: e-NAM facilitates electronic payments directly to farmers' bank accounts, reducing delays and ensuring that farmers are not subjected to the risks associated with cash transactions.
5. **Real-Time Market Data**: The platform provides farmers real-time information about prices, demand, and supply trends, empowering them to decide where and when to sell their produce.
6. **Interstate Trading**: One of the most transformative features of e-NAM is its facilitation of interstate trading, allowing farmers to sell their produce across state boundaries without physical barriers. This reduces regional price disparities and ensures a more equitable distribution of agricultural resources.

Impact of E-NAM on Farmers: The introduction of e-NAM has had a profound impact on the agricultural sector in India, particularly in empowering farmers and improving their economic outcomes:

1. **Fair Pricing**: By eliminating intermediaries and ensuring transparent price discovery, e-NAM has significantly improved the prices that farmers receive for their produce. Studies have shown that e-NAM farmers often secure better prices than traditional mandi transactions.

2. **Market Access**: e-NAM has expanded market access for farmers, particularly those in remote and underserved regions. By connecting them to a larger pool of buyers, the platform reduces their reliance on local mandis and enables them to tap into markets with higher demand and better prices.
3. **Reduced Transaction Costs**: The digital nature of e-NAM reduces transaction costs associated with traditional trading, such as transportation to distant markets, commissions paid to intermediaries, and time spent in negotiations.
4. **Encouragement of Best Practices**: The emphasis on quality testing and transparency in pricing has incentivized farmers to adopt better farming practices, such as improved post-harvest handling and grading of produce, which can enhance their marketability.
5. **Timely Payments**: The electronic payment system ensures that farmers receive payments directly and promptly, reducing the risk of fraud and delays associated with cash transactions.
6. **Financial Inclusion**: By linking farmers to the formal banking system, e-NAM promotes financial inclusion and empowers farmers to access credit and other financial services.

Defies in the Implementation of E-NAM: Despite its successes, the implementation of e-NAM has faced several challenges that highlight the complexities of transforming traditional agricultural markets:

1. **Limited Awareness and Adoption**: Many farmers, particularly smallholders, remain unaware of e-NAM or lack the technical skills to use the platform effectively. Bridging this gap requires targeted awareness campaigns and capacity-building programs.
2. **Infrastructure Gaps**: The success of e-NAM depends on reliable infrastructure, such as internet connectivity, electricity, and quality testing facilities. Many rural areas in India still lack these essential resources, limiting the platform's reach.
3. **Resistance from Traditional Stakeholders**: Intermediaries, who have traditionally dominated agricultural markets, often resist the adoption of e-NAM as it threatens their economic interests. Overcoming this resistance requires strong policy support and stakeholder engagement.
4. **Integration of Mandis**: While e-NAM has made significant progress in integrating mandis, the process is complex and time-consuming. Variations in mandi regulations across states and the lack of standardization in trading practices pose significant challenges.

5. **Language and Usability Barriers**: Many farmers are uncomfortable using digital platforms, particularly those available only in English or Hindi. Developing multi-lingual, user-friendly interfaces is crucial to ensuring inclusivity.
6. **Price Volatility and Risk Management**: While e-NAM facilitates transparent price discovery, it does not address price volatility or provide risk management tools such as hedging mechanisms, which are critical for protecting farmers from market fluctuations.

E-NAM as a Model for Global Digital Marketplaces: E-NAM serves as a model for other countries seeking to develop digital marketplaces for their agricultural sectors. Its focus on transparency, inclusivity, and efficiency provides valuable lessons for addressing the challenges of traditional agricultural markets. Key takeaways from the e-NAM experience include:

1. **Policy Support and Stakeholder Engagement**: Strong government support and collaboration with stakeholders, including farmers, traders, and technology providers, are critical for the success of digital marketplaces.
2. **Infrastructure Development**: Investing in rural infrastructure, such as internet connectivity and quality testing facilities, is essential for ensuring digital platforms reach their full potential.
3. **Capacity Building**: Providing training and support to farmers is crucial for ensuring they can effectively use digital platforms and benefit from their features.
4. **Scalability and Interoperability**: Designing scalable and interoperable platforms with other systems, such as financial services and logistics networks, enhances their utility and impact.

Case Study

Doubling farmer income has been a central goal of agricultural policies globally, particularly in developing nations like India, where agriculture sustains a significant portion of the population. The mission to achieve this ambitious target by leveraging digital initiatives has gained momentum in recent years, driven by technological advancements and government interventions. This case study delves into the multifaceted role of digital tools in enhancing farmer income, focusing on key initiatives, their impact, and the challenges ahead. It explores how digital platforms, mobile apps, precision farming technologies, and e-commerce solutions collectively empower farmers to achieve higher incomes by optimizing resources, improving market access, and minimizing risks.

Context – Challenges to Farmer Income: Farmers, particularly smallholders, face persistent challenges that limit their incomes. These include fluctuating crop prices, fragmented supply chains, lack of access to timely credit, dependency on intermediaries, and climate-induced risks such as droughts or floods. Additionally, inefficiencies in resource use, such as over-reliance on fertilizers and water, further erode profitability. The traditional systems of agriculture and marketing often fail to address these issues comprehensively. Against this backdrop, digital initiatives emerge as transformative tools to bridge knowledge, connectivity, and market access gaps.

Key Digital Initiatives Driving Farmer Income Growth

1. **E-NAM (Electronic National Agriculture Market):** As discussed earlier, e-NAM provides a unified platform integrating mandis across India. By enabling transparent price discovery and eliminating intermediaries, e-NAM empowers farmers to access broader markets and secure better prices for their produce. Farmers participating in e-NAM have reported up to 20% price increases compared to traditional mandi sales.
2. **Direct Benefit Transfer (DBT) for Subsidies:** Government schemes for subsidizing fertilizers, seeds, and other inputs are now linked with digital Direct Benefit Transfer (DBT) mechanisms. This ensures timely and transparent disbursal of subsidies directly into farmers' accounts, reducing leakages and administrative inefficiencies.
3. **Kisan Suvidha and PM Kisan Mobile Apps:** Mobile applications such as **Kisan Suvidha** provide real-time weather forecasts, market prices, and pest management advice, while platforms like **PM Kisan** facilitate direct income support to farmers. These apps bridge the information gap and help farmers plan their activities more effectively.
4. **Digital Agriculture Platforms for Credit and Insurance:** Digital lending platforms, powered by Artificial Intelligence (AI) and blockchain, assess farmers' creditworthiness using alternative data sources such as transaction histories, satellite imagery, and weather data. Platforms like Aadhaar-enabled Payment System (AePS) have made credit more accessible. Additionally, digital insurance schemes offer timely compensation for crop failures, reducing income volatility.
5. **E-Commerce for Agricultural Produce:** Startups like DeHaat, AgroStar, and Ninjacart connect farmers directly with consumers, institutional buyers, and exporters, bypassing traditional supply chains. These platforms fetch higher prices and provide logistical support and demand forecasting tools.

6. **Precision Agriculture Technologies:** Internet of Things (IoT)-enabled sensors, drones, and satellite imagery help farmers monitor crop health, soil conditions, and water usage in real-time. These tools enable efficient use of inputs, reducing costs and increasing yields. For instance, farmers using IoT-driven irrigation systems have reported up to 40% water savings and 20-30% yield improvements.

7. **Smartphone Penetration and Digital Literacy Campaigns:** With smartphone penetration growing exponentially, even in rural areas, farmers now have access to a wealth of information and digital services. Government and NGO-led digital literacy programs are crucial in equipping farmers with the skills to utilise these tools effectively.

Case Studies of Success Stories

1. **Digital Marketplaces in Maharashtra:** Farmers participating in digital marketplaces like e-NAM and private e-commerce platforms have reported significant income growth in Maharashtra. For example, a group of onion farmers in Nashik leveraged e-NAM to sell directly to buyers in Gujarat and Karnataka, realizing a price increase of 30%.

2. **Karnataka's Precision Farming Project:** The Karnataka government's precision agriculture initiative, in partnership with agritech startups, deployed drones and IoT sensors to monitor paddy and maize fields. Farmers reduced their fertilizer usage by 25% and achieved yield improvements of 15%, leading to a 40% increase in net income.

3. **Agri-Tech Startup Success: DeHaat:** DeHaat, a prominent agri-tech startup, provides end-to-end agricultural solutions, from input supplies to market linkages, via a digital platform. In Bihar, farmers using DeHaat services reported income growth of 50% due to improved productivity and better market prices.

4. **Digital Irrigation in Tamil Nadu:** Farmers in Tamil Nadu's delta regions adopted IoT-driven smart irrigation systems under a government-sponsored pilot project. These systems optimized water use, enabling farmers to cultivate a second crop during the dry season, doubling their annual income.

Impact of Digital Initiatives on Farmer Income

1. **Increased Market Access:** Digital platforms have dismantled geographical barriers, allowing farmers to access larger, more competitive markets. This has led to better price realization and reduced dependency on exploitative local intermediaries.

2. **Optimized Resource Use:** IoT, drones, and precision agriculture tools reduce input costs by optimizing fertilizers, water, and pesticides. This efficiency translates into higher profitability for farmers.

3. **Risk Mitigation:** Weather forecasting apps and crop insurance schemes enabled by digital platforms help farmers mitigate risks associated with climate variability. Timely alerts and financial protection ensure stability in income levels.

4. **Time Savings and Operational Efficiency:** Automating processes like loan applications, subsidy disbursals, and quality testing saves farmers time, allowing them to focus on core agricultural activities.

5. **Empowerment and Financial Inclusion:** Digital initiatives empower farmers by giving them control over their decisions, enhancing their bargaining power. Integrating digital payment systems fosters financial inclusion, enabling farmers to access credit and save for future needs.

Way Forward

- **Policy Interventions:** Governments must prioritize investments in rural digital infrastructure and create supportive regulatory frameworks for agritech startups.
- **Public-Private Partnerships:** Collaborations between governments, private sector players, and NGOs can drive innovation and scale solutions.
- **Localized Solutions:** Digital initiatives must be tailored to the specific needs and contexts of different regions, considering factors like crop types, climate, and socioeconomic conditions.
- **Integration of Emerging Technologies:** Blockchain for traceability, AI for predictive analytics, and machine learning for customized recommendations can further enhance the effectiveness of digital initiatives.
- **Capacity Building:** Equipping farmers with digital skills and fostering an ecosystem of agritech service providers will accelerate the adoption of digital tools.

Digital initiatives hold immense potential to double farmer income by transforming how agriculture is practiced and marketed. Platforms like e-NAM, mobile apps, precision agriculture tools, and digital payment systems have already demonstrated their impact in enhancing productivity, reducing costs, and improving market access. However, achieving scale and inclusivity remains a challenge. By addressing infrastructure gaps, fostering digital literacy, and promoting innovation, policymakers and stakeholders can unlock the full potential of digital initiatives, ensuring that every farmer benefits from this technological revolution. This case study underscores the critical role of technology in achieving sustainable and equitable growth in agriculture, serving as a model for global efforts to uplift farming communities.

Future Scenario

The future of agriculture in Bharat (India) is poised for a transformation driven by the convergence of artificial intelligence (AI), machine learning (ML), autonomous farming systems, and the Internet of Things (IoT). When integrated effectively, these technologies can revolutionize the agricultural sector, ensuring food security and striving toward the ambitious goal of Zero Hunger by 2047. With AI-powered predictive crop yield models, autonomous farming systems, and IoT-enabled monitoring and control mechanisms, agriculture in Bharat could witness a quantum leap, making the sector more sustainable, efficient, and resilient to climate change. This chapter explores the pivotal role these innovations will play in reshaping Indian agriculture and the journey toward achieving food security.

AI for Predictive Crop Yield Models: The application of AI in predictive crop yield models represents a critical advancement in agricultural planning and management. Traditional farming techniques often relied on historical data, weather patterns, and expert knowledge to forecast crop yields. However, these methods were prone to inaccuracies due to agricultural environments' complex and dynamic nature. AI, specifically machine learning algorithms, can overcome these challenges by analyzing vast amounts of data, including weather forecasts, soil conditions, satellite imagery, and historical crop performance. Through advanced pattern recognition, these models can predict crop yields with remarkable precision, enabling farmers to make data-driven decisions.

In the future, AI models will not only predict yields but will also optimize farming practices. By integrating real-time data from sensors in the field, AI can provide insights into optimal planting times, irrigation needs, and pest management strategies tailored to specific geographic locations. Such personalized recommendations will enhance productivity and reduce resource wastage, making farming more efficient and environmentally sustainable. For instance, AI can detect plant stress or disease patterns early, enabling timely interventions and minimizing crop loss. The predictive models of the future will incorporate climate change projections, ensuring that farming systems can adapt to the increasingly unpredictable weather patterns caused by global warming. This will be critical in mitigating the impacts of droughts, floods, and other climate-related events that often lead to crop failures in India.

Autonomous Farming Systems: The Role of Drones and IoT: One of the most exciting prospects for the future of agriculture is the widespread use of autonomous farming systems, particularly those integrating drones and IoT technologies. Autonomous drones have already made their presence felt in modern farming by being used for aerial surveillance, crop monitoring, precision spraying, and fertilization tasks. As these technologies advance, drones will become more intelligent, capable of performing complex tasks autonomously without human intervention. In the future, drones will be equipped with advanced sensors, including multispectral and hyperspectral cameras, to capture a wide range of data about crop health, soil quality, and moisture levels. They will be able to monitor crops in real-time, detect early signs of diseases or pests, and apply targeted treatments, reducing the need for broad-spectrum pesticide use and lowering input costs.

In parallel, IoT devices will play an increasingly critical role in autonomous farming systems. IoT-enabled sensors in fields and greenhouses will collect real-time data on soil moisture, temperature, humidity, pH levels, and nutrient availability. This data will be fed into AI algorithms that can analyze and interpret the information to optimize irrigation, fertilization, and crop management practices. IoT sensors can also track livestock health and well-being, improving farm management efficiency. Together with AI, IoT will allow farmers to make smarter, more informed decisions, improving their yields while minimizing resource usage.

Autonomous tractors, harvesters, and robotic systems are also on the horizon, potentially revolutionising labor-intensive tasks. These machines can plant, cultivate, and harvest crops with minimal human involvement, significantly reducing labor costs and improving productivity. By utilizing AI and IoT data, these autonomous systems will perfectly synchronise with the environment, optimizing their movements and operations to reduce energy consumption and environmental impact. This paradigm shift will allow farmers to scale up their operations while addressing labor shortages and the rising cost of farm labor.

Ensuring Zero Hunger in Bharat by 2047: The ambitious goal of ensuring Zero Hunger in Bharat by 2047 aligns with the nation's aspirations for self-sufficiency in food production and eliminating malnutrition. AI and automation will play a pivotal role in ensuring food is produced more efficiently and distributed equitably. Farmers can optimize their production capacity, reduce losses, and enhance food security by leveraging AI-driven predictive crop models. Predicting the outcomes of various agricultural strategies and weather conditions will allow farmers to prepare for fluctuations in supply and demand, avoiding crises like food shortages.

The role of AI in facilitating efficient food distribution networks cannot be overstated. One of the critical challenges India faces in achieving food security is inefficiency in the food supply chain, which leads to large-scale post-harvest losses. AI-based systems can revolutionize supply chain management by predicting demand, optimizing transportation routes, and ensuring food products promptly reach consumers. AI will also help monitor food safety standards, reduce contamination risks, and ensure that food products meet quality standards throughout the supply chain. Additionally, AI can aid in the targeted delivery of food to areas suffering from food insecurity. By analyzing regional data on crop yields, weather patterns, and consumption rates, AI can identify areas where food scarcity is likely to occur and preemptively distribute resources to mitigate hunger. Blockchain, coupled with AI, could further enhance the transparency and traceability of food distribution, ensuring that food reaches those who need it the most.

IoT and autonomous farming systems will play a role in improving yields and reducing wastage at various stages of production and distribution. Smart irrigation systems, powered by IoT sensors, will reduce water wastage, while IoT-enabled temperature control systems will ensure that perishable foods like fruits, vegetables, and dairy products maintain their freshness during transport. This will minimize food losses, ensuring that more produce is available for consumption rather than wasted due to inefficient handling.

Path Ahead: Despite the immense potential of AI, IoT, and autonomous systems in achieving Zero Hunger in Bharat by 2047, several challenges must be overcome. Adopting these technologies requires significant infrastructure investment, training, and education for farmers, particularly those in rural and underserved areas. Ensuring that smallholder farmers have access to AI tools and IoT devices will be critical to achieving equitable growth in the agricultural sector. The digital divide must be addressed, as many rural areas still lack reliable internet access and technical support.

The government and private stakeholders will need to invest in creating an ecosystem that supports the development and deployment of these technologies. This includes building digital literacy, providing financial support for farmers to adopt new technologies, and ensuring the technological infrastructure is robust and scalable. Public-private partnerships will be crucial in driving these innovations forward and ensuring they benefit farmers nationwide, particularly those in economically disadvantaged regions. Finally, policies and regulations must be developed to ensure that AI and autonomous systems are used responsibly and ethically in agriculture. This includes addressing data privacy concerns, ensuring AI algorithms' transparency, and minimising technological advancements' environmental impact.

The future of agriculture in Bharat is inextricably linked to the adoption of AI, IoT, and autonomous farming systems. By harnessing the power of predictive crop yield models, drones, and IoT devices, India can enhance its agricultural productivity, optimize resource usage, and pave the way for achieving Zero Hunger by 2047. While significant challenges remain, the potential benefits are immense: a more sustainable, efficient, and equitable agricultural system that feeds a growing population. By embracing technological advancements and investing in infrastructure, Bharat can lead toward a future where hunger is eradicated and food security is ensured for all.

12

Urban Sustainability and IT

Introduction

Urban sustainability has become a critical focus as the global population increasingly concentrates in cities. With over 56% of the world's population residing in urban areas, cities are significant drivers of economic growth and innovation but also contribute disproportionately to environmental degradation, resource depletion, and social inequities. Urban sustainability aims to balance economic, social, and environmental dimensions, ensuring that cities are livable, resilient, and capable of meeting the needs of current and future generations. Information Technology (IT) is pivotal in advancing urban sustainability by enabling smarter resource management, reducing environmental footprints, and fostering inclusive and equitable development.

Challenge of Urban Sustainability: Cities face numerous sustainability challenges, including:

1. **Resource Inefficiencies**: High energy consumption, water scarcity, and inefficient waste management are pressing issues in urban centers.
2. **Environmental Degradation**: Urban areas are responsible for 75% of global carbon emissions and are hotspots for air and water pollution.
3. **Social Inequities**: Rapid urbanization often leads to disparities in housing, healthcare, and access to basic services.
4. **Climate Resilience**: Urban areas are increasingly vulnerable to climate-induced risks such as flooding, heatwaves, and rising sea levels.
5. **Infrastructure Strain**: Aging infrastructure and inadequate urban planning exacerbate the strain on resources and services.

Smart Cities – IT as a Catalyst for Sustainable Urbanization

1. **Efficient Energy Management**: Smart grids and energy management systems leverage IoT (Internet of Things) sensors and AI (Artificial Intelligence) to optimize energy distribution and reduce consumption. IT-enabled renewable energy integration into urban grids promotes a shift toward clean energy, reducing greenhouse gas emissions. For example,

smart metering systems allow households to monitor and control energy use, contributing to demand-side management.

2. **Water Conservation and Management**: IT solutions such as GIS (Geographic Information Systems) and IoT sensors monitor water distribution networks, detect leaks, and optimize usage. Smart irrigation systems in urban green spaces and AI-driven water demand forecasting ensure efficient water resource management. Cities like Singapore have adopted IT-driven water reclamation and recycling technologies, setting global benchmarks in water sustainability.

3. **Urban Mobility and Transportation**: IT-enabled intelligent transportation systems (ITS) improve urban mobility while reducing emissions. Real-time traffic management, optimisation of public transit, and promoting shared mobility platforms like ride-hailing and bike-sharing contribute to sustainable urban transport. Electric vehicle (EV) integration with smart charging infrastructure also benefits from IT-driven innovations.

4. **Waste Management and Circular Economy**: IT facilitates efficient waste collection, segregation, and recycling through IoT-enabled smart bins and waste-tracking systems. Digital platforms connecting consumers, recyclers, and manufacturers promote a circular economy, reducing landfill dependency and resource extraction.

5. **Urban Planning and Monitoring**: Big data analytics and urban modeling tools allow city planners to simulate development scenarios, optimize land use, and design green infrastructure. GIS mapping aids in identifying urban heat islands and planning interventions to mitigate their impact. Additionally, drones and satellite imagery monitor urban growth, helping policymakers address encroachment and habitat destruction.

IT and Environmental Sustainability

1. **Air Quality Monitoring and Management**: IoT-enabled air quality sensors deployed across cities provide real-time data on pollutant levels. AI algorithms analyze this data to identify pollution sources and recommend mitigation measures. Publicly accessible air quality indexes encourage community engagement in reducing emissions.

2. **Green Infrastructure and Biodiversity**: IT supports designing and maintaining green spaces, urban forests, and biodiversity corridors. Satellite-based monitoring and GIS tools track vegetation health and guide afforestation efforts. Urban sustainability initiatives increasingly use IT to balance urban development with ecological preservation.

3. **Carbon Footprint Reduction**: IT solutions help businesses and individuals track and reduce their carbon footprints. Smart buildings, powered by IoT sensors, monitor energy use and optimize heating, cooling, and lighting systems to reduce emissions. Blockchain technology enables transparent carbon credit trading systems, incentivizing emission reductions.

Social Equity and IT-Driven Urban Inclusion

1. **Digital Platforms for Citizen Engagement**: IT fosters participatory governance through mobile applications and digital platforms that allow citizens to report issues, access services, and contribute to decision-making processes. Initiatives like e-governance portals ensure transparency and accountability in urban administration.

2. **Affordable Housing and Smart Solutions**: IT-enabled platforms connect urban dwellers with affordable housing options, ensuring equitable access to shelter. AI-driven predictive analytics guide policymakers in addressing housing shortages and improving living conditions in informal settlements.

3. **Healthcare and Education Accessibility**: Telemedicine platforms and mobile health applications bridge healthcare gaps in underserved urban areas. Similarly, IT-enabled digital classrooms and online education tools improve access to quality education, fostering inclusivity in urban development.

Building Climate-Resilient Cities

1. **Disaster Prediction and Management**: AI-driven predictive models analyze climate patterns and urban vulnerabilities to forecast extreme weather events, guiding proactive disaster preparedness. Early warning systems and real-time data dissemination reduce human and economic losses during disasters.

2. **Urban Cooling Solutions**: IT supports the development of urban cooling strategies to combat heatwaves. GIS mapping identifies vulnerable areas, while IoT-enabled cooling systems optimize energy use in cooling urban spaces.

3. **Flood Management and Resilience**: IT tools like flood modeling software and IoT-enabled monitoring systems help cities design effective drainage systems and manage stormwater. These technologies reduce flood risks in low-lying urban areas.

Role of Emerging Technologies in Urban Sustainability

1. **Artificial Intelligence (AI)**: AI enables predictive analytics, automated decision-making, and optimization across urban systems, from energy grids to waste management. AI-driven urban analytics tools help policymakers prioritize sustainability initiatives.

2. **Internet of Things (IoT)**: IoT devices collect and transmit real-time data from urban systems, enabling efficient resource management. IoT forms the backbone of urban sustainability initiatives from smart meters to connected vehicles.

3. **Blockchain Technology**: Blockchain fosters transparency and accountability in urban resource management. Its applications in energy trading, waste recycling, and carbon credit systems ensure equitable and efficient resource distribution.

4. **Big Data and Cloud Computing**: Big data analytics and cloud platforms process vast amounts of urban data, providing actionable insights for sustainability planning. These technologies enable scalable and cost-effective urban solutions.

Challenges in IT-Driven Urban Sustainability: Despite its transformative potential, IT-driven urban sustainability faces challenges such as:

1. **Digital Divide**: Unequal access to digital technologies may exclude vulnerable populations from the benefits of IT solutions.

2. **Cybersecurity Risks**: Increased reliance on digital systems exposes urban infrastructure to cyber threats.

3. **Data Privacy Concerns**: The collection and use of urban data must comply with stringent privacy regulations to build public trust.

4. **High Costs of Implementation**: Deploying IT solutions in urban infrastructure requires significant investment, which may strain municipal budgets.

IT is a cornerstone of urban sustainability, enabling smarter resource management, enhancing environmental protection, and promoting social equity. By integrating cutting-edge technologies like AI, IoT, and blockchain, cities can address complex challenges, from resource inefficiencies to climate risks. However, ensuring inclusivity, affordability, and data security is crucial to realizing the full potential of IT-driven urban sustainability. With sustained investment, policy support, and community engagement, IT can transform cities into hubs of sustainable innovation, resilience, and prosperity for all.

Current Scenario

The current era of rapid urbanization has made the concept of smart cities an essential component of sustainable urban development. Smart cities leverage Information Technology (IT) to enhance the quality of life for citizens, optimize resource use, and improve governance. Globally, over 500 cities have adopted smart technology to address resource inefficiency, traffic congestion, pollution, and inadequate public services. In India, the Smart Cities Mission, launched in 2015, envisions the transformation of 100 cities through IT-enabled solutions, aiming for sustainable urbanization and improved living standards. However, while IT plays a transformative role in smart city initiatives, significant issues arise due to inadequate data integration, undermining the efficacy of urban planning and management.

IT in Smart City Initiatives: Smart city initiatives worldwide emphasize the integration of IT across various urban domains, including energy, water, transportation, waste management, and public services. The following are key IT-enabled components in smart cities:

Smart Infrastructure

- **IoT-Enabled Systems**: The Internet of Things (IoT) allows for collecting real-time data from urban infrastructure, such as water pipelines, electricity grids, and transportation networks. Sensors embedded in infrastructure provide actionable insights for predictive maintenance and efficient resource management.

- **Smart Buildings**: IT-powered smart buildings optimize energy consumption, enhance occupant comfort, and reduce operational costs through systems that adjust lighting, heating, and cooling based on real-time conditions.

Urban Mobility and Transportation

- Intelligent Transportation Systems (ITS) use sensors, cameras, and GPS data to improve traffic flow, manage public transit schedules, and reduce congestion. Real-time traffic data shared through mobile apps enhances commuting efficiency.

- IT solutions also support electric vehicle (EV) infrastructure, enabling smart charging and reducing the carbon footprint of urban transport systems.

Governance and Citizen Engagement

- E-governance platforms powered by IT promote transparency, accountability, and citizen participation. Mobile apps and online portals allow residents to report grievances, access services, and participate in policy discussions.

- Geographic Information Systems (GIS) and data visualization tools support decision-making by mapping urban assets, planning land use, and assessing environmental impacts.

Waste and Resource Management

- IT solutions such as IoT-enabled waste bins monitor and optimize waste collection schedules. Smart recycling systems track material flows, promoting circular economy practices.
- Energy management systems integrate renewable energy sources and manage grid loads to ensure efficient energy use.

Public Safety and Emergency Management

- Smart surveillance systems, powered by AI and IoT, enhance public safety by monitoring high-risk areas and enabling rapid response during emergencies.
- Early warning systems for disasters like floods and earthquakes use real-time data and predictive analytics to minimize loss of life and property.

Issues in Urban Planning Due to Inadequate Data Integration: Despite these advancements, inadequate data integration remains a critical challenge in urban planning and smart city implementation. The inability to effectively collect, integrate, and analyze data across multiple urban domains creates bottlenecks in achieving the goals of smart cities.

Fragmented Data Silos

- Data generated by various city departments, private entities, and IoT devices often exist in isolated silos. A lack of interoperability and standardized protocols prevents the seamless exchange of information, limiting holistic decision-making.
- For instance, transportation data may not be integrated with air quality monitoring systems, making it difficult to design comprehensive solutions for reducing urban pollution.

Incomplete and Inaccurate Data

- In many cities, the datasets required for urban planning are either incomplete or outdated. This is particularly problematic in rapidly growing urban areas, where real-time data is essential for dynamic planning.
- For example, outdated maps of water pipelines can result in inefficient maintenance schedules and frequent disruptions to water supply.

Technical and Organizational Challenges

- Integrating diverse data sources requires advanced IT infrastructure, which is often lacking in many urban centers, especially in developing countries.
- Institutional inertia and resistance to change also hinder the adoption of integrated data systems. Different departments may be reluctant to share data due to concerns over jurisdiction, accountability, or loss of control.

Lack of Skilled Workforce

- Urban planning departments often lack IT expertise to manage, analyze, and interpret complex datasets. This skills gap slows the adoption of modern technologies and reduces the effectiveness of planning processes.

Data Privacy and Security Concerns

- Integrating large volumes of citizen data raises significant privacy and security concerns. The absence of robust regulatory frameworks and data protection measures exacerbates the risks of unauthorized access and misuse.
- For example, smart surveillance systems can inadvertently infringe on citizens' privacy if not managed with strict data governance policies.

Challenges in Real-Time Data Utilization

- Real-time data is critical for addressing dynamic urban challenges, such as traffic congestion or disaster response. However, many cities struggle to implement systems capable of efficiently processing and acting on real-time data.
- The delay in leveraging real-time data often results in reactive rather than proactive urban management.

Impact of Inadequate Data Integration on Urban Planning

Inefficient Resource Allocation

- Urban planners do not understand resource demand and supply patterns comprehensively without integrated data systems. This results in inefficient allocation of resources, such as water, energy, and public transit.

Suboptimal Infrastructure Development

- Urban infrastructure projects require accurate data to assess future demand and environmental impact. Inadequate data integration leads to underutilized or overburdened infrastructure, affecting urban sustainability.

Missed Opportunities for Optimization

- Cities fail to capitalize on optimization opportunities when data from different domains is not combined. For instance, linking energy consumption data with weather forecasts could enhance energy efficiency, but such integration is rarely achieved.

Inability to Address Complex Urban Challenges

- Urban challenges such as traffic congestion, pollution, and housing shortages require multi-dimensional solutions based on integrated datasets. Fragmented data systems hinder the development of such solutions.

Reduced Citizen Trust and Participation

- Inconsistent or incomplete data undermines the credibility of urban planning initiatives, reducing citizen trust and participation in governance processes.

Addressing Data Integration Challenges: To overcome the challenges of inadequate data integration, cities must adopt a comprehensive approach that includes the following strategies:

Standardization and Interoperability

- Developing standardized data collection, storage, and sharing protocols ensures system interoperability. Open data platforms can facilitate collaboration between stakeholders.

Investments in IT Infrastructure

- Upgrading IT infrastructure, including data centers, cloud computing systems, and high-speed internet connectivity, is essential for managing large datasets and enabling real-time data analysis.

Capacity Building

- Training urban planners and administrators in data analytics, GIS, and other IT tools is crucial for effective implementation. Partnerships with academic institutions and private sector organizations can bridge the skills gap.

Data Governance Frameworks

- Establishing robust data governance policies ensures privacy, security, and ethical use. Regulatory frameworks must define clear guidelines for data sharing and accountability.

Public-Private Partnerships (PPPs)

- Collaborations between government bodies, technology providers, and research organizations can accelerate the adoption of integrated data systems.

Citizen-Centric Design

- Engaging citizens in the planning process through participatory platforms fosters trust and ensures that urban solutions align with community needs.

IT has immense potential to transform urban planning and management, laying the foundation for smarter and more sustainable cities. However, inadequate data integration remains a significant obstacle, leading to inefficiencies and missed opportunities. Addressing these challenges requires concerted efforts to standardize data systems, invest in IT infrastructure, build capacity, and establish robust governance frameworks. With the right strategies, cities can harness the power of IT to overcome data integration barriers and achieve their smart city ambitions, ultimately improving the quality of life for all urban residents.

IT Solutions for Urban Planning

Integrating Information Technology (IT) into urban planning has revolutionized how cities are designed, managed, and optimized. Urban areas face complex challenges, including rapid population growth, resource scarcity, and environmental degradation, which demand innovative, data-driven approaches to planning and governance. Geographic Information Systems (GIS), smart waste management systems, and IT-enabled resource efficiency strategies are at the forefront of modern urban planning solutions. These technologies facilitate efficient land use, streamline resource allocation, and promote sustainable urban development.

Geographic Information Systems (GIS) for Efficient Urban Design: GIS has emerged as a powerful tool for urban planners, providing spatial insights and enabling the visualization, analysis, and interpretation of data in geographic contexts. Its role in urban planning is multifaceted:

Land Use Planning and Zoning

- GIS helps planners identify optimal land use patterns by overlaying layers of demographic, environmental, and economic data.
- It enables zoning that balances residential, commercial, industrial, and recreational needs, fostering well-organized urban growth.

Infrastructure Planning

- GIS aids in designing transportation networks, utilities, and public facilities by analyzing spatial connectivity and population density.

- For instance, transit-oriented development (TOD) plans leverage GIS to optimize public transit access and reduce car dependency.

Environmental Impact Assessment

- By integrating ecological data, GIS tools enable planners to evaluate the environmental impact of urban projects.
- Planners can identify critical habitats, flood-prone zones, and air pollution hotspots, guiding sustainable development decisions.

Disaster Management and Resilience

- GIS facilitates risk assessment and disaster preparedness by mapping vulnerabilities like flood zones, seismic activity areas, and heat islands.
- It also supports emergency response planning, enabling faster and more effective disaster mitigation strategies.

Community Engagement and Decision-Making

- GIS-based platforms allow citizens to visualize proposed plans and participate in urban design discussions.
- This fosters transparency, inclusivity, and trust in urban governance.

Smart Waste Management Systems: Waste management is a critical aspect of urban sustainability, and IT-driven smart waste management systems are addressing inefficiencies in traditional practices. These systems leverage IoT, GIS, and data analytics to enhance waste collection, segregation, recycling, and disposal.

IoT-Enabled Smart Bins

- IoT sensors in waste bins monitor fill levels and send real-time data to centralized systems, enabling optimized collection routes.
- This reduces fuel consumption and operational costs while minimizing the environmental impact of waste collection vehicles.

Waste Segregation and Recycling

- AI-powered sorting systems use image recognition and robotic automation to separate recyclable materials from mixed waste streams.
- Smart recycling kiosks incentivize citizens to deposit recyclable items by offering rewards, promoting community participation in recycling programs.

GIS for Waste Management

- GIS tools optimize waste collection routes, ensuring efficient allocation of resources and minimizing delays.
- By mapping waste generation patterns, planners can identify underserved areas and deploy additional infrastructure where needed.

Circular Economy Integration

- IT platforms connect waste generators with recyclers and manufacturers, facilitating the reuse of materials and reducing landfill dependency.
- Blockchain technology ensures transparency and traceability in material flows, building stakeholder trust.

Data-Driven Decision-Making

- Smart waste management systems generate actionable insights from historical data, helping municipalities predict waste generation trends and plan accordingly.

GIS-Based Design for Resource – Efficient Cities: Urban areas are resource-intensive, consuming vast energy, water, and raw materials. GIS-based design approaches aim to optimize resource use while minimizing waste and environmental degradation.

Energy Efficiency and Renewable Integration

- GIS tools identify locations suitable for renewable energy installations, such as solar farms and wind turbines.
- Energy demand mapping guides planners in designing efficient distribution networks and reducing transmission losses.

Water Resource Management

- GIS-based hydrological modeling helps planners optimize water supply networks, minimize leaks, and manage stormwater.
- By mapping water demand and availability, GIS ensures equitable distribution and sustainable use of water resources.

Green Infrastructure Planning

- GIS facilitates the design of urban green spaces, parks, and green roofs by analyzing spatial data on land availability, heat islands, and air quality.
- Green infrastructure reduces urban heat effects, improves air quality, and promotes biodiversity.

Transportation and Mobility

- GIS supports the development of sustainable transportation systems by mapping population density, travel patterns, and existing transit networks.
- Planners can optimize public transit routes, reduce traffic congestion, and promote non-motorized transport options like cycling and walking.

Waste-to-Energy and Resource Recovery

- GIS identifies sites for waste-to-energy plants, factoring in waste generation patterns, proximity to residential areas, and environmental constraints.
- These plants convert waste into usable energy, reducing reliance on fossil fuels and landfills.

Urban Heat Island Mitigation

- GIS mapping of temperature variations across a city guides interventions such as planting trees, increasing albedo of surfaces, and designing reflective rooftops.
- These measures lower urban temperatures, improving comfort and reducing energy consumption for cooling.

Future Directions: To fully realize the potential of IT in urban planning, future efforts should focus on:

Data Standardization and Integration

- Establishing common data standards and promoting system interoperability will enable seamless integration across urban domains.

Capacity Building

- Training programs for urban planners and decision-makers can bridge the skills gap and foster widespread adoption of IT tools.

Public-Private Partnerships

- Collaborations with technology providers and research institutions can reduce implementation costs and bring cutting-edge solutions to cities.

Citizen-Centric Design

- Engaging citizens in planning ensures that IT solutions address community needs and gain public support.

Regulatory Frameworks

- Comprehensive data governance, privacy, and cybersecurity policies will build trust and facilitate wider adoption of IT-driven systems.

The application of IT in urban planning, particularly through GIS, smart waste management systems, and resource-efficient designs, offers transformative potential for sustainable urban development. These technologies enable data-driven decisions, optimize resource use, and enhance the quality of life for urban residents. However, addressing challenges such as data integration, high costs, and skill gaps is essential for their successful implementation. By adopting a holistic approach that combines technical innovation, capacity building, and citizen engagement, cities can harness IT to build resilient, efficient, and inclusive urban environments.

Challenges in Urban Growth and Resource Management

Urban growth presents opportunities and challenges, particularly in managing critical resources like housing, waste, and transportation systems. As cities expand due to population growth and economic opportunities, the strain on urban infrastructure and resources intensifies. The challenges of inadequate housing, inefficient waste management, and persistent traffic congestion hinder sustainable urban development and compromise the quality of life for urban residents. Addressing these interconnected issues requires a comprehensive understanding of their causes, implications, and potential solutions.

Housing Challenges in Rapid Urban Growth

Urbanization and Housing Demand

- Urban population growth, fueled by migration from rural areas, leads to skyrocketing demand for housing. Cities in developing countries are particularly affected, as they often lack the infrastructure to accommodate rapid influxes of people.
- The UN-Habitat estimates that nearly 1 billion people globally live in slums, where overcrowding and substandard living conditions are prevalent.

Affordable Housing Shortages

- Rising land prices and construction costs make affordable housing scarce, forcing low-income populations into informal settlements or inadequate housing.
- Policies for affordable housing often fall short due to limited funding, inefficient implementation, and competing land use priorities.

Environmental Impact of Housing

- Expanding urban areas often encroach on natural ecosystems, leading to deforestation, loss of biodiversity, and increased vulnerability to climate risks such as flooding.

- Energy-inefficient housing contributes to urban carbon footprints, exacerbating climate change.

Infrastructure Deficiencies

- Informal settlements typically lack access to basic services such as clean water, sanitation, and electricity. This exacerbates health risks and perpetuates cycles of poverty.

Policy and Planning Gaps

- Urban planning often fails to anticipate long-term housing needs, leading to haphazard development. Poor coordination between government agencies and private developers further compounds the issue.

Waste Management Challenges in Growing Urban Areas

Rising Waste Generation

- Urbanization and rising living standards increase waste generation, particularly in high-density areas. According to the World Bank, cities generate over 2 billion tons of solid waste annually, a figure expected to double by 2050.
- The composition of urban waste is becoming more complex, with growing volumes of e-waste, plastics, and hazardous materials.

Inefficient Collection and Disposal Systems

- In many cities, waste collection systems are underfunded and poorly managed, resulting in significant portions of waste being left uncollected. This contributes to pollution, blocked drainage systems, and health hazards.
- Open dumping and unregulated landfills are common, releasing greenhouse gases like methane and contaminating soil and water.

Lack of Recycling Infrastructure

- Recycling rates in most cities remain low due to inadequate infrastructure, public awareness, and limited markets for recycled materials.
- Informal waste pickers play a significant role in recycling but often lack social and economic protections.

Economic and Institutional Barriers

- Waste management services are costly, and many municipalities face budget constraints.

- Poor governance and corruption can undermine efforts to improve waste systems, while fragmented agency responsibilities lead to inefficiencies.

Environmental and Public Health Impacts

- Improper waste disposal contributes to air, water, and soil pollution. For instance, burning waste in open dumps releases toxic pollutants that harm respiratory health.
- Unmanaged waste attracts pests and pathogens, increasing the risk of diseases like cholera and dengue fever.

Traffic Congestion as a Barrier to Urban Growth

Increasing Vehicle Ownership

- Rising incomes and inadequate public transportation systems drive private vehicle ownership, leading to severe traffic congestion in urban areas.
- By 2050, the number of vehicles on the road will double, with cities in developing nations experiencing the fastest growth.

Inadequate Public Transit Systems

- Many urban transit systems are outdated, overcrowded, and poorly connected, failing to meet the mobility needs of growing populations.
- Investment in public transportation is often inadequate, prioritizing road infrastructure for private vehicles instead.

Urban Sprawl and Land Use Inefficiencies

- Poor urban planning contributes to sprawling cities with residential areas far from employment hubs. This increases commute times and dependence on private vehicles.
- Mixed land use, which could reduce travel distances, is rarely implemented effectively.

Economic Costs of Congestion

- Traffic congestion imposes substantial economic costs, including lost productivity, increased fuel consumption, and higher business logistics costs.
- For example, congestion in Indian cities is estimated to cost the economy nearly $22 billion annually.

Environmental and Health Impacts

- Prolonged idling in traffic contributes to air pollution, with transportation being a major source of greenhouse gas emissions.
- Poor air quality in congested cities increases respiratory and cardiovascular health risks, particularly for vulnerable populations.

Policy and Governance Gaps

- Fragmented transportation policies and lack of coordination among city, regional, and national authorities hinder effective congestion management.
- Resistance to congestion pricing or carpooling incentives further exacerbates the issue.

Interconnected Nature of Urban Challenges: The challenges of housing, waste management, and traffic congestion are deeply interconnected, creating a feedback loop that exacerbates urban inefficiencies:

- **Housing and Congestion**: Lack of affordable housing near city centers forces people to live in distant suburbs, increasing commuting distances and traffic congestion.
- **Waste Management and Housing**: Informal settlements often lack organized waste collection, leading to health and environmental hazards that further marginalize their residents.
- **Congestion and Waste**: Traffic congestion impedes waste collection vehicles, delaying operations and increasing costs.

Strategies for Addressing Urban Growth Challenges

Integrated Urban Planning

- Adopting holistic planning approaches considering housing, transportation, and waste management can enhance resource efficiency and equity.
- Tools like GIS and urban modeling software can provide insights into land use optimization and infrastructure needs.

Affordable Housing Initiatives

- Public-private partnerships can mobilize resources for affordable housing projects.
- Policies like inclusionary zoning and subsidies for low-income households can make housing accessible to marginalized groups.

Innovative Waste Management Solutions

- Smart waste management systems, including IoT-enabled bins and AI-powered sorting technologies, can improve collection efficiency and recycling rates.
- Circular economy practices prioritising waste reduction, reuse, and recycling should be mainstreamed.

Sustainable Transportation Systems

- Expanding and modernizing public transit systems can reduce dependence on private vehicles.
- Promoting active transportation modes like cycling, walking, carpooling, and shared mobility services can alleviate congestion.

Policy and Governance Reforms:

- Clear regulatory frameworks, improved coordination among agencies, and citizen engagement are essential for effective urban management.
- Data-driven decision-making and transparent monitoring mechanisms can enhance accountability and trust.

Technology Integration

- Leveraging IT solutions such as GIS, IoT, and big data analytics can optimize resource allocation and provide real-time insights for urban governance.
- Smart city initiatives can serve as platforms for piloting innovative solutions to housing, waste, and transportation challenges.

Case Study

Smart cities have gained significant traction globally as a solution to rapid urbanization, resource scarcity, and sustainability challenges. Smart city initiatives aim to integrate information technology, data analytics, and citizen engagement to improve urban areas' quality of life, economic growth, and environmental sustainability. This case study examines the implementation of smart city projects, focusing on key technologies, governance models, challenges, and success stories. Examples from global cities, including Singapore, Barcelona, and India's Smart Cities Mission, highlight how technology-driven approaches transform urban management.

Definition and Objectives of Smart City Projects: Smart city projects are characterized by the use of advanced technologies, such as the Internet of Things (IoT), Geographic Information Systems (GIS), Artificial Intelligence (AI), and Big Data analytics, to address urban challenges. The objectives typically include:

1. Enhancing urban infrastructure and services, such as transportation, energy, waste management, and water supply.
2. Promoting sustainability by reducing resource consumption and minimizing environmental impact.
3. Fostering economic growth through innovation and the creation of technology-driven ecosystems.
4. Improving governance through transparency, accountability, and citizen participation.
5. Ensuring inclusivity by addressing the needs of diverse demographic groups, including marginalized communities.

Key Components of Smart City Projects

Smart Governance

- E-governance platforms enable seamless interaction between citizens and municipal authorities, improving service delivery and reducing bureaucratic inefficiencies.
- Examples include online grievance redressal systems, real-time public service updates, and participatory planning platforms.

Smart Infrastructure

- IoT-enabled sensors and actuators monitor and manage urban systems in real-time.
- Applications include smart grids for energy management, automated traffic signals, and sensor-based water distribution systems.

Smart Mobility

- Intelligent transportation systems integrate data from various sources to optimize public transit, reduce traffic congestion, and enhance accessibility.
- Shared mobility services, such as bike-sharing and ride-hailing platforms, promote sustainable transportation.

Smart Environment

- Air quality sensors, waste segregation systems, and energy-efficient buildings contribute to environmental sustainability.
- Renewable energy integration, such as solar rooftops and wind farms, reduces dependency on fossil fuels.

Smart Economy

- Innovation hubs and technology parks within smart cities foster entrepreneurship and attract investments.
- Digital marketplaces enable local businesses to reach wider audiences and streamline supply chains.

Global Examples of Smart City Implementation

Singapore

- Singapore's Smart Nation initiative emphasizes data-driven urban management and citizen-centric solutions.
- Key projects include autonomous public transport trials, smart waste management systems, and MyResponder, an app that connects citizens with emergency medical responders.
- Challenges include balancing privacy concerns with the extensive use of surveillance technologies and maintaining inclusivity for digitally less-savvy populations.

Barcelona, Spain

- Barcelona employs IoT networks to manage energy consumption, reduce traffic congestion, and monitor environmental parameters.
- Smart streetlights with sensors adjust brightness based on real-time foot traffic, significantly reducing energy costs.
- The city's focus on participatory governance through the "Decidim Barcelona" platform ensures that citizens contribute to policymaking.

India's Smart Cities Mission

- Launched in 2015, India's Smart Cities Mission aims to develop 100 cities as sustainable and inclusive urban development models.
- Projects include integrated command and control centers (ICCCs), solar-powered infrastructure, and GIS-based urban planning tools.

- Challenges include funding constraints, limited technical expertise at local levels, and delays in project execution due to bureaucratic hurdles.

Case Study – Pune Smart City, India: Pune, one of the cities selected under India's Smart Cities Mission, provides a comprehensive example of smart city implementation:

Key Projects

- **Integrated Traffic Management System (ITMS)**: Traffic signals are synchronized using AI and IoT, reducing congestion and improving safety.
- **24x7 Water Supply System**: GIS-based modeling and IoT sensors monitor water distribution and minimize leaks.
- **Smart Lighting**: Energy-efficient LED streetlights with motion sensors reduce energy consumption.
- **Citizen Engagement**: The Pune Connect app allows residents to report issues, access city services, and participate in governance.

Outcomes

- A reduction in water losses by 15% and energy savings of 20% in street lighting.
- Improved response times for citizen grievances and increased transparency in service delivery.

Challenges

- Limited public awareness about smart city initiatives initially led to low citizen participation.
- Integration of legacy infrastructure with new technologies required significant investments and technical expertise.

Challenges in Smart City Implementation

Funding and Financial Constraints

- Smart city projects are capital-intensive, and municipalities often struggle to secure sufficient funding.
- Public-private partnerships (PPPs) have been instrumental but challenge aligning private profit motives with public welfare goals.

Technology Integration

- Integrating new technologies with existing urban infrastructure can be complex and expensive.

- Ensuring interoperability among different systems and platforms is a persistent challenge.

Privacy and Security Concerns

- The extensive use of data and surveillance technologies raises concerns about privacy and data misuse.
- Cybersecurity threats, including hacking and data breaches, pose significant risks to critical urban systems.

Digital Divide

- In many regions, disparities in digital literacy and access to technology exclude marginalized populations from benefiting fully from smart city initiatives.

Institutional and Governance Barriers

- Weak institutional capacities, fragmented responsibilities, and resistance to change hinder the effective implementation of smart city projects.
- Lack of skilled personnel at municipal levels further exacerbates these issues.

Strategies for Effective Implementation

Robust Policy Frameworks

- Clearly defined goals, timelines, and responsibilities ensure accountability and coordination among stakeholders.
- Policies should emphasize inclusivity, sustainability, and resilience in urban planning.

Capacity Building

- Training programs for municipal staff and urban planners can bridge skill gaps and enhance technical expertise.
- Partnerships with academic institutions and research organizations can provide additional support.

Citizen Engagement

- Ensuring that citizens are active participants in planning and implementation fosters trust and ensures that projects address community needs.
- Digital platforms for participatory governance can enhance transparency and accountability.

Public-Private Partnerships (PPPs)

- PPPs can mobilize private sector expertise and funding while ensuring public welfare through well-defined contractual agreements.

Focus on Scalability and Interoperability

- Solutions should be designed to scale and integrate with other urban systems, maximizing efficiency and cost-effectiveness.

Future Directions: The future of smart cities lies in advancing technologies such as AI, blockchain, and 5G, which can enhance urban management and citizen engagement:

1. **AI and Machine Learning**: Predictive analytics can optimize traffic flows, energy consumption, and emergency response.
2. **Blockchain**: Transparent and tamper-proof systems can facilitate efficient service delivery and resource allocation.
3. **5G Networks**: High-speed connectivity will enable seamless communication among IoT devices, enhancing smart city functionalities.

The implementation of smart city projects demonstrates the transformative potential of technology in addressing urban challenges. While successes like Singapore, Barcelona, and Pune highlight the benefits of integrating IT into urban management, significant challenges remain, particularly in funding, governance, and inclusivity. Addressing these challenges requires coordinated efforts among governments, private stakeholders, and citizens, emphasizing transparency, innovation, and equity. With continued technological advancements and best practices, smart cities can be models for sustainable, efficient, and inclusive urban living.

Future Scenario

The future of urban development lies in autonomous urban management systems powered by advanced Artificial Intelligence (AI) technologies. These systems promise to revolutionize city governance, resource allocation, and mobility by addressing persistent urban challenges such as overcrowding, inefficient resource use, and traffic congestion. With the integration of AI-enabled crowd and traffic flow management, urban areas are poised to become hubs of efficiency, sustainability, and enhanced quality of life. This chapter delves into the potential advancements, underlying technologies, and transformative impacts of such systems, alongside the challenges and ethical considerations that must be addressed.

Autonomous Urban Management Systems: Autonomous urban management systems represent a paradigm shift in how cities are operated. By leveraging AI, the Internet of Things (IoT), and real-time data analytics, these systems automate decision-making processes, optimize resource allocation, and ensure seamless urban functioning. Key features include:

1. **Real-Time Data Integration**: Autonomous systems aggregate data from IoT sensors, cameras, and connected devices to provide a comprehensive, real-time overview of urban dynamics.
2. **Predictive Analytics**: Machine learning algorithms analyze historical and real-time data to forecast demand patterns for water, energy, and transportation resources.
3. **Self-Regulating Mechanisms**: Systems adjust automatically to changing conditions, such as rerouting traffic during peak hours or managing energy distribution based on consumption trends.
4. **Interconnectivity**: Urban subsystems, such as transportation, utilities, and waste management, are interconnected, enabling coordinated responses to complex scenarios.

AI-Enabled Crowd and Traffic Flow Management: Crowd and traffic flow management are among the most critical applications of AI in autonomous urban systems. AI algorithms, combined with IoT-enabled devices and 5G connectivity, allow cities to optimize mobility and resource distribution:

Crowd Flow Management

- **Smart Surveillance**: AI-powered cameras monitor crowd density in public spaces such as transit hubs, shopping centers, and event venues.
- **Dynamic Space Allocation**: Algorithms predict crowd movements and dynamically allocate spaces to prevent overcrowding and ensure safety.
- **Emergency Response**: In emergencies, AI systems analyze crowd behavior in real-time, identifying evacuation routes and guiding people to safety.

Traffic Flow Optimization

- **Adaptive Traffic Signals**: AI algorithms adjust traffic light timings based on real-time vehicle flow, reducing congestion and minimizing travel times.
- **Autonomous Vehicles (AVs)**: Coordination among AVs using AI reduces road accidents and improves traffic efficiency by optimizing vehicle speeds and routes.

- **Public Transit Integration**: AI integrates data from buses, trains, and other transit modes to synchronize schedules, reduce waiting times, and enhance accessibility.

Event Management

- AI systems analyze crowd and traffic data during large-scale events to manage entry points, parking, and transit options, minimizing disruptions.
- Real-time updates through mobile apps provide users with personalized navigation and parking suggestions.

Efficient Resource Utilization: The integration of autonomous systems and AI in urban management leads to the efficient use of resources across multiple domains:

Energy Management

- AI-powered smart grids predict energy demand and balance supply from renewable and conventional sources, minimizing waste and ensuring reliability.
- Demand-side management systems incentivize energy-saving behaviors among residents and businesses.

Water Resource Optimization

- IoT-enabled water distribution networks monitor leaks, predict demand fluctuations, and ensure equitable water supply.
- AI models predict droughts or flooding, enabling proactive water resource management and disaster preparedness.

Waste Management

- Autonomous waste collection systems streamline recycling and disposal, including sensor-equipped bins and robotic waste sorters.
- Predictive analytics optimize collection schedules, reducing operational costs and environmental impact.

Land Use and Housing

- AI systems analyze population density and land usage patterns to recommend optimal zoning policies.
- Autonomous construction technologies, including 3D printing and robotics, enable cost-effective and sustainable housing solutions.

Enabling Technologies: The realization of autonomous urban systems and AI-enabled crowd and traffic flow management relies on a convergence of several advanced technologies:

1. **5G Connectivity**: High-speed, low-latency networks are essential for real-time communication between IoT devices and AI systems.
2. **IoT Networks**: Sensors embedded in infrastructure collect data on parameters such as traffic density, air quality, and resource usage.
3. **Cloud and Edge Computing**: Cloud platforms store and process large volumes of urban data, while edge devices handle localized, real-time computations.
4. **Machine Learning and Computer Vision**: These technologies enable AI to analyze data, identify patterns, and make decisions with minimal human intervention.

Transformative Impacts: The implementation of these systems has profound implications for urban sustainability and quality of life:

1. **Environmental Benefits**: Reduced traffic congestion lowers emissions, while optimized energy and water management decrease waste and resource depletion.
2. **Economic Gains**: Efficient urban systems reduce operational costs, attract investments, and stimulate innovation.
3. **Improved Safety and Resilience**: Real-time monitoring and predictive analytics enhance disaster preparedness and response, protecting lives and property.
4. **Enhanced Accessibility**: Optimized public transportation systems ensure equitable mobility for all residents, including the elderly and disabled.

Global Examples and Future Projections: Cities worldwide are pioneering elements of these future systems:

1. **Singapore**: AI-driven traffic management and crowd monitoring systems are integrated into the city's Smart Nation initiative.
2. **Copenhagen, Denmark**: IoT-enabled waste management and energy-efficient smart grids reduce the city's carbon footprint.
3. **Pune, India**: AI-enhanced command centers optimize traffic flow, emergency response, and resource management.

Autonomous urban management systems and AI-enabled crowd and traffic flow management hold transformative potential for urban sustainability and efficiency. By integrating advanced technologies into city governance, these systems promise to optimize resource use, reduce environmental impact, and enhance quality of life. However, their successful implementation hinges on addressing privacy, equity, and technological integration challenges. With strategic investments and ethical governance, the vision of smart, autonomous cities can become a cornerstone of sustainable urban futures.

13

IT and Waste Management

Introduction

Integrating Information Technology (IT) into waste management has revolutionized how societies address the growing challenges of waste generation, collection, segregation, recycling, and disposal. Rapid urbanisation and industrialisation have escalated the global waste burden, necessitating efficient and sustainable solutions. IT is pivotal in enabling smart waste management systems through real-time monitoring, data analytics, and automation. This chapter explores the transformative impact of IT on waste management, highlighting key technologies, their applications, challenges, and prospects.

Growing Need for IT in Waste Management: The generation of municipal solid waste (MSW) globally has reached alarming levels, driven by population growth, changing consumption patterns, and inadequate infrastructure. Traditional waste management systems, characterized by manual processes and limited oversight, struggle to cope with waste's increasing volume and complexity. Key challenges include:

1. **Inefficient Collection and Transportation**: Conventional waste collection methods often result in underutilized resources, fuel wastage, and higher operational costs.
2. **Limited Segregation at Source**: The lack of effective segregation leads to lower recycling rates and increased environmental pollution.
3. **Insufficient Recycling and Recovery**: Poor tracking and management systems hinder the recovery of valuable materials, exacerbating landfill overflow.
4. **Environmental and Health Impacts**: Unmanaged waste contributes to air, soil, and water pollution, posing risks to ecosystems and public health.

Key IT – Driven Waste Management Technologies

Internet of Things (IoT)

- IoT-enabled sensors embedded in waste bins monitor fill levels and transmit real-time data to waste collection systems.

- Smart bins notify collection teams when they reach capacity, optimizing collection routes and reducing operational costs.
- Applications include monitoring hazardous waste to ensure safe handling and disposal.

Geographic Information Systems (GIS)

- GIS facilitates spatial analysis and mapping for efficient route planning and waste facility siting.
- Cities like Amsterdam and Singapore utilize GIS to design optimal waste collection and recycling networks, reducing fuel consumption and emissions.

Artificial Intelligence (AI) and Machine Learning (ML)

- AI algorithms analyze waste composition to automate sorting processes in recycling plants, improving accuracy and efficiency.
- Predictive models forecast waste generation trends, enabling proactive management and policy formulation.

Blockchain Technology

- Blockchain ensures transparency and traceability in waste recycling and disposal.
- It is particularly effective in tracking electronic waste (e-waste) and ensuring compliance with environmental regulations.

Mobile Applications and Cloud Platforms

- Mobile apps enable citizens to report waste-related issues, locate recycling centers, and participate in waste segregation programs.
- Cloud-based systems centralize data collection and analysis, facilitating real-time decision-making and stakeholder coordination.

Applications in Waste Management

Smart Waste Collection Systems

- IT enables dynamic waste collection schedules, minimizing missed pickups and unnecessary trips.
- GPS-enabled vehicles and route optimization algorithms reduce fuel usage and emissions.

Waste Segregation and Recycling

- Automated Material Recovery Facilities (MRFs) use AI and robotics for precise waste sorting, enhancing recycling rates.
- IT systems categorize recyclable materials by type and quality, ensuring efficient resource recovery.

Hazardous Waste Management

- IoT sensors monitor chemical and biological waste to prevent leaks and ensure safe disposal.
- IT platforms track hazardous waste from generation to disposal, maintaining regulatory compliance.

E-Waste Management

- IT solutions streamline the collection and recycling of electronic waste, addressing the growing challenge of obsolete gadgets and devices.
- Blockchain-based systems trace the lifecycle of e-waste, promoting accountability among manufacturers and recyclers.

Public Awareness and Engagement

- Digital platforms educate citizens about waste segregation, recycling practices, and the environmental impact of improper disposal.
- Gamification and reward-based systems encourage community participation in waste reduction initiatives.

Global Examples of IT – Driven Waste Management

South Korea

- South Korea employs RFID (Radio Frequency Identification) tags on waste bags to monitor and charge households based on waste generation.
- The initiative has significantly reduced waste generation and increased recycling rates.

Pune, India

- The Pune Municipal Corporation leverages mobile apps and GIS platforms to manage waste collection, monitor waste bins, and engage with citizens.
- Segregation at source and community-driven initiatives have improved waste management efficiency.

Sweden

- Sweden's smart waste-to-energy plants use IoT and AI to optimize waste conversion into electricity and heat.
- The country's digital platforms promote citizen awareness and participation in recycling programs.

Future Prospects and Innovations

Circular Economy Integration

- IT solutions will be critical in transitioning to a circular economy, where waste is minimized, and materials are reused or recycled.

AI-Enhanced Predictive Systems

- Advanced AI models will predict waste generation patterns, enabling preemptive measures to manage waste surges during festivals or natural disasters.

Smart Waste Markets

- Blockchain-enabled platforms will facilitate trade in recyclable materials, promoting resource efficiency and economic growth.

Zero-Waste Cities

- Cities will deploy fully integrated IT systems to achieve zero-waste goals, leveraging IoT, AI, and GIS for seamless waste management.

Global Collaboration

- IT-enabled platforms will foster international collaboration on waste management best practices, technology sharing, and policy development.

The application of IT in waste management represents a transformative approach to addressing the global waste crisis. By enabling real-time monitoring, efficient resource allocation, and citizen engagement, IT systems contribute to cleaner, greener, and more sustainable urban environments. Despite challenges such as high costs and technical gaps, the continued advancement of technologies like IoT, AI, and blockchain promises a future where waste is managed proactively and sustainably. As cities worldwide adopt IT-driven waste management practices, the vision of a circular economy and zero-waste society becomes increasingly attainable.

Current Scenario in IoT in Waste Tracking and Management

The rapid urbanization, growing population, and increased consumption of goods have led to an escalating waste crisis worldwide, making waste management one of the most critical environmental challenges of the 21st century. Traditional waste management systems rely heavily on manual processes and have often struggled to keep pace with waste's growing volume and complexity, particularly in urban areas. Integrating Information and Communication Technologies (ICT), particularly the Internet of Things (IoT), has revolutionized waste tracking and management. IoT provides real-time data collection and analysis, optimizing various aspects of waste management, from collection to recycling and disposal. However, despite the transformative potential of IoT, several challenges persist, particularly in electronic waste (e-waste) recycling, hazardous waste handling, and overall recycling inefficiencies, all of which contribute to the overuse of landfills.

IoT in Waste Tracking and Management: The application of IoT in waste management is enabling significant advancements in tracking, collection, and disposal. IoT enables waste management systems to become smarter and more efficient by using sensors, smart bins, and data analytics platforms. These technologies facilitate real-time monitoring of waste levels, enabling timely collection and reducing operational inefficiencies. Several key applications of IoT in waste management include:

1. **Smart Bins and Sensors**: IoT-enabled waste bins are embedded with sensors that detect fill levels. These sensors transmit data to a central management system, optimizing waste collection routes and schedules. In smart cities like San Francisco and Barcelona, sensors in waste bins alert waste collection trucks when full, ensuring that waste is picked up promptly without overfilling or excessive trips. This reduces fuel consumption, reduces carbon emissions, and increases operational efficiency.

2. **Route Optimization**: GPS-enabled vehicles and real-time traffic data allow waste management fleets to optimize their routes. This reduces unnecessary trips, improves fuel efficiency, and contributes to a reduction in greenhouse gas emissions. For example, Seoul uses IoT to manage its waste collection routes, reducing transportation costs and congestion.

3. **Waste Composition Analysis**: IoT technologies can be integrated with waste sorting facilities to assess waste composition in real-time. Automated sensors and imaging technologies enable waste sorting systems to separate recyclable materials from non-recyclable waste, improving the efficiency of recycling efforts. These systems, which are increasingly being deployed in European countries, can automatically detect recyclable materials such as plastic, metal, and paper, making recycling faster and more efficient.

4. **Real-time Monitoring and Reporting**: With IoT, cities can monitor waste management activities in real-time, ensuring that data on waste generation, collection, and disposal are up-to-date and accurate. This data can generate

insights into waste patterns, optimize operations, and create reports for better policy decision-making. Real-time data can also help in compliance with environmental regulations by ensuring that waste is disposed of properly and that recycling goals are being met.

Gaps in E - Waste Recycling: Electronic waste, or e-waste, is one of the fastest-growing waste streams worldwide. The rapid pace of technological advancement and planned obsolescence in the electronics industry results in a constant influx of discarded devices. According to the Global E-Waste Monitor, approximately 53.6 million metric tons of e-waste were generated globally in 2019, which is expected to continue rising. E-waste presents unique challenges in waste management due to its complex composition, which includes valuable materials such as gold, silver, and copper and toxic substances like lead, mercury, and cadmium.

1. **Lack of Effective Collection Infrastructure**: While IoT technologies are advancing in general waste management, their integration into e-waste collection remains limited. One of the main challenges in e-waste management is the lack of formalized collection channels. Many consumers dispose of electronic devices through informal channels, such as throwing them away in regular trash bins or selling them to unregulated markets, bypassing formal recycling systems. Although IoT can help monitor e-waste generation and streamline collection, its widespread adoption has been slow due to limited infrastructure and public awareness.

2. **E-Waste Segregation**: A major issue in e-waste recycling is improper segregation. IoT can facilitate the categorization of e-waste, ensuring that valuable materials are separated from harmful substances. However, the complex nature of e-waste, with its diverse materials and components, makes segregation challenging. While some automated sorting technologies are in use, they are still imperfect and require constant refinement to efficiently separate recyclable from non-recyclable components.

3. **Toxicity and Environmental Hazard**: E-waste contains hazardous materials, such as heavy metals and toxic chemicals, that pose serious risks to human health and the environment if not disposed of properly. IoT-based tracking systems can help monitor the disposal process, but significant gaps exist in the safe handling and disposal of hazardous e-waste materials. The absence of regulations and inadequate recycling infrastructure in many parts of the world further exacerbate the problem.

4. **Lack of Global Standards and Regulations**: While some countries have established regulations for e-waste management, the global lack of harmonized standards has made it difficult to develop a coordinated approach

to e-waste recycling. IoT can aid in creating transparent supply chains and recycling processes, but it requires widespread regulatory support to ensure that e-waste is disposed of safely and effectively across borders.

Challenges in Hazardous Waste Handling: Hazardous wastecomprising industrial chemicals, batteries, medical waste, and radioactive substancesposes significant health and environmental risks. While IoT technologies have made advancements in managing hazardous waste, the sector still faces several significant challenges:

1. **Tracking and Traceability**: Safely handling and disposing hazardous waste requires stringent tracking and traceability. IoT can improve these processes by providing real-time location data and condition monitoring. However, many countries lack the technological infrastructure to implement such systems on a large scale. The complexity of hazardous waste regulations complicates the tracking process, as different materials require different handling and disposal protocols.
2. **Improper Disposal and Illegal Dumping**: Hazardous waste is often illegally dumped in unregulated areas due to inadequate enforcement of regulations. Although IoT solutions can help detect illegal dumping by monitoring landfills and waste collection sites, they are not yet widely implemented worldwide. The failure to monitor hazardous waste disposal and insufficient resources to combat illegal dumping pose a significant challenge to sustainable waste management.
3. **Treatment and Recycling Efficiency**: The recycling of hazardous waste is a highly complex and specialized process. IoT can enhance the efficiency of these operations by providing real-time data on the waste's composition, quantity, and condition. However, many recycling facilities still rely on outdated technologies not fully optimized for handling hazardous materials. The lack of investment in modern infrastructure and technologies limits the effectiveness of IoT in improving recycling rates for hazardous waste.

Recycling Inefficiencies and Landfill Overuse: Despite advancements in recycling technologies, inefficiencies in recycling and the overuse of landfills remain significant global challenges. The reasons for these inefficiencies include:

1. **Poor Waste Segregation**: In many regions, improper waste segregation at the source results in recyclable materials mixed with non-recyclables, making recycling difficult and costly. IoT-enabled sensors in waste bins can help improve segregation, but this requires public participation and awareness, which is often lacking.

2. **Limited Recycling Infrastructure**: While many cities have recycling systems, the infrastructure is often underdeveloped, inadequate, or poorly managed. Advanced sorting technologies such as AI and robotics, which can improve recycling efficiency, are not yet deployed at a large scale, leaving many valuable materials unrecovered.
3. **Overreliance on Landfills**: Landfills remain the default disposal method in many countries, especially in low-income regions, leading to the overuse of landfill space. Landfills release harmful gases, contaminate soil and water, and contribute to climate change. IoT can provide data on landfill usage, waste composition, and environmental impact, but without sufficient waste diversion and recycling programs, the problem of landfill overuse persists.

Integrating IoT into waste management can transform how waste is tracked, collected, recycled, and disposed of. Through real-time data collection, route optimization, waste segregation, and resource tracking, IoT can significantly improve the efficiency and sustainability of waste management systems. However, several critical gaps remain, particularly in e-waste recycling, hazardous waste handling, and addressing recycling inefficiencies. Effective solutions require technological advancements, greater infrastructure investments, regulatory improvements, and public awareness. As IoT technology continues to evolve and its integration deepens, it can play a central role in reducing landfill overuse, improving recycling rates, and ultimately contributing to a more sustainable and circular economy.

Future Scenario

The future of waste management is increasingly intertwined with advanced technologies such as Artificial Intelligence (AI) and Blockchain, which are poised to radically transform waste segregation, recycling processes, and overall waste disposal practices. With the escalating challenges of global waste generation, resource depletion, and environmental degradation, integrating these technologies into waste management systems is a matter of technological advancement and a critical necessity for sustainable development. As we look to the future, AI-based segregation and recycling systems and Blockchain-based incentives for responsible waste disposal present a paradigm shift in how waste can be managed, recycled, and valued within the global economy.

AI-Based Segregation and Recycling Systems: In the future, AI technologies are expected to play an increasingly central role in improving the efficiency and effectiveness of waste segregation and recycling systems. As urban populations grow and the volume of waste generated by households, industries, and businesses increases, traditional waste sorting systems, which rely on manual labor or basic mechanical processes, will become unsustainable. AI, with its ability to process vast amounts of data, learn from patterns, and automate complex processes, holds immense potential to enhance waste management systems, especially of material recovery facilities (MRFs) and waste treatment plants.

AI-Powered Waste Segregation: Waste segregation is one of the most critical and challenging aspects of recycling. Improper segregation of waste at the sourcesuch as mixing recyclables with non-recyclablesoften leads to significant inefficiencies and contamination, reducing the effectiveness of recycling processes. Future AI systems will employ advanced computer vision, machine learning, and robotics to identify and sort waste materials with far greater accuracy and efficiency than current systems.

AI systems will utilize real-time data from sensors and cameras in waste bins and collection trucks to determine waste composition as it is generated and collected. These systems will detect the physical attributes of materials (e.g., size, shape, and color) and chemical compositions, allowing for a more nuanced approach to waste sorting. For example, AI could detect different plastics based on polymer types, identifying those suitable for recycling and those that should be directed to other disposal or treatment routes. Machine learning algorithms will continuously improve sorting accuracy, adapting to new materials and evolving waste types. Robotics integrated with AI will automate the sorting process in recycling plants, allowing for faster and more accurate separation of materials. Robotic arms, equipped with AI-powered image recognition, can work alongside human operators or independently, significantly increasing throughput while minimizing errors and contamination. AI-powered machines can sort common recyclables such as plastics, paper, and metals and more complex materials like electronic waste (e-waste) and composite packaging materials that are typically difficult to handle with conventional methods.

Enhanced Recycling Efficiency: AI-based systems will also significantly improve recycling operations' efficiency. AI-powered predictive analytics will help recycling plants forecast waste volumes and types, enabling better planning and resource allocation. This data-driven approach will optimize recycling, reducing downtime, energy consumption, and operational costs. Additionally, AI will assist in detecting and identifying hazardous or non-recyclable materials before they enter the recycling stream, preventing contamination and ensuring that only suitable materials undergo the recycling process.

One of the breakthroughs AI will bring to recycling is its ability to continuously monitor the quality and quantity of recyclables. AI-based systems can track the recyclable materials' lifecycle, monitoring the recycling process's efficiency at every stage. This will provide valuable insights that can be used to fine-tune operations, identify bottlenecks, and ensure that the recycling process is as efficient and sustainable as possible.

AI for Waste-to-Energy (WTE) Optimization: Beyond segregation and recycling, AI is also expected to revolutionize the Waste-to-Energy (WTE) technologies field. AI algorithms will optimize waste conversion into energy by predicting and controlling combustion, maximizing energy recovery and minimizing emissions. AI could monitor the chemical composition of waste fed into incinerators, adjusting the operating parameters to ensure complete combustion and reduce harmful byproducts. AI will enable smart energy management in WTE plants, ensuring that energy generation from waste aligns with grid demand and supply.

Blockchain for Incentivizing Responsible Waste Disposal: While AI will be central to the operational efficiency and technological advancement of waste management systems, Blockchain technology holds the potential to address one of the key challenges in waste management: encouraging and ensuring responsible waste disposal practices at the individual and corporate levels. The decentralized and immutable nature of Blockchain makes it an ideal tool for incentivizing waste segregation, tracking waste generation and disposal, and creating transparent and traceable systems for recycling and waste management.

Blockchain for Waste Tracking and Transparency: Blockchain technology can provide a transparent and immutable record of waste generation, collection, recycling, and disposal. Blockchain can use smart contracts and decentralized applications (dApps) to track every step of the waste management process, from when waste is generated to its final disposal or recycling. This traceability can significantly reduce illegal dumping, waste mismanagement, and fraud, as all transactions and waste-related activities are recorded on the Blockchain ledger, which is publicly available and immutable. For example, Blockchain can track e-waste from its point of generation (e.g., consumer electronics) to its ultimate disposal or recycling. Smart contracts can automatically trigger payments to recycling facilities or waste handlers when certain conditions are met, ensuring that waste is disposed of in an environmentally responsible manner. Blockchain can also certify recyclable materials, ensuring they are processed under environmental regulations and standards.

Tokenization of Waste for Incentives: Blockchain can be used to tokenize waste, offering incentives for responsible waste disposal and recycling. By implementing a Blockchain-based reward system, individuals and businesses can earn digital tokens or cryptocurrency as a reward for proper waste segregation and recycling. For instance, when a household or a business correctly segregates recyclable materials and places them in designated bins, recycling could be verified using IoT sensors and recorded on the Blockchain. These verifiable actions could then be rewarded with tokens, which could be exchanged for discounts, cash, or other rewards in a gamified manner. This system encourages greater participation in waste segregation and recycling efforts. Blockchain's transparency ensures that rewards are distributed fairly, based on verified and traceable actions, preventing fraudulent activities. In countries like Switzerland and Germany, where citizens are already incentivized to return bottles and containers for recycling, a Blockchain-based system could expand the concept to broader waste streams, such as paper, plastic, and electronic waste.

Blockchain for Circular Economy and Extended Producer Responsibility (EPR): Blockchain can also play a key role in facilitating the transition to a circular economy. Through smart contracts and transparent supply chains, Blockchain can enable Extended Producer Responsibility (EPR), whereby manufacturers are held accountable for the lifecycle of their products, including their disposal. Blockchain can create a decentralized record that tracks products from the point of sale to disposal, ensuring that manufacturers take responsibility for the recycling and proper disposal of their products once they reach the end of their useful life. For instance, a company that produces electronics could be required to participate in a Blockchain-enabled EPR system, where their products are tracked throughout the supply chain and then responsibly recycled at the end of their life. The Blockchain ledger would record each step, ensuring that the product is recycled in an environmentally friendly way and that the producer is financially incentivized or penalized accordingly.

Blockchain and Public-Private Partnerships in Waste Management: Blockchain can also facilitate collaboration between governments, waste management companies, and other stakeholders by providing a common platform for information sharing and accountability. For example, municipalities could partner with private waste management firms to implement Blockchain-based systems for tracking waste collection and recycling activities. The data collected on the Blockchain would be publicly accessible, providing real-time insights into waste management operations and ensuring that all stakeholders adhere to environmental standards.

Future Considerations: While AI-based segregation systems and Blockchain-based incentives hold immense promise, there are still several challenges to their widespread adoption. First, there is a need for significant investment in the necessary infrastructure to implement these technologies at scale. In many regions, particularly low-income countries, the financial and technological barriers to implementing such systems may be prohibitive. Additionally, regulatory frameworks must evolve to address the complexities of Blockchain and AI integration in waste management, ensuring that privacy concerns, data security, and environmental standards are upheld. The success of these technologies will depend on public participation. Implementing AI and Blockchain in waste management requires significant behavioral change, especially in developing nations, where awareness and understanding of these technologies may be limited.

The future of waste management lies in the successful integration of AI and Blockchain technologies, which promise to revolutionize how waste is segregated, tracked, recycled, and disposed of. AI-based systems will bring unparalleled efficiency to the recycling process, while Blockchain will provide transparency, traceability, and incentives for responsible waste disposal. Together, these technologies will be crucial in transitioning to a circular economy, where waste is minimized, resources are efficiently recycled, and sustainability is prioritized. However, overcoming the infrastructural, regulatory, and behavioral challenges will be essential to realizing the full potential of these innovations.

IT for Waste Segregation and Recycling

In the quest for more sustainable waste management practices, integrating advanced Information Technology (IT) solutions, particularly Artificial Intelligence (AI) and robotics, has emerged as a game-changer for waste segregation and recycling. As global waste generation continues to rise, driven by increased consumption, urbanization, and industrial activities, the traditional waste sorting and recycling methods have proven inefficient, labor-intensive, and resource-draining. Conventional recycling methods, which often rely on human labor or basic mechanical sorting systems, have struggled to keep pace with the sheer volume and complexity of modern waste streams, especially with the advent of more diverse and intricate materials like composite packaging, e-waste, and mixed plastics. This has led to increased contamination in recyclable materials, lower recycling rates, and higher landfill diversion costs. However, the emergence of AI and robotics has the potential to revolutionize waste segregation, offering a more efficient, cost-effective, and environmentally friendly approach to managing waste.

AI and Robotics: Waste sorting is the critical first step in recycling, and its efficiency directly influences the success of the entire recycling process. Traditionally, waste sorting has relied on manual labor, with workers tasked with identifying and separating recyclable materials from non-recyclable waste. While some automation technologies, such as conveyor belts and basic mechanical sorters, have been employed in recycling plants, they fall short when dealing with today's society's increasingly complex waste streams. AI and robotics, however, offer a level of precision and speed that far surpasses these conventional methods, enabling a more efficient and automated waste segregation process.

AI-Powered Waste Sorting Systems: Computer vision and machine learning are at the heart of AI-based waste segregation. AI algorithms, powered by large datasets of labeled waste materials, can identify and classify different types of waste based on their visual attributes. Using high-definition cameras and sensors, AI can capture images of waste materials on conveyor belts, analyze their composition, and determine their recyclability. This process involves using deep learning models, a subset of machine learning, which train AI systems to recognize patterns in the waste material. These models can distinguish between plastics, metals, paper, glass, organic waste, and non-recyclable or hazardous materials.

The ability of AI to learn from experience makes it highly adaptable. As waste streams evolve, AI systems can be retrained and updated with new data, allowing them to handle emerging waste materials more effectively. The precision of AI enables it to identify not only basic categories of waste but also more complex materials, such as composite materials that contain a mixture of recyclable and non-recyclable components. For example, packaging made from multiple layers of different materials can be efficiently identified and separated by AI systems, reducing contamination in the recycling process.

AI-based sorting systems also optimize the speed and efficiency of waste segregation. Unlike human workers, who can become fatigued or make errors due to repetitive tasks, AI systems can operate continuously, processing large volumes of waste with minimal downtime. Additionally, AI can help recycle materials that were previously considered difficult or impossible to process, increasing the overall recycling rate. Through AI algorithms, recycling facilities can automatically adjust the sorting process in real-time, optimizing the system's performance based on the type and quantity of processed materials.

Robotics and Automation in Waste Sorting: Robotics complement AI in waste segregation. While AI provides the intelligence and decision-making capabilities for identifying and classifying materials, robotics allows for physically handling and sorting these materials. Robotic arms, equipped with advanced AI-based vision systems, can quickly and accurately pick and sort waste materials from conveyor belts or bins, significantly reducing the need for manual labor and improving the overall efficiency of the recycling process.

One of the most promising robotics applications in waste sorting is robotic arms with vacuum grippers or suction systems, which can pick up and handle a wide range of waste materials, from light plastics to heavier metals. These robotic systems are often equipped with AI-powered cameras and sensors that allow them to "see" and "feel" the waste, making real-time decisions on grabbing and separating materials based on size, shape, and material type. Robots are sometimes designed to work with human workers, performing repetitive tasks while humans handle more complex or delicate materials. Robotics can help address labor shortages and reduce the safety risks associated with manual labor in waste sorting facilities. Workers in recycling plants are often exposed to hazardous materials, sharp objects, and other safety risks. By automating many hazardous tasks, robotic systems can improve worker safety and reduce the overall labor costs in waste management.

Optimizing Resource Recovery: Integrating AI and robotics in waste sorting enhances the process's efficiency and improves resource recovery. In traditional waste sorting systems, significant amounts of recyclable materials are often lost due to contamination, inefficient sorting, or the inability to identify specific materials. AI and robotics address these issues by improving the accuracy of material identification and reducing contamination rates. This, in turn, helps recycling facilities recover a higher proportion of valuable materials, such as metals, plastics, and paper, which can be reused in manufacturing processes.

AI can also track the lifecycle of materials within the recycling process, ensuring that they are processed according to established environmental standards and regulations. By providing real-time data on the amount and type of recycled materials, AI systems can enable better decision-making regarding which materials should be prioritized for recovery. With its ability to rapidly and accurately process waste, robotics allows for faster recovery of valuable materials, reducing the need for landfilling or incineration. AI and robotics have the potential to enhance the circular economy by enabling closed-loop recycling systems. In a circular economy, products and materials are reused, refurbished, and recycled indefinitely, minimizing waste and resource depletion. Through AI-based sorting systems, recycling plants can more effectively close the loop on material recovery, ensuring that valuable resources are continually reintroduced into the supply chain.

Future Considerations: While AI and robotics offer significant advantages for waste segregation and recycling, several challenges must be addressed to realize their full potential. One of the major challenges is the high cost of implementing AI and robotic systems, particularly in developing countries or smaller-scale recycling operations. The upfront investment required for AI-powered sorting systems, robotics, and the necessary infrastructure can be prohibitively expensive for many municipalities and waste management companies. The complexity of these systems requires specialized training and expertise, which can add to the operational costs. Another challenge is the need for large, high-quality datasets to train AI models. For AI-based sorting systems to be effective, they must be trained on diverse waste materials, ensuring they can accurately identify and classify a wide array of materials. However, collecting and labeling sufficient data to train AI models can be time-consuming and resource-intensive. In some cases, the variability in waste streams across different regions may require the development of region-specific AI models, adding layer of complexity to the implementation process.

Additionally, while robotics is highly effective in automating many aspects of waste sorting, certain materials may still present challenges for robotic systems. For example, materials that are difficult to grasp, such as small or irregularly shaped items, may require advanced robotic systems with specialized grippers or suction mechanisms. In some cases, hybrid systems that combine robotic and human labor may still be required to handle particularly challenging materials for robots. Finally, there are concerns about the potential environmental impact of robots themselves. While robotics and AI can contribute to more efficient recycling, robots and AI systems' production, operation, and disposal carry an environmental footprint. It will be important for the waste management industry to consider the lifecycle environmental impact of these technologies, ensuring that they contribute positively to sustainability goals rather than exacerbating environmental problems.

Future Vision

As one of the world's fastest-growing economies, India faces significant challenges in waste generation, resource depletion, and environmental degradation. Rapid urbanization, industrialization, and consumption patterns have led to an exponential increase in waste, with the country's landfills growing at an unsustainable rate. The conventional linear take-make-dispose model, which has historically driven industrial and consumer behavior, has resulted in massive inefficiencies and resource wastage. In response to these pressing challenges, India is increasingly adopting a circular economy (CE) to promote sustainable resource use, waste reduction, and environmental protection. Circular economy models, which emphasize reducing, reusing, recycling, and regenerating resources, offer a transformative approach that aligns with India's sustainable development goals and its vision of achieving a cleaner, more resilient economy.

Understanding Circular Economy: A circular economy (CE) is an alternative economic model that contrasts with the traditional linear economy. In a linear economy, raw materials are extracted, transformed into products, used, and discarded as waste. By contrast, a circular economy is designed to keep products, materials, and resources in use for as long as possible, extracting the maximum value from them while in use. The model also involves the recovery and regeneration of materials at the end of each lifecycle, thereby minimizing waste and reducing the consumption of finite resources. This systemic approach to production and consumption aims to create a closed-loop system, where waste is minimized, products are reused, and natural resources are preserved. Transitioning to a circular economy holds immense potential to address the country's resource-intensive growth trajectory, reduce environmental harm, and create sustainable economic opportunities. The Indian government, businesses, and civil society have started to recognize the importance of circular economy principles in addressing the challenges of over-exploitation of resources and the opportunity to foster a new, sustainable economic growth model.

India's Transition to a Circular Economy: India's transition to a circular economy is still in its early stages, but there has been significant progress in recent years. Several factors are driving the momentum toward circular models, including:

1. **Waste Management Challenges**: India generates over 60 million tons of municipal solid waste annually, which is expected to increase by 5-6% annually. While India has made strides in waste management through initiatives like Swachh Bharat Abhiyan (Clean India Mission), the linear waste disposal model still predominates, leading to overflowing landfills, improper waste segregation, and resource wastage. A shift to circular economy practices, especially regarding waste management, could significantly reduce landfill usage, prevent resource leakage, and enhance material recovery.
2. **Resource Scarcity and Over-extraction**: India's rapid industrialization and economic growth have increased the demand for natural resources, including water, minerals, energy, and agricultural inputs. This has led to concerns over the depletion of finite resources and the environmental impact of resource extraction. Circular economy models aim to address these concerns by promoting resource efficiency, reducing the dependency on virgin materials, and encouraging recycling and reuse.
3. **Environmental Sustainability and Climate Change**: With the growing threat of climate change, India faces the dual challenge of promoting economic growth while mitigating environmental impacts. Circular economy principles can help decouple economic growth from environmental degradation by minimizing waste and promoting sustainable production and consumption patterns. By reducing waste, improving energy efficiency, and minimizing carbon emissions, circular economy models can contribute to achieving India's climate goals.
4. **Government Initiatives and Policy Support**: The Indian government has increasingly supported circular economy initiatives. The National Resource Efficiency Policy (2019), launched by the Ministry of Environment, Forest, and Climate Change (MoEFCC), outlines resource efficiency and waste management strategies. Introduction of extended producer responsibility (EPR) regulations, particularly for plastic waste, has incentivized industries to adopt circular practices. The government's push for circular economy models has also been reflected in initiatives like the Atal Mission for Rejuvenation and Urban Transformation (AMRUT), which aims to make cities more sustainable and resource-efficient.
5. **Private Sector Innovation**: Indian businesses, particularly in sectors such as manufacturing, agriculture, and technology, are beginning to explore circular models in their operations. Companies increasingly recognise the potential for cost savings, resource optimization, and market differentiation

by adopting circular economy principles. Many companies are implementing closed-loop systems for recycling and upcycling materials, redesigning products for longer life cycles, and using waste as a resource.

Future Vision: The future of circular economy models in India lies in their widespread adoption across industries, urban centers, and rural areas. As the country moves forward, there are several key areas where IT-enabled innovations, policy reforms, and collaborative efforts can drive the shift to a circular economy. Some of the key strategies for scaling circular economy models in India are:

1. **Circular Design in Manufacturing**: One of the critical areas where circular economy models can be integrated is product design and manufacturing. Adopting circular design principles, such as designing products for durability, disassembly, and reuse, will enable businesses to reduce waste and resource consumption. For example, product manufacturers in India can shift from the traditional "take-make-dispose" model to a circular design approach where products are created with materials that are easy to repair, upgrade, or recycle at the end of their life cycle. This will reduce the need for new resources, enhance the longevity of products, and minimize waste generation.

2. **Improved Waste Management and Recycling Infrastructure**: Efficient waste management and recycling are at the core of a circular economy. India needs to strengthen its waste segregation, collection, and recycling infrastructure to realise the full potential of recycling and resource recovery. This includes developing material recovery facilities, e-waste collection centers, and advanced sorting technologies powered by artificial intelligence (AI) and robotics. Incentivizing decentralized recycling initiatives at the community level will allow for the recovery of valuable materials from waste streams, further enhancing the circular economy. IT-enabled solutions, such as mobile waste tracking and management applications, can help improve waste collection efficiency and reduce contamination in recycling streams.

3. **Circular Supply Chains in Agriculture**: Agriculture plays a vital role in India's economy, contributing significantly to employment and GDP. However, the sector is also one of the largest consumers of natural resources, particularly water, energy, and soil. Circular economy models can play a transformative role in agricultural sustainability by promoting resource-efficient practices, such as precision farming, agro-waste recycling, and water conservation. Using technology like Internet of Things (IoT)-enabled sensors, AI, and big data analytics, farmers can optimize water usage, monitor soil health, and improve crop yields while reducing environmental impact. Agricultural waste, such as crop residues, can be repurposed for

energy generation, composting, or animal feed, creating closed-loop systems within agricultural supply chains.

4. **Urban Circular Economy Models**: As urbanization accelerates, India's cities face increasing pressure on their infrastructure, resources, and waste management systems. Transitioning to circular economy models in urban areas can significantly alleviate these challenges. Smart cities in India can integrate circular principles by adopting energy-efficient building designs, waste-to-energy technologies, and sustainable transportation solutions. For example, urban areas can invest in biogas plants that convert organic waste into renewable energy and create systems for recycling construction and demolition waste. Further, IT solutions like GIS (Geographic Information Systems) can map and manage resources, enabling more efficient urban planning and resource allocation.
5. **Policy and Regulatory Support for Circular Economy**: To drive the adoption of circular economy models, the Indian government must continue strengthening its regulatory frameworks. This includes developing policies that incentivize resource efficiency, waste reduction, and recycling and creating a robust legal and financial framework that supports circular business models. Extended Producer Responsibility (EPR) regulations can be expanded to cover more industries, ensuring that producers are held accountable for the end-of-life management of their products. Additionally, providing financial incentives and subsidies to businesses that adopt circular economy practices can accelerate the transition to a more sustainable economic model.
6. **Public Awareness and Capacity Building**: A successful transition to a circular economy also requires a shift in societal attitudes toward consumption and waste. Public awareness campaigns can play a key role in educating citizens about the benefits of circular economy models, such as reducing waste, conserving resources, and supporting sustainable development. In addition, capacity-building efforts for businesses, local governments, and communities can help them implement circular practices more effectively. Training programs focused on sustainable production, waste management, and resource recovery will equip stakeholders with the knowledge and skills to drive the circular economy forward.
7. **Collaboration Between Stakeholders**: The successful implementation of circular economy models in India requires collaboration across various sectors, including government, industry, academia, and civil society. Collaborative platforms can foster innovation, create synergies, and provide

a space for sharing best practices. By working together, stakeholders can co-develop solutions that address the specific challenges faced by different industries, regions, and communities.

14

IT-Driven Social Sustainability

Introduction

Social sustainability has gained significant attention in the contemporary development, especially in information technology (IT). Social sustainability refers to the ability of a society to maintain and improve social well-being, equity, and quality of life for its present and future generations. It involves fostering an environment where all individuals, irrespective of their socioeconomic backgrounds, have equal access to resources, opportunities, and services. Information technology (IT) has emerged as a transformative tool in achieving social sustainability by addressing various social challenges such as inequality, lack of access to education, healthcare, and basic services. It also fosters community engagement, empowerment, and participation in governance.

Bridging the Digital Divide: One of the most significant ways IT contributes to social sustainability is by bridging the digital divide, which refers to the gap between those who have access to information and communication technologies (ICT) and those who do not. Technological access has exacerbated social inequalities in many parts of the world, particularly in developing countries. By providing equitable access to the internet and digital platforms, IT can help level the playing field, allowing marginalized populations to access essential services such as education, healthcare, and employment opportunities. For instance, mobile phones and internet connectivity have become essential tools for accessing information and services. In rural areas, where access to traditional infrastructure may be limited, mobile-based solutions are being used for a wide range of purposes, from accessing government services to receiving agricultural advice. By improving connectivity and digital literacy, IT can empower individuals in underserved areas, allowing them to participate more fully in economic, educational, and social activities, ultimately contributing to greater social equity.

Improving Access to Education: Education is a critical component of social sustainability, as it plays a key role in reducing poverty, promoting social inclusion, and enhancing individual well-being. IT-driven solutions, such as online learning platforms, digital classrooms, and e-learning tools, have revolutionized education by expanding access to knowledge and learning opportunities. In many parts of the world, particularly in rural or remote areas, traditional education systems are often inadequate or unavailable due to a lack of resources, infrastructure, or qualified teachers. Information technology can fill this gap by providing access to high-quality educational content, regardless of geographical location. For example, Massive Open Online Courses (MOOCs) and online degree programs have enabled individuals to access educational resources from top universities and institutions across the globe. In countries like India, where a large proportion of the population lives in rural areas, mobile-based education apps deliver basic education and vocational training, providing opportunities for skill development and employment. These IT-enabled educational platforms increase access to education and promote lifelong learning and the development of skills critical for the future workforce.

Enhancing Healthcare Access and Equity: Healthcare is another area where IT has had a profound impact on social sustainability. IT in healthcare, often called eHealth or digital health, has revolutionized how healthcare services are delivered, making them more accessible, affordable, and efficient. In many developing countries, healthcare services are often scarce, with rural and marginalized communities facing significant barriers to access. IT solutions, such as telemedicine, mobile health applications, and electronic health records, are overcoming these barriers by bringing healthcare services directly to people in need.

Telemedicine, for example, allows patients in remote or underserved areas to consult with doctors and specialists through video conferencing, thereby overcoming the need for travel and reducing healthcare costs. Similarly, mobile health apps enable individuals to monitor their health, access medical advice, and receive reminders for medication or appointments, enhancing health outcomes and promoting preventative care. By improving healthcare access and quality, IT reduces health disparities and fosters social sustainability.

Promoting Social Inclusion and Participation: Social inclusion is a core element of social sustainability, as it ensures that all individuals can participate fully in society. IT promotes social inclusion by providing platforms for individuals and communities to connect, collaborate, and participate in social, political, and economic processes. Social media, for instance, has become an important tool for marginalized groups to voice their concerns, share their experiences, and engage in advocacy and activism. Digital platforms provide spaces for social movements, grassroots organizations, and individuals to raise awareness about issues such as gender equality, poverty, and human rights.

Additionally, IT-enabled platforms can enhance citizen engagement in governance and decision-making processes. E-governance platforms, which provide access to government services, information, and grievance redressal systems, enable citizens to participate more actively in the political process. By facilitating transparency, accountability, and participation, IT empowers individuals and communities, ensuring that all voices are heard and considered in decision-making. This fosters a more inclusive and equitable society, where social sustainability is promoted through active civic engagement.

Improving Employment Opportunities: Employment and economic empowerment are essential components of social sustainability, and IT has played a pivotal role in expanding access to job opportunities for gig economy. Digital platforms such as job portals, freelancing websites, and online marketplaces have made it easier for individuals to find employment, connect with potential employers, and work remotely. These platforms are particularly beneficial for individuals who face barriers to traditional employment, such as women, people with disabilities, or those living in remote areas. IT enables the development of new industries and sectors, such as the tech industry, which has become a significant driver of economic growth and job creation. By providing access to digital skills training and promoting entrepreneurship through online platforms, IT helps individuals develop the skills needed to succeed in the modern economy. The rise of online businesses and startups has opened up new avenues for economic empowerment, especially for individuals in developing countries who may not have access to traditional financial systems or capital.

Addressing Climate Change and Environmental Sustainability: Social sustainability cannot be achieved without addressing environmental sustainability, as environmental degradation disproportionately impacts vulnerable populations. IT solutions can potentially address climate change and environmental sustainability, supporting social sustainability efforts. For example, IT can monitor environmental conditions, track resource usage, and optimize energy consumption. Smart grids, energy management systems, and IoT devices allow individuals and businesses to reduce energy waste, lower carbon emissions, and contribute to sustainable resource management.

In addition, IT-driven solutions, such as climate modeling and data analytics, can help predict and mitigate the effects of climate change, such as natural disasters or extreme weather events. Early warning systems, powered by data analytics and IoT sensors, enable governments and communities to respond more effectively to environmental crises, minimizing their impact on vulnerable populations. Using IT in sustainable agriculture, including precision farming techniques and crop monitoring systems, helps optimize resource use, reduce waste, and promote environmentally friendly practices.

Ensuring Equity in Resource Distribution: IT also plays a crucial role in ensuring the equitable distribution of resources, particularly regarding financial services, energy access, and food security. Financial inclusion has become a significant focus in many developing countries, where large population segments remain unbanked. Digital financial services, including mobile banking and digital payment systems, have provided millions of individuals with access to financial resources, enabling them to save, invest, and access credit. This has particularly benefited women and rural populations, who often face challenges accessing traditional banking services. Similarly, IT can help address energy poverty by enabling smart grid systems that optimize energy distribution and reduce wastage. Solar power systems, powered by IoT devices and AI algorithms, can provide affordable and sustainable energy to off-grid communities. In food security, IT solutions such as digital marketplaces, supply chain optimization tools, and agricultural data platforms can improve food distribution systems, reduce food waste, and ensure that food reaches to needy.

Current Scenario

The role of Information Technology (IT) in enhancing social equity and community welfare has been evolving over the last few decades, increasingly becoming an integral part of efforts to achieve social justice, inclusivity, and economic development. IT provides solutions that help bridge the gaps in access to education, healthcare, finance, and social services, among other areas. It can empower marginalized and underserved communities by improving access to essential services, promoting social mobility, and facilitating civic engagement. However, despite these positive developments, challenges persist in utilizing IT systems effectively to address local needs, particularly in rural, remote, and underserved urban areas. These limitations prevent the full realization of IT's potential in promoting true social equity.

IT as a Tool for Enhancing Social Equity: In the contemporary landscape, IT has been leveraged to reduce disparities in access to vital services and resources. By increasing access to information, education, healthcare, and financial services, IT systems have helped empower individuals, especially in marginalized communities, and reduce inequalities often stem from socioeconomic and geographical factors.

1. **Education**: In education, IT has been a transformative tool for enhancing social equity. Digital learning platforms, online resources, and e-learning applications have opened up educational opportunities to millions who previously lacked access to quality education. For example, MOOCs (Massive Open Online Courses) allow students from developing regions to access high-quality courses from top institutions worldwide. In India, digital education initiatives such as SWAYAM (Study Webs of Active Learning for Young Aspiring Minds) and DIKSHA (Digital Infrastructure for Knowledge Sharing) aim to provide free, online education to millions of students, especially in rural and remote regions. Mobile learning apps designed in local languages enhance literacy and skill-building among people who cannot access traditional forms of education.

2. **Healthcare**: Telemedicine and mobile health apps are bridging the healthcare access gap for rural and underserved communities. Digital health platforms enable remote consultations, diagnosis, and treatment, which are especially vital in regions with a shortage of healthcare infrastructure and professionals. Platforms such as India's eSanjeevani, which offers teleconsultation services, are making healthcare more accessible to populations that would otherwise need to travel long distances for basic medical care. Similarly, health monitoring systems using wearable devices connected to the internet are helping individuals in remote areas track their health and receive medical advice in real-time.

3. **Financial Inclusion**: Another significant area IT has impacted social equity is financial inclusion. In many developing countries, a large portion of the population is excluded from formal financial services. However, the advent of mobile banking platforms such as M-Pesa in Kenya, Google Pay, Paytm in India, and other fintech solutions has significantly improved access to banking services. These platforms enable individuals in remote or underserved areas to make digital payments, save money, and access micro-loans without needing traditional banking infrastructure. This, in turn, reduces the inequality in access to financial resources and services, empowering individuals economically and socially.

4. **Civic Engagement**: IT has also facilitated the creation of digital platforms for civic participation, enabling citizens to have a more active role in governance. E-government initiatives have been implemented to streamline access to government services and provide citizens with a platform to raise grievances, make suggestions, and engage in policy discussions. In India, platforms like MyGov and the Digital India initiative provide citizens with tools to communicate directly with government authorities and participate in decision-making processes. Promoting transparency and accountability, these platforms help enhance the democratic process and ensure that underrepresented groups can voice their concerns.

Limitations of Current IT Systems in Addressing Local Needs: Despite the remarkable advancements in using IT to foster social equity, significant limitations persist in addressing local needs, especially in areas with weak infrastructure, low digital literacy, and a lack of resources. These limitations highlight the challenges in implementing IT solutions that enhance community welfare and equity.

1. **Digital Divide and Accessibility**: The digital divide is one of the most significant limitations in leveraging IT for social equity. While urban areas tend to have better access to digital infrastructure, rural and remote areas are often left behind. In India, for example, despite the government's efforts to

improve internet access through initiatives like BharatNet, rural areas still face connectivity issues, limiting individuals' ability to access IT-enabled services such as online education, telemedicine, and e-governance. Poor internet infrastructure, unreliable electricity, and lack of hardware such as smartphones or computers continue to hinder IT-based interventions in these regions.

2. **Digital Literacy**: Even when infrastructure is available, a lack of digital literacy limits the ability to use IT systems effectively. Many people in rural areas, women, and the elderly face challenges using digital tools, accessing online services, and engaging with digital platforms. Digital illiteracy is a significant barrier to fully realising IT's potential in promoting social equity. The gap between those who are digitally literate and those who are not is growing, exacerbating existing social and economic divides. For example, while e-commerce platforms offer opportunities for small businesses to grow, many rural artisans and farmers lack the skills to use these platforms, thus missing out on potential markets and income opportunities.

3. **Cultural and Linguistic Barriers**: A further challenge in adopting IT solutions is the lack of localized content and language support. While English is the dominant language online, many underserved communities speak regional languages and dialects. Platforms that do not support these languages or that are tailored to urban-centric cultural contexts are less likely to be effective in fostering social equity. For example, online educational content in regional languages and culturally appropriate formats is still limited, especially in areas where dialects differ significantly from the language used in mainstream education.

4. **Fragmented Data Systems**: While many governments have developed digital platforms to offer services such as healthcare, education, and welfare, these systems are often fragmented and lack integration. This fragmentation makes it difficult to provide holistic solutions and to track individuals' needs across various service sectors. For instance, a person receiving government welfare benefits might also need healthcare services, but due to the lack of data integration between these services, they might not be able to access both seamlessly. Integrated IT solutions must provide a more comprehensive approach to social welfare and address local needs.

5. **Infrastructure Constraints**: Even in urban areas, the lack of reliable digital infrastructure poses a significant challenge. While there is greater access to digital technologies in cities, issues such as inadequate bandwidth, network congestion, and system downtimes can undermine the effectiveness of IT

solutions. For example, during the COVID-19 pandemic, many cities saw a surge in online education and telemedicine services, but infrastructure issues, including network overloads, led to disruptions, affecting the smooth delivery of services.

6. **Inefficiency in Service Delivery**: Even when IT solutions are in place, inefficiency in service delivery can hinder their ability to address local needs. Many digital platforms, especially those created by governments, are slow, non-intuitive, or poorly designed. For example, in India, the PMGDISHA (Pradhan Mantri Gramin Digital Saksharta Abhiyan) aimed to empower rural populations digitally, but challenges such as inadequate infrastructure, poor internet speeds, and insufficient training for users have made it difficult to achieve its goals effectively. Additionally, there is a lack of post-implementation support to ensure citizens can continue using these platforms effectively.
7. **Privacy and Security Concerns**: Another critical limitation concerns privacy and data security. The extensive use of digital platforms raises significant questions about personal and sensitive data security. Vulnerable populations are often less aware of cybersecurity risks and may fall victim to fraud or exploitation. The lack of robust data protection frameworks in many developing countries has made individuals wary of using IT solutions, particularly for financial or healthcare services.
8. **Sustainability of IT Interventions**: Many IT-driven initiatives, particularly those focused on community welfare, are often short-term projects that lack long-term sustainability. While a government or NGO may roll out an IT solution, these systems become outdated or ineffective without proper maintenance, capacity building, and support investment. For example, a rural digital literacy program may teach people how to use basic internet services, but without sustained follow-up, the benefits diminish over time, and the technology becomes underutilized.

Bridging Inequities through IT

Information Technology (IT) has emerged as a transformative tool that can potentially reduce socioeconomic disparities across various dimensions, particularly in gender, caste, and regional inclusion. In developing countries, particularly in regions like India, these structural inequities have persisted for centuries, impacting access to resources, opportunities, and services. However, with the advent of IT, there are growing opportunities to bridge these gaps by providing access to education, healthcare, financial services, and civic engagement platforms. This shift towards digital inclusion is vital in creating equitable societies that ensure fair participation of marginalized communities in economic, social, and political spheres.

Gender Inequality and Empowerment: Gender inequality remains a significant challenge in many societies, with women and girls disproportionately affected by limited access to education, healthcare, economic opportunities, and decision-making power. IT has been recognized as a powerful tool to address gender-based disparities and empower women by improving access to essential services, providing platforms for education, and enabling financial inclusion.

1. **Access to Education**: The digital revolution has enabled unprecedented access to educational resources for women and girls, particularly in rural and underserved areas where traditional schooling systems may be inadequate or unavailable. Through digital platforms and e-learning tools, women can access high-quality educational content and skill-building courses without leaving their homes. Initiatives like the "Beti Bachao Beti Padhao" scheme in India, which aims to promote girls' education and provide digital learning opportunities, have significantly impacted. Online learning platforms such as Coursera, edX, and Udemy offer courses in multiple languages, free or low-cost, allowing women to upskill and gain employable skills.

2. **Economic Empowerment and Financial Inclusion**: IT is pivotal in enhancing financial inclusion for women. Mobile banking platforms, digital payment systems, and fintech services have revolutionized how women access financial services. In rural areas, where women may not have access to traditional banking infrastructure, mobile money services like M-Pesa in Kenya or Paytm and PhonePe in India have enabled women to manage their finances, make transactions, and even access microcredit. This financial autonomy fosters economic empowerment, allowing women to engage in entrepreneurship, improve their households' financial well-being, and contribute to economic growth. The ability to send remittances digitally has also enabled migrant women workers to provide for their families, reducing the control traditionally exerted by men over women's financial resources.

3. **Healthcare Access**: IT has also revolutionized women's healthcare access, particularly in rural or remote areas where access to medical facilities and professionals may be limited. Telemedicine services and mobile health apps have enabled women to consult with healthcare professionals remotely, receive timely diagnoses, and access maternal and child healthcare services. For example, India's "eSanjeevani" telemedicine platform provides women in remote areas with virtual consultations, reducing the burden of traveling long distances for healthcare services. Mobile health apps like "Portea" and "HealthifyMe" also help women track their health metrics, access

information on reproductive health, and receive mental health counsellingall of which contribute to a healthier and more empowered female population.

4. **Social and Political Participation**: IT can also bridge gender inequalities in political and social participation. Platforms for civic engagement, such as India's MyGov portal, provide women a platform to voice their concerns, participate in discussions, and engage in policymaking processes. Social media platforms like Facebook, Twitter, and Instagram allow women to amplify their voices, share their stories, and organize campaigns for social change. These platforms have been instrumental in movements such as #MeToo, where women globally have used IT to raise awareness about gender-based violence and inequality. Digital literacy programs targeted at women have empowered them to participate in social movements and become active political and social discourse contributors.

IT and Caste – Based Inequality for Bridging Divides: Caste-based discrimination, deeply embedded in certain societies, especially in India, continues to marginalize a large section of the population. People from lower castes, particularly Dalits and Adivasis, have historically faced systemic exclusion from educational, economic, and social opportunities. While addressing caste discrimination is a multifaceted challenge requiring structural reforms, IT can play a crucial role in bridging caste-based inequities by increasing access to education, enhancing employment opportunities, and improving social mobility.

1. **Access to Education and Skill Development**: IT has provided marginalized caste groups access to educational resources. The widespread availability of digital learning platforms has democratized education by allowing anyone, irrespective of caste, to learn and develop new skills. Government initiatives like the "National Digital Literacy Mission" aim to provide digital literacy to people from marginalized communities, including Dalits, Adivasis, and backward castes. These initiatives ensure that lower caste communities are not excluded from the digital economy and that they can harness the potential of IT to enhance their knowledge and skill sets.

2. **Employment and Economic Opportunities**: One of the most critical ways IT helps reduce caste-based inequalities is by enhancing access to employment opportunities. Digital platforms such as LinkedIn, Upwork, and various freelance job portals enable individuals from marginalized caste backgrounds to access jobs and clients beyond their immediate social circles. E-commerce platforms like Amazon and Flipkart also provide opportunities for entrepreneurs from lower castes to sell products, including handicrafts, textiles, and agricultural produce, to national and international markets. Additionally, IT systems can promote affirmative action in hiring

by ensuring that caste-based reservations are respected and that lower-caste individuals are not discriminated against in the hiring process.

3. **Social Inclusion and Awareness**: Digital platforms can also serve as a means of social inclusion for marginalized caste groups by providing a space for them to share their stories and raise awareness about caste-based discrimination. Platforms like Facebook, Twitter, and YouTube have allowed Dalit activists and organizations to raise their voices against systemic oppression, mobilize support for social justice initiatives, and educate others about the importance of equality and human rights. Social media movements like #DalitLivesMatter and the #StandWithDalit campaigns have gained momentum, helping to bring caste-based discrimination into the public consciousness and pushing for social change.

4. **Policy Advocacy**: IT also plays a significant role in advocating for policies that address caste-based inequities. E-government platforms allow marginalized caste groups to petition for their rights, engage with policymakers, and access government schemes to uplift their socioeconomic status. For instance, Dalit and Adivasi communities can access information about government scholarships, subsidies, and housing schemes through digital platforms, which helps them avail of benefits that would have otherwise been inaccessible due to bureaucratic hurdles or caste-based discrimination.

Regional Inequities and IT Solutions for Inclusive Development: India's vast geographical diversity, including urban-rural divides and disparities between regions, presents significant challenges regarding equitable development. While urban centres often have greater access to IT infrastructure, rural areas face difficulties due to poor connectivity, lack of awareness, and insufficient resources. IT solutions can play a critical role in bridging regional inequalities by facilitating access to resources, promoting regional development, and ensuring no region is left behind in the digital economy.

1. **Bridging the Urban-Rural Divide**: One of the key challenges in India is the stark contrast between urban and rural areas regarding access to infrastructure, services, and opportunities. IT can bridge this divide by bringing services like e-education, telemedicine, and e-governance to rural areas. The government's "Digital India" initiative aims to provide high-speed internet access to rural areas through projects like BharatNet, which aims to connect rural India through optical fiber networks. By increasing internet penetration and digital infrastructure in rural regions, these initiatives ensure that rural populations can access the same services as their urban counterparts.

2. **Supporting Regional Development**: IT is also key in promoting regional development, particularly in economically backward states and regions. By leveraging IT tools for better governance and resource management, these regions can attract investment, improve public services, and stimulate economic growth. For example, the "e-Choupal" initiative by ITC Limited connects farmers in rural India with markets, providing them with accurate weather data, crop prices, and agricultural advice via the internet. Such initiatives empower farmers to make informed decisions, improving agricultural productivity and reducing the economic disparity between regions.

3. **Decentralized Governance and Local Solutions**: IT also enables decentralized governance, allowing local governments to design region-specific solutions to address their unique challenges. IT systems like GIS (Geographical Information Systems) and data analytics tools enable governments to identify regional issues such as water scarcity, land degradation, or inadequate healthcare access and design targeted interventions. For example, GIS-based solutions can help identify areas that lack essential services such as roads, electricity, and sanitation, facilitating more efficient allocation of resources.

4. **Addressing Regional Disparities in Health and Education**: Regional disparities in access to healthcare and education remain a challenge in many parts of India. IT-based platforms can ensure equitable access to these services by bridging regional gaps. Telemedicine platforms such as eSanjeevani provide remote consultations to populations in underserved regions, while online education platforms like SWAYAM allow students from less-developed regions to access quality educational content. These IT-driven solutions ensure that individuals from all regions, regardless of socioeconomic status or geographical location, can access the same opportunities for growth and development.

Digital Tools for Social Empowerment

The digital transformation, fueled by rapid technological advancements such as Artificial Intelligence (AI), has opened new pathways for social empowerment, particularly for marginalized communities. In an era where economic growth is increasingly driven by knowledge, digital tools offer transformative opportunities for skill development and employment generation. AI, alongside other digital technologies, is playing a pivotal role in creating accessible platforms for skill-building, job placement, and entrepreneurship, thereby helping to bridge the socioeconomic divide. Empowering marginalized groups, including women, rural populations, minorities, and individuals from lower socioeconomic backgrounds, is crucial for inclusive development. The integration of AI in skill-building platforms and employment portals has the potential to drive social mobility, reduce unemployment, and foster sustainable livelihoods. This section explores the current landscape of digital tools for skill development and employment generation, particularly focusing on AI-driven solutions and how they contribute to social empowerment.

AI and Skill Development: AI has emerged as a powerful tool in revolutionizing education and skill-building. The potential of AI to cater to personalized learning experiences, adapt to individual learning styles, and offer real-time feedback is especially beneficial for marginalized groups. Traditional education and skill development systems, especially in underserved areas, often fail to provide the necessary education and training, limiting access to valuable skills in demand in the job market. AI-enabled platforms can overcome these barriers by providing affordable, scalable, and tailored solutions to skill development, thus enhancing the employability of individuals from marginalized communities.

1. **Personalized Learning Platforms**: AI-powered learning platforms such as Coursera, Udemy, and edX have democratized access to knowledge and skills. These platforms offer courses on various subjects, from technical skills like coding, data analysis, and AI, to soft skills such as communication, leadership, and critical thinking. AI algorithms track learners' progress, identify their strengths and weaknesses, and tailor the learning experience to suit their needs. This personalized approach helps marginalized individualswho may not have access to formal education or resourcesgain skills directly aligned with job market demands.

2. **Adaptive Learning Systems**: In regions where educational infrastructure is lacking, AI-driven adaptive learning systems have the potential to bridge the education gap. These systems use AI to continuously assess learners' performance and adapt content delivery based on their progress. For instance, AI systems in platforms like Khan Academy or Duolingo provide real-time feedback to students, allowing them to learn at their own pace. This technology is particularly effective for learners from rural or disadvantaged backgrounds who may face challenges accessing quality educational resources. AI-based platforms can enhance learning outcomes by tailoring content to suit the learner's context, such as their language proficiency, pace of learning, and educational background.

3. **Gamification and Virtual Training**: AI-driven gamification techniques transform skill development into a more engaging and interactive experience. Platforms such as Duolingo for language learning or Google's AI-based games for coding offer learners fun, game-like experiences that

encourage sustained engagement. For marginalized groups, especially youth, gamification offers an innovative way to keep them motivated and focused on acquiring new skills. Additionally, virtual training platforms powered by AI, such as simulators and virtual labs, provide practical hands-on experience in various domains, ranging from agriculture to digital marketing. These tools allow individuals in remote or underserved areas to acquire practical skills without access to physical infrastructure or expensive resources.

Digital Skill-Building for Marginalized Groups: Marginalized groups often face several barriers when it comes to skill-building, including limited access to quality education, gender bias, caste-based discrimination, and geographical isolation. Digital tools, particularly AI and related technologies, have the potential to break down these barriers by providing equal access to education and job opportunities. As such, skill-building initiatives that leverage digital technologies are crucial for promoting inclusivity and social equity.

1. **Targeted Digital Literacy Programs**: One of the first steps in empowering marginalized groups through digital tools is ensuring basic digital literacy. Many government and non-government organizations implement digital literacy programs to teach marginalised communities basic computing and internet skills. In India, initiatives like "Pradhan Mantri Gramin Digital Saksharta Abhiyan" (PMGDISHA) focus on providing digital literacy to rural populations, particularly women and low-income groups. These initiatives provide access to basic digital skills and help individuals in marginalized communities build confidence in using technology for everyday tasks, such as accessing government services, applying for jobs, or participating in online education programs.

2. **Bridging Gender Gaps in Skill Development**: Women, especially in rural and low-income areas, face a gender gap in access to education and economic opportunities. IT-based skill-building platforms have the potential to bridge this gap by offering women the flexibility to learn at their own pace, without the constraints of traditional education systems. Programs like the "Women's Digital Literacy Program" (WDLP) in India focus on providing training in digital literacy and tech skills specifically for women. Similarly, organizations like Women Who Code, Girls Who Code, and TechSheCan offer online coding, AI, and digital marketing training to empower women to enter the tech industry. AI-based platforms further enable personalized learning experiences for women by catering to their learning styles and needs, ensuring they are not left behind in the digital age.

3. **Inclusion of Rural and Remote Populations**: Access to physical training institutes or colleges is often a major barrier to skill development for marginalised rural communities. AI-driven platforms help overcome this barrier by offering online courses that are accessible from anywhere with an internet connection. Localized content delivery in regional languages has expanded opportunities for rural populations to access skill-building programs. Initiatives like "Skill India" aim to provide online training in areas like entrepreneurship, manufacturing, and digital literacy, making it possible for rural individuals to gain valuable skills from the comfort of their homes.

4. **Breaking Caste and Ethnic Barriers**: In countries like India, caste-based discrimination continues to limit access to education and employment for marginalized groups. Digital platforms, particularly those offering AI-based content, can help break down these barriers by offering impartial, merit-based learning and employment opportunities. For example, AI-driven job portals and skill development platforms can reduce human bias in recruitment processes by focusing on candidates' skills, experience, and qualifications rather than their social background. By ensuring that marginalized communities have access to skill-building programs that prioritize merit and potential, digital tools help level the playing field for all, regardless of caste or ethnicity.

AI – Driven Employment Portals and Platforms: As skill-building accelerates, the next challenge is ensuring that individuals from marginalized groups can access meaningful employment opportunities. AI-based employment portals have the potential to revolutionize the job market by making job search processes more efficient, transparent, and accessible. These portals are designed to match individuals with jobs that suit their skills, experience, and preferences while removing traditional barriers such as geographic location, caste, or gender.

1. **Personalized Job Matching**: AI-based job portals, such as LinkedIn, Indeed, and Upwork, use advanced algorithms to analyze users' profiles, skills, and career interests to match them with job opportunities. These platforms improve job search by providing personalized recommendations, automated job alerts, and relevant networking opportunities. In addition, platforms like "Naukri.com" and "Monster" in India have started using AI to offer job matching services tailored specifically for women and underrepresented communities, helping them find jobs that might otherwise be inaccessible through traditional job search methods.

2. **AI-Powered Resume Screening and Interview Preparation**: One of the main barriers for marginalized groups in the job market is the bias in traditional

recruitment processes. AI can help mitigate this by offering unbiased resume screening and interview preparation tools. For example, AI-based platforms like HireVue use algorithms to assess candidates' interviews objectively, focusing on skills and competencies rather than factors such as appearance, accent, or background. This reduces discrimination based on gender, caste, or regional origins, ensuring that candidates from marginalized groups are evaluated fairly based on their qualifications and potential.

3. **Gig Economy and Freelance Opportunities**: The rise of the gig economy has opened new avenues for employment, particularly for individuals who may not have access to full-time, traditional jobs. AI-driven platforms such as TaskRabbit, Freelancer, and Fiverr provide freelancers with the tools and opportunities to find jobs that match their skills and interests. These platforms offer a flexible and inclusive alternative to traditional employment for marginalised groups, enabling individuals to work remotely and manage their schedules. These platforms also provide mechanisms to evaluate and rate workers, ensuring that the gig economy operates fairly and accountable.
4. **Training and Upskilling for Employment**: AI-driven portals like Coursera, Skillshare, and Pluralsight provide access to courses and offer certification programs recognized by employers globally. By integrating AI into these platforms, marginalized individuals can access high-quality, globally relevant training programs in data science, artificial intelligence, digital marketing, and software development. These platforms also provide job placement support, helping graduates connect with potential employers who are specifically looking for skills in emerging technologies.

Case Study

Women's empowerment through digital tools has become a central theme in India's journey towards achieving inclusive economic growth. With the advent of the Digital India initiative, which was launched in 2015, the country has seen a profound transformation in its digital ecosystem. The initiative aims to promote digital literacy, facilitate e-governance, improve digital infrastructure, and create an enabling environment for technological innovations. The rise of women entrepreneurs leveraging digital platforms presents a compelling case study for understanding how technology can bridge gender gaps, facilitate economic participation, and promote gender equity. This section explores the impact of Digital India on women entrepreneurs, highlighting the role of technology in transforming their businesses and addressing the unique challenges they face.

Digital India Initiative and Its Impact on Women Entrepreneurs: The Digital India initiative focuses on empowering citizens by providing them with the tools and access to digital technology, and women, particularly in rural and underserved areas, are a significant beneficiary of these efforts. By improving access to the internet, expanding digital literacy programs, and promoting digital entrepreneurship, Digital India has paved the way for women to enter and thrive in the entrepreneurial landscape. Traditionally, women have faced several barriers to entrepreneurship, including limited access to capital, information, networks, and markets and social and cultural barriers that hinder their participation in economic activities. However, digital tools have allowed women to overcome many obstacles by giving them new opportunities to scale their businesses, access global markets, and participate in the digital economy.

1. **Access to Information and Market Reach**: One of the key ways digital tools have empowered women entrepreneurs is through the democratization of information. The internet has bridged these gaps in rural areas, where traditional communication and market access were limited. Platforms such as Google, YouTube, and WhatsApp have become critical tools for women entrepreneurs, enabling them to access tutorials, business advice, market trends, and even government schemes for financing and business development. Women can now run businesses from the comfort of their homes, selling handmade crafts, promoting services, or marketing agricultural products. Through e-commerce platforms like Amazon, Flipkart, and Etsy, women entrepreneurs can reach customers globally, circumventing geographic barriers that previously constrained their business opportunities.

2. **Skill Development and Capacity Building**: With the availability of digital platforms for skill development, women entrepreneurs can acquire new skills crucial for growing their businesses. Platforms such as Coursera, Udemy, and LinkedIn Learning offer training programs in various areas, such as digital marketing, financial management, e-commerce, and entrepreneurship. Many of these courses are free or affordable, making them accessible to women who may not have formal education or training. Additionally, government-led initiatives such as the "Pradhan Mantri Kaushal Vikas Yojana" (PMKVY) have focused on providing skill development programs that cater to women, particularly in areas like technology, sewing, and beauty services. These skills, combined with the increasing access to the internet, empower women to start their businesses and scale them effectively.

3. **E-Governance and Support Systems**: The Digital India initiative has also facilitated greater transparency, ease of business, and access to governmental support for entrepreneurs. Several government schemes and portals have been created to provide financial and regulatory support to

women entrepreneurs. For example, the "Stand Up India" scheme provides loans to women entrepreneurs to start new businesses, and platforms like "Udyam Registration" make it easier for women to register their businesses online. Women can also access information about taxation, funding, and business policies through government portals such as the Ministry of Micro, Small, and Medium Enterprises (MSME) website and the National Bank for Agriculture and Rural Development (NABARD).

4. **Networking and Mentorship**: The digital ecosystem also provides women entrepreneurs with opportunities to network and find mentors who can guide them through the challenges of entrepreneurship. Platforms like LinkedIn, Facebook, and Twitter enable women to connect with potential customers, collaborators, and business partners and find mentors who can provide guidance. Digital networks have helped women entrepreneurs break through traditional gender norms and restrictions, providing them with access to a global community of professionals and business leaders. Women can now learn from others' experiences and expertise, thus enhancing their capacity to grow their businesses effectively.

Challenges Faced by Women Entrepreneurs in the Digital Era: While digital tools and platforms have undoubtedly opened up new opportunities for women entrepreneurs, the path to success has its challenges. Women, particularly in rural or conservative areas, face significant barriers when fully realizing their entrepreneurial potential. These challenges must be addressed to ensure that the benefits of Digital India reach all women, particularly those from marginalized communities.

1. **Digital Literacy and Skills Gap**: Despite the widespread availability of digital technologies, digital literacy remains challenging for many women, especially in rural and underserved areas. A significant portion of the female population in these areas lacks basic digital literacy skills, which hinders their ability to leverage the full potential of digital tools for business development. While initiatives like the "Pradhan Mantri Gramin Digital Saksharta Abhiyan" (PMGDISHA) have made strides in improving digital literacy among rural populations, more needs to be done to ensure that women are equipped with the skills required to navigate the digital ecosystem and build successful businesses.

2. **Access to Finance**: Financial inclusion remains a major barrier for women entrepreneurs in India. While initiatives like the "Mudra Yojana" have aimed to provide micro-financing to small and medium entrepreneurs, women still face discrimination when accessing loans and funding from traditional financial institutions. The lack of a robust credit history and the

conservative approach of many financial institutions towards lending to women limits their access to capital. Though digital platforms like Peer-to-Peer (P2P) lending platforms and crowdfunding sites are helping to overcome this issue, access to finance remains a significant challenge for many women entrepreneurs.

3. **Cultural and Social Barriers**: In many parts of India, traditional gender norms significantly constrain women's ability to pursue entrepreneurial ventures. Women in rural or conservative areas often face societal pressure to prioritize household responsibilities over their business aspirations. The social stigma associated with women stepping outside the home to run a business or participate in the public sphere can limit their ability to succeed. Digital platforms allow women to work from home and run businesses online, which helps bypass some of these social restrictions. However, these cultural barriers persist and continue limiting women's participation in the entrepreneurial landscape.

4. **Gender Bias and Stereotyping**: Even in the digital world, gender bias remains a barrier to the success of women entrepreneurs. Digital platforms, whether social media, e-commerce websites, or business forums, can perpetuate traditional gender stereotypes. For instance, women entrepreneurs may face skepticism or discrimination when their products or services are marketed online, especially in traditionally male-dominated industries such as technology or manufacturing. There is also the challenge of dealing with harassment or trolling, particularly on social media platforms, which can discourage women from actively engaging in the digital economy.

5. **Inadequate Infrastructure in Rural Areas**: Despite significant improvements in digital infrastructure under the Digital India initiative, rural areas still face challenges regarding internet connectivity, power supply, and basic infrastructure. Poor connectivity in remote areas hampers women entrepreneurs' ability to use digital platforms for business expansion fully. In many instances, unreliable internet access or power outages disrupt business operations and online transactions, making it difficult for women entrepreneurs in rural areas to scale their businesses effectively.

Successful Case Studies of Women Entrepreneurs in the Digital India Ecosystem: Despite these challenges, several women entrepreneurs have leveraged the power of digital tools to build successful businesses. These case studies exemplify how women navigate the digital landscape and break barriers to become leaders in their respective fields.

1. **Rupali Sharma (Founder of "Craftsvilla")**: Rupali Sharma, the founder of Craftsvilla, an online marketplace for handmade products, has used digital tools to empower artisans, particularly women, by connecting them to a global customer base. Craftsvilla is one of India's leading e-commerce platforms for ethnic products, and it has enabled thousands of rural artisans, many of whom are women, to sell their goods online. Sharma's success demonstrates how digital platforms can provide a scalable solution for women entrepreneurs, creating a marketplace that connects rural producers to global buyers, and creating livelihoods for women artisans nationwide.
2. **Kiran Mazumdar-Shaw (Founder of Biocon)**: Although Biocon was founded before the digital revolution in India, Kiran Mazumdar-Shaw, a renowned entrepreneur in the biotechnology sector, has embraced digital technologies to revolutionize Biocon's operations. The company's use of AI and machine learning for drug discovery and manufacturing is a testament to how women-led businesses embrace the digital ecosystem to scale globally. Mazumdar-Shaw's leadership has encouraged other women in the pharmaceutical and biotech sectors to use digital tools for innovation and growth.
3. **Vandana Luthra (Founder of VLCC)**: Vandana Luthra, the founder of VLCC, a beauty and wellness brand, has successfully integrated digital platforms into her business model. Luthra's business has expanded its reach beyond India's borders through an online presence, including e-commerce, social media engagement, and digital marketing. Her success highlights the role of digital marketing, e-commerce platforms, and social media in empowering women entrepreneurs in the beauty and wellness industry.

The Digital India initiative has created a fertile ground for women entrepreneurs to emerge and flourish in India's rapidly growing digital economy. By improving access to information, resources, and markets, digital tools have provided women with the means to overcome traditional barriers such as limited access to finance, skill gaps, and socio-cultural constraints. However, challenges related to digital literacy, financial access, and gender bias persist. To ensure that more women can benefit from the digital revolution, there needs to be a concerted effort to address these gaps through targeted interventions, including skill development programs, financial inclusion initiatives, and policies that promote gender equity in the digital space. With the right support, women entrepreneurs in India have the potential to become powerful drivers of economic growth and social change.

Future Scenario

The future of conflict resolution and social impact assessment is poised for a radical transformation driven by advancements in artificial intelligence (AI) and virtual reality (VR) technologies. These tools promise to reshape how societies understand, engage with, and address social conflicts and their broader implications of inclusive community engagement. As the world grapples with political unrest, environmental degradation, and social inequality, integrating AI and VR into conflict resolution frameworks and social impact assessments offers new hope for finding more effective, transparent, and inclusive solutions. In this future scenario, AI tools will be instrumental in identifying, analyzing, and mitigating conflicts, while VR platforms will provide innovative ways to engage communities, especially marginalized groups, in decision-making processes and policy development.

AI Tools for Conflict Resolution: The application of AI in conflict resolution is rapidly gaining momentum, as the technology has the potential to offer data-driven insights into complex, multifaceted social issues. One of the primary benefits of AI in conflict resolution lies in its ability to process large datasets from various sourcessuch as social media, news outlets, and historical conflict dataenabling the identification of patterns that may otherwise remain unnoticed. This ability to analyze vast amounts of information in real-time allows AI systems to predict the onset of conflicts, understand the underlying causes, and recommend strategies for intervention. AI can be employed in various conflict scenarios, from political and ethnic conflicts to environmental and resource-based disputes.

AI-based predictive models can forecast potential conflicts by analyzing early-warning indicators such as rising tensions, economic inequality, resource scarcity, or social unrest. For instance, machine learning algorithms can process data from online platforms to detect shifts in public sentiment, which could indicate an impending conflict. By incorporating diverse indicators from economic variables to social media discourseAI systems can provide conflict prevention agencies with actionable intelligence, helping them intervene before conflicts escalate into violence or instability. This predictive capability is particularly valuable for governments, NGOs, and international peacebuilding and conflict prevention organisations.

AI can be used to develop mediation and negotiation tools that facilitate dialogue between opposing groups. Virtual agents powered by natural language processing (NLP) can simulate conversations and offer solutions based on conflict resolution models. These AI-driven platforms can help mediators identify common ground between conflicting parties, propose compromises, and reduce tensions. AI-powered chatbots, for instance, can mediate negotiations, ensuring that both sides are heard and understood, ultimately fostering a more collaborative approach to resolving disputes. Additionally, AI tools can be employed to enhance post-conflict recovery efforts. In societies recovering from conflict, AI can assist in rebuilding trust and fostering reconciliation. For example, AI-driven sentiment analysis can monitor public opinions and gauge the effectiveness of peace agreements or reconciliation processes. By tracking how communities react to post-conflict policies, AI systems can provide insights that help policymakers better adjust strategies to address affected populations' needs and concerns.

Social Impact Assessment Through AI: Social impact assessment (SIA) is a crucial process for understanding the effects of policies, projects, and developments on communities and the environment. As development initiatives become increasingly complex, the need for robust tools to assess their social impact becomes more pressing. AI can play a transformative role in improving social impact assessments' accuracy, efficiency, and inclusiveness. Through machine learning algorithms, AI can analyze data from various sourcesincluding socioeconomic data, environmental metrics, and community feedbackto evaluate proposed projects' short-term and long-term impacts.

AI-driven impact assessment tools can integrate vast amounts of qualitative and quantitative data, allowing for a more comprehensive analysis of social and environmental consequences. For instance, AI can process geographic information system (GIS) data to assess the spatial distribution of impacts, helping policymakers understand how a project will affect different regions or communities. AI can analyze demographic data to predict how specific population groupssuch as women, children, or marginalized communitiesmay disproportionately affect a particular development initiative.

One of the key advantages of using AI in social impact assessments is its ability to perform real-time analysis. For example, in large-scale infrastructure projects, AI can continuously monitor the environmental and social impact as the project progresses. By analyzing data from sensors, satellite imagery, and local reports, AI can detect early negative impactssuch as pollution, displacement, or social unrestand recommend corrective actions to mitigate these effects. This dynamic approach to impact assessment allows for adaptive management, where policies and strategies can be adjusted as new data becomes available, leading to more sustainable and socially responsible development outcomes.

AI can improve the inclusivity of social impact assessments by enabling greater participation from diverse stakeholders. Traditional methods of gathering community feedback often rely on surveys, focus groups, or public consultations, which accessibility issues, language barriers, or biases can limit. AI tools can help overcome these challenges by analyzing digital content, such as social media posts, online forums, and digital petitions. Natural language processing algorithms can assess sentiment and identify emerging concerns within communities, providing a more comprehensive and inclusive view of the social impact of a project. This allows for integrating voices that might otherwise be marginalized in traditional assessment processes.

Virtual Reality for Inclusive Community Engagement: In parallel with the rise of AI, virtual reality (VR) is emerging as a powerful tool for enhancing community engagementfor inclusive decision-making. VR offers an immersive experience that can transport individuals into simulated environments, allowing them to interact with complex scenarios in a highly engaging and intuitive way. This has significant implications for community participation, as VR can enable marginalized and underrepresented groups to engage in discussions and decision-making processes that directly affect their lives.

One of the most compelling applications of VR in community engagement is the creation of immersive simulations that allow stakeholders to experience the potential outcomes of different policies, projects, or interventions. For instance, in urban planning, VR platforms can simulate the impact of a new housing development, infrastructure project, or environmental policy. Community members can "walk through" the proposed changes in a virtual environment, providing them with a better understanding of how these initiatives will affect their daily lives. By experiencing these changes firsthand, citizens can offer more informed and meaningful feedback, ensuring that development is more aligned with the needs and aspirations of the community.

VR also has the potential to bridge geographical and social divides, enabling people from different parts of the world to engage in dialogue and collaboration. In conflict zones or areas with high levels of social tension, VR can be used to create safe spaces for dialogue and understanding. For example, VR can be used to facilitate peacebuilding workshops, where individuals from conflicting groups can experience the perspectives and lived experiences of others in an immersive and empathetic way. This can foster greater understanding and empathy, paving the way for more constructive discussions and conflict resolution. VR can also be employed to visualize social impact assessments in a way that makes the results more accessible and understandable to diverse audiences.

VR platforms can be used for training and capacity-building in communities, particularly where traditional education or training methods are inaccessible. For instance, VR simulations can train community members in disaster preparedness, healthcare, or conflict resolution skills. These interactive, hands-on experiences can be more effective than traditional classroom-based learning, especially in communities with limited educational resources.

Future of AI and VR in Conflict Resolution and Social Impact Assessment: As AI and VR technologies evolve, their potential to transform conflict resolution and social impact assessment will only increase. In the future, AI systems will likely become more sophisticated, capable of processing even larger datasets and providing deeper insights into the underlying causes of conflicts and their potential solutions. AI's ability to predict and prevent conflicts will become more accurate, helping to prevent violence and instability before they occur. Similarly, AI-driven social impact assessments will become more dynamic, continuously adapting to new data and adjusting policies to mitigate negative impacts on communities and the environment. With advancements in haptic feedback, real-time interaction, and enhanced realism, VR will continue to evolve. These improvements will make VR simulations more immersive, allowing for more meaningful community engagement. Integrating AI and VR technologies will enable even more powerful platforms for social impact assessment and conflict resolution, where virtual environments are used for engagement and real-time data analysis, predictive modelling, and scenario testing. These platforms will empower communities, providing them with the tools and insights to make informed decisions and actively participate in shaping their future.

15

Economic Transformation Through IT

Introduction

The economic transformation through information technology (IT) has emerged as one of the most influential drivers of modern economic development across the globe. In the 21st century, IT has catalyzed sweeping changes in industrial processes, global markets, and the overall economic structure, impacting everything from productivity to business models, labor markets, and governmental policy. The proliferation of IT has led to profound shifts in economic paradigms, influencing sectors ranging from agriculture and manufacturing to services and finance. The integration of IT into these various domains has spurred growth by fostering innovation, reducing inefficiencies, improving resource management, and creating new opportunities for economic development.

Productivity Gains and Economic Growth: At the core of IT-driven economic transformation is the significant increase in productivity it has enabled. IT tools and infrastructure, such as automation, big data analytics, and cloud computing, have drastically improved business operations and service delivery across multiple sectors. In industries like manufacturing, the integration of IT has led to the rise of smart factories, where the application of sensors, Internet of Things (IoT) technologies, and robotics has enabled real-time monitoring, predictive maintenance, and optimized production processes. These innovations have reduced operational costs, improved product quality, and increased output efficiency, enhancing productivity levels across economies.

IT has revolutionized business models in the services sector by enabling digital platforms that connect consumers with service providers in real-time. The advent of e-commerce platforms, such as Amazon and Alibaba, has transformed retail by allowing businesses to reach global markets without needing physical storefronts. Similarly, IT-enabled financial platforms, such as mobile banking, peer-to-peer lending, and blockchain technologies, have democratized access to financial services, allowing for more inclusive economic participation. These digital services have not only improved consumer convenience but also reduced transaction costs and expanded market access, fueling economic growth and creating new avenues for entrepreneurship and job creation.

The widespread adoption of IT has facilitated the shift from traditional industries to knowledge-based economies. The advent of digital technologies such as artificial intelligence (AI), machine learning, and automation has given rise to entirely new industries and business models. For instance, the rise of the gig economy, powered by IT platforms such as Uber, Airbnb, and Upwork, has transformed labor markets, enabling individuals to monetize their skills and assets in ways previously unimaginable. This shift towards knowledge-intensive sectors has led to higher-value-added industries, which are typically more resilient to economic shocks and are characterized by greater capacity for innovation and high levels of economic productivity.

Enhancing Global Trade and Connectivity: One of the most significant effects of IT on the global economy has been the dramatic transformation of international trade and connectivity. The internet, in particular, has eliminated geographical boundaries, enabling businesses to engage in global trade relatively easily. E-commerce and online marketing have enabled even small and medium-sized enterprises (SMEs) to access international markets, offering products and services to customers across borders. The expansion of digital payment systems has made cross-border transactions simpler, cheaper, and faster, removing many traditional barriers to international trade.

IT has also enabled the development of global supply chains that rely on digital tools for coordination, logistics management, and real-time communication. Enterprises are now able to manage complex supply chains across different continents with unprecedented levels of efficiency. Technologies such as blockchain and IoT have played a critical role in improving the traceability and security of these supply chains, enhancing the trust and reliability of international trade. By improving the efficiency of global supply chains, IT has enabled businesses to access a broader range of suppliers, reduce inventory costs, and improve product availability, leading to more competitive pricing and overall economic benefits. The proliferation of digital platforms and social media has also facilitated the globalization of ideas, cultural exchange, and innovation. IT has created an interconnected global community where individuals, businesses, and governments can share knowledge, collaborate, and innovate in ways previously limited by time and distance. This democratization of knowledge has led to a more open, competitive, and dynamic global economy, where new ideas and innovations can spread quickly, enabling rapid economic development in emerging economies.

Job Creation and Labor Market Transformation: While IT-driven economic transformation has resulted in job creation, it has also led to significant shifts in labor markets and employment patterns. Integrating IT into various sectors has created new job categories requiring advanced skills in software development, data science, cybersecurity, and digital marketing. These high-skill, high-wage jobs drive economic transformation, especially in emerging economies investing heavily in IT education and infrastructure. The rise of the digital economy has created opportunities for individuals to engage in the labor market in new and flexible ways, such as through remote work or freelance platforms.

The gig economy is a direct result of IT, as platforms like Uber, Lyft, and TaskRabbit have enabled individuals to offer their services on a flexible, on-demand basis. This has led to the creation of millions of jobs globally, particularly in sectors where traditional employment models have been slow to evolve. The flexibility provided by digital platforms has not only transformed the labor market but has also increased job access for marginalized groups, including women, youth, and those in rural areas, by eliminating geographical barriers to employment. IT's influence on the labor market has also raised concerns about job displacement, particularly in sectors that rely heavily on manual labor or routine tasks. The automation of processes through robotics and AI has led to the reduction of certain low-skilled jobs, particularly in manufacturing and administrative sectors. As a result, many workers face the challenge of reskilling and upskilling to remain competitive in a rapidly changing job market. The rise of IT-driven industries has highlighted the need for a robust educational infrastructure to provide workers with the skills necessary to thrive in the digital economy.

Challenges and Economic Inequality: While IT has contributed to significant economic transformation, it has also introduced new challenges, particularly regarding economic inequality. The digital dividethe gap between those with access to technology and those withoutremains a significant issue in developed and developing countries. In rural areas and low-income communities, lack of access to affordable technology and internet connectivity can limit opportunities for individuals and businesses to participate in the digital economy. This disparity can exacerbate existing inequalities in income, education, and job opportunities, leaving certain populations disadvantaged in the rapidly evolving global economy.

The fast-paced nature of technological change has led to disparities in how countries and regions benefit from IT-driven economic transformation. While developed nations have been quick to capitalize on new technologies, many developing countries face challenges such as inadequate infrastructure, limited access to education, and regulatory hurdles that hinder their ability to embrace digital transformation fully. These challenges can further entrench economic inequality, creating a situation where countries and regions that are already economically disadvantaged find themselves left behind in the digital age.

Governments and international organizations must prioritize addressing these inequalities to ensure that the benefits of IT-driven economic transformation are broadly shared. This can be achieved through investments in digital infrastructure, affordable internet access, and education systems that equip individuals with the skills needed for the digital economy. Policies promoting digital inclusion, such as providing subsidies for broadband access or supporting digital literacy programs, can help bridge the digital divide and ensure that all individuals can benefit from the economic opportunities offered by IT.

Sustainability and Green Transformation: IT *also plays* a key role in promoting economic sustainability, particularly in environmental conservation and resource management. Smart technologies, such as IoT sensors and data analytics, are being used to monitor energy consumption, optimize supply chains, and reduce waste, contributing to more efficient use of natural resources. In agriculture, IT solutions *enable* precision farming, where sensors and AI algorithms help optimize irrigation, fertilization, and pest control, leading to higher crop yields with fewer resources. Also, IT is driving the shift towards green technologies and renewable energy sources. Digital platforms enable the efficient integration of renewable energy systems into national grids, optimizing the distribution of solar, wind, and hydroelectric power. Smart grids, powered by IT, allow for real-time monitoring and management of energy usage, enabling utilities to balance supply and demand more effectively and reduce carbon emissions.

Current Scenario in Context of IT in Driving Business Innovation and Financial Inclusion

Information Technology (IT) has become a pivotal driver of business innovation and financial inclusion, shaping the modern economic landscape. As the world embraces digital transformation, IT continues influencing how businesses innovate, operate, and interact with customers, leading to significant shifts in economic models, organizational structures, and consumer behaviors. In parallel, financial inclusion, traditionally a challenge for underserved and marginalized communities, has been revolutionized by deploying IT solutions, particularly in developing regions. Integrating IT into financial systems has enabled broader access to banking services, credit, and insurance, while driving business models prioritising efficiency, inclusivity, and scalability.

Business Innovation through IT: Business innovation powered by IT, has fundamentally reshaped industries across the globe, from traditional sectors like manufacturing and retail to newer domains such as tech, finance, and healthcare. The rapid advancements in cloud computing, big data analytics, and artificial intelligence (AI) have created an environment where businesses can rapidly innovate, scale operations, and develop new products and services to meet emerging market demands. Cloud computing has, in particular, facilitated the digitalization of business models, allowing companies to move away from traditional infrastructure and adopt more agile, cost-effective systems. This shift has enabled small and medium-sized enterprises (SMEs) to access the same tools and technologies as large corporations, leveling the playing field and encouraging greater competition and innovation across sectors.

In manufacturing, the advent of Industry 4.0, characterized by integrating cyber-physical systems, IoT, and AI, has transformed how goods are produced, distributed, and consumed. These innovations have introduced smarter supply chains, enabling companies to manage inventory, logistics, and production in real-time. For example, companies are now using AI algorithms to optimize production schedules, predict maintenance needs, and even enhance the design of products based on consumer preferences and market trends. The fusion of IT with business operations has increased efficiency and led to new business models, such as the subscription economy, digital platforms, and service-based revenue models.

The retail industry, in particular, has experienced a radical transformation due to IT. E-commerce platforms have revolutionized how consumers shop, enabling businesses to reach a global audience and offer various products. The success of companies like Amazon, Alibaba, and Flipkart illustrates how businesses can leverage IT to create highly scalable, personalized, and efficient sales channels. Digital marketing tools, such as targeted ads, customer relationship management (CRM) systems, and data analytics, have further enhanced the ability of businesses to engage with customers and tailor their offerings to meet the specific needs and preferences of different market segments.

IT has enabled businesses to enhance their customer service and communication capabilities. With the rise of chatbots, AI-driven customer support systems, and social media engagement platforms, companies can provide more responsive and personalized service. This improves the overall customer experience and helps businesses build stronger, more loyal relationships with their clientele. Therefore, IT in business innovation goes beyond introducing new products; it is also about creating smarter, more effective ways to operate, communicate, and engage with customers, leading to sustained business growth and competitiveness in a digital-first world.

IT in Financial Inclusion: Parallel to driving business innovation, IT has played a transformative role in advancing financial inclusion, particularly in developing economies. Traditional banking systems, often limited by physical infrastructure and regulatory barriers, have left millions without access to essential financial services such as savings accounts, credit, and insurance. However, the rise of mobile banking, digital wallets, and fintech innovations has expanded access to financial services in underserved and remote areas, overcoming many traditional barriers to financial inclusion.

Mobile banking and digital wallets, such as M-Pesa in Kenya and Paytm in India, have promoted financial inclusion by allowing individuals to store, transfer, and receive money using a mobile phone. These mobile platforms have dramatically reduced the cost of financial transactions, particularly for low-income populations who would otherwise be excluded from formal banking systems. M-Pesa, for instance, has brought millions of unbanked individuals into the financial system, offering services such as micro-lending, bill payments, and remittances previously unavailable to them. Mobile banking has become an especially important tool for women, rural populations, and small-scale entrepreneurs, enabling them to access financial services without needing a bank account or physical branch. In addition to mobile banking, the rise of digital lending platforms and peer-to-peer (P2P) lending has expanded access to credit for underserved populations. Traditional credit systems often fail to provide loans to individuals without credit history or collateral. However, fintech platforms like LendingClub, Kiva, and Credit Karma use alternative data, such as mobile phone usage, social media activity, and transaction history, to assess creditworthiness. This approach allows individuals excluded from traditional credit systems to access affordable loans and other financial services, facilitating economic participation and upward mobility.

Blockchain technology is another critical innovation in the field of financial inclusion. Blockchain can radically alter how financial services are delivered by providing a secure, transparent, and decentralized way of recording transactions. In addition to improving security and reducing fraud, blockchain can reduce the costs associated with cross-border payments and remittances. In regions with limited access to traditional financial institutions, blockchain-based platforms can enable individuals to send and receive money across borders with minimal fees and delays. Blockchain's ability to create decentralized financial services (DeFi) means that individuals can access financial products without relying on traditional intermediaries, further increasing the reach and accessibility of financial services. Fintech innovations have paved the way for microinsurance models protecting low-income households from financial shocks. Traditional insurance systems have long been out of reach for those in poverty, who often cannot afford the premiums or meet the conditions of conventional policies. However, microinsurance platforms are designed to be more affordable and accessible to low-income populations, providing coverage for health, agriculture, and property-related risks. By leveraging IT to distribute and manage insurance products, these platforms can reach more people, offering them a safety net in times of need.

Despite the promise of digital financial services, many individuals in low-income communities remain skeptical of new technologies or lack trust in the systems they use. Financial literacy remains a significant barrier to widespread adoption, as many people do not fully understand how digital financial services work or how to use them safely and effectively. To overcome this, it is essential to provide education and awareness programs that help individuals understand the benefits and risks of digital finance and how to navigate these systems securely.

Digital Transformation in the Indian Economy

India is undergoing an unprecedented digital transformation, and the interplay of FinTech, e-commerce, and digital payments has become central to this evolution.

Over the past decade, these sectors have drastically reshaped the Indian economy by enhancing financial inclusion, enabling new business models, and creating a more efficient and transparent economy. This digital shift has transformed how individuals access and use financial services and revolutionized how businesses operate, connect with customers, and scale. At the heart of India's digital transformation is the rapid adoption of technology, the rise of innovative startups, and a supportive government framework to foster economic growth through digital empowerment.

FinTech – Revolutionizing Financial Services: Financial Technology, or FinTech, has emerged as a cornerstone of India's digital transformation, fundamentally altering the landscape of financial services. The Indian FinTech ecosystem is among the largest and most dynamic in the world, fueled by a combination of factors, including high smartphone penetration, expanding internet connectivity, and a young, tech-savvy population. FinTech encompasses various services, including digital payments, mobile banking, peer-to-peer lending, robo-advisory, insurance technology (InsurTech), and blockchain-based innovations. Together, these services have greatly improved access to financial products and services for most of the population, especially those previously excluded from the formal financial system. A key milestone in India's FinTech journey has been the Jan Dhan Yojana (PMJDY) introduction, which opened millions of bank accounts, many of which were linked to digital payment platforms. This initiative created a fertile ground for the widespread adoption of digital financial services. Government-backed schemes such as Aadhaar (a biometric identification system) and the Unified Payments Interface (UPI) have been crucial in driving digital inclusion, particularly in rural areas where traditional banking infrastructure was minimal.

The UPI, launched by the National Payments Corporation of India (NPCI), has been a game-changer in simplifying digital payments. It enables instantaneous and low-cost peer-to-peer transactions, making it easier for individuals to transfer money, pay bills, and make online purchases. UPI has become the backbone of India's digital payment infrastructure, experiencing exponential growth since its inception in 2016. It has made India a global leader in digital transactions, surpassing even developed economies in the volume of digital payments processed. Additionally, digital wallets such as Paytm, PhonePe, and Google Pay have become ubiquitous, offering consumers a seamless and convenient way to manage their finances. These platforms provide payment services and a range of financial products, including insurance, mutual funds, and gold investments, making it easier for consumers to diversify their financial portfolios.

The growth of digital payments and FinTech services has also fueled the rise of alternative lending platforms. Online lenders such as Lendingkart, Capital Float, and Indifi Technologies have democratized access to credit, especially for small and medium-sized enterprises (SMEs) and individuals with limited or no credit history. These platforms leverage big data, AI, and machine learning to assess the creditworthiness of borrowers using alternative data sources, such as mobile phone usage patterns and transaction history, instead of traditional credit scores. As a result, they can offer faster, more flexible loan products, empowering millions of underserved individuals and businesses to access credit.

Blockchain technology is another area in which FinTech has made significant strides. Blockchain's promise of secure, transparent, and decentralized financial transactions can potentially overhaul many aspects of India's financial ecosystem, from remittances to cross-border payments. While blockchain adoption in India is still nascent, its potential to provide greater security, reduce fraud, and cut costs associated with traditional financial transactions is undeniable. The Reserve Bank of India (RBI) and other regulatory bodies are exploring how blockchain can be utilized to improve the efficiency of India's banking system, particularly in the realm of digital payments and financial inclusion.

E-commerce – Transforming the Retail Landscape: The rapid growth of e-commerce in India is another crucial pillar of the country's digital transformation. E-commerce has fundamentally reshaped how Indians shop, access goods and services, and interact with businesses. The Indian e-commerce market has experienced explosive growth, driven by increased internet penetration, affordable smartphones, and a burgeoning middle class with rising disposable incomes. According to a report by the Indian Brand Equity Foundation (IBEF), India's e-commerce industry is expected to reach $200 billion by 2026, up from $38.5 billion in 2017, making it one of the fastest-growing e-commerce markets in the world.

A significant driver of this growth has been the rise of platforms like Flipkart, Amazon India, and Snapdeal, making it easier for consumers to purchase products from the comfort of their homes. These platforms offer a vast range of products, including electronics, clothing, groceries, and personal care items, creating a one-stop-shop experience for consumers. The convenience of doorstep delivery, competitive pricing, and frequent discount offers have accelerated e-commerce adoption among Indian consumers. In addition to retail, e-commerce platforms have expanded into other sectors such as food delivery, travel, and services. Companies like Swiggy, Zomato, and Ola have become household names, offering digital-first solutions that have transformed consumer behavior. Food delivery services, for example, have not only catered to the growing demand for convenience but have also provided a much-needed avenue for small restaurants and local food vendors to reach a broader customer base. Similarly, ride-hailing platforms like Uber and Ola have redefined transportation in urban centers, contributing to the digitalization of the mobility sector.

The increasing adoption of digital payment systems has also facilitated the rise of e-commerce in India. As more consumers use online platforms to purchase products, digital wallets and UPI-based payments have become essential tools for seamless transactions. This convergence of e-commerce and digital payments has created a positive feedback loop, further propelling the growth of India's digital economy. However, while e-commerce has transformed the retail landscape, it also presents challenges such as logistical inefficiencies, concerns over data privacy, and the dominance of a few large players in the market. Despite these challenges, the sector remains one of the most dynamic and innovative in the Indian economy, with continued growth expected in the coming years.

Digital Payments – Fueling India's Cashless Economy: Digital payments have been at the core of India's economic transformation, with the government's push for a *cashless economy* gaining significant momentum over the past few years. The demonetization exercise 2016, which aimed to curb black money and encourage digital transactions, marked a watershed moment in the country's financial history. The move accelerated the adoption of digital payments, leading to a surge in UPI transactions, digital wallet usage, and mobile banking services.

The government has also introduced several initiatives to promote digital payments. The Digital India campaign, launched in 2015, aims to transform India into a digitally empowered society and knowledge economy. Through this initiative, the government has made significant investments in digital infrastructure, such as increasing internet connectivity in rural areas, promoting digital literacy, and enhancing the security of online transactions. Bharat Interface for Money (BHIM) mobile payment application based on UPI, has provided a simple and secure platform for making payments and transfers. In addition to UPI, the growth of digital payments in India has been fueled by the rise of mobile wallets, point-of-sale (POS) systems, and QR code-based payment systems. The adoption of mobile wallets, such as Paytm, PhonePe, and Google Pay, has been especially rapid, with millions of users leveraging these platforms for everyday transactions and more complex financial services. These wallets offer a range of services, from bill payments and money transfers to online shopping and ticket bookings, making them indispensable in the daily lives of many Indians.

The success of digital payments in India has not only been driven by technology but also by robust government regulation. The Reserve Bank of India (RBI) has implemented several measures to ensure the safety and security of digital payment systems, including mandating two-factor authentication for online transactions and enhancing fraud detection mechanisms. These regulations have helped build trust in digital payment platforms, paving the way for widespread adoption. Despite the rapid growth of digital payments, challenges remain, including cybersecurity risks, the need for greater financial literacy, and ensuring the seamless integration of digital payments across diverse sectors and regions. Nevertheless, the Indian government's push for a cashless economy, coupled with increasing technological advancements, positions India on the cusp of a digital payments revolution.

Future-Proofing India's Economy

As India continues its rapid march toward becoming one of the world's largest and most dynamic economies, ensuring its future economic resilience becomes paramount. Key to this resilience is the integration of advanced technologies such as Artificial Intelligence (AI) and Blockchain, which hold transformative potential for various sectors, including economic modelling and the empowerment of micro, small, and medium enterprises (MSMEs). When strategically applied, these technologies can safeguard India's economy from future shocks and unlock new opportunities for growth, efficiency, and inclusivity. This section explores the role of AI-based economic modelling and Blockchain technology in future-proofing India's economy, particularly by enhancing decision-making, forecasting, and supporting MSMEs in their journey towards digital transformation.

AI-Based Economic Modelling: The integration of AI in economic modelling presents a groundbreaking opportunity for policymakers and businesses in India. Traditional economic models rely on historical data and statistical methods to make predictions and guide decision-making. However, these models are typically limited in accounting for complex, dynamic, and interconnected factors that influence economic trends. This is where AI, specifically machine learning (ML) and deep learning, comes into play. By leveraging vast amounts of real-time data, AI-based models can provide more accurate and nuanced predictions, allowing for more informed and adaptive policy decisions.

In India, AI-based economic modelling can significantly enhance forecasting capabilities for key economic indicators such as GDP growth, inflation, employment rates, and sectoral performance. For instance, AI algorithms can analyze data from multiple sourcesfrom financial markets and industrial production to social media trends and satellite imageryto predict shifts in demand, supply chain disruptions, and potential economic downturns. This real-time capability would be especially valuable in the Indian economy, where challenges such as inflation, unemployment, and supply chain inefficiencies are ever-present. AI can thus act as a dynamic tool for economic forecasting, moving beyond static models to real-time predictions and scenario-based analysis.

The application of AI-based economic models extends to assessing the impact of various economic policies, including taxation, subsidies, and fiscal interventions. Machine learning models can simulate different policy scenarios, accounting for immediate and long-term effects, enabling policymakers to evaluate trade-offs and anticipate unintended consequences. For example, India's ongoing tax reforms or agricultural policy changes, AI models can help forecast their economic impacts on various sectors, regions, and demographics. These models would enable better allocation of resources, more effective regulation, and, ultimately, a more resilient economy.

AI can also play a significant role in financial and economic risk management. The ability to predict financial crises, exchange rate volatility, or commodity price fluctuations can empower businesses and government bodies to implement preemptive strategies. With India's diverse economic composition and regional disparities, this predictive capability can be crucial in avoiding or mitigating financial instability. AI can assist in monitoring and responding to external economic shocks, such as changes in global trade dynamics or disruptions caused by climate change, which have a particularly pronounced impact on India's economyhowever, the successful deployment of AI in economic modelling hinges on overcoming several challenges. First, the quality of the data used in AI models is critical; inaccurate or incomplete data can lead to erroneous predictions. Second, there is a need for robust infrastructure and computational power to handle the massive datasets required for AI-based analysis. Third, building human capacity in AI and data science is essential for interpreting and translating the results into actionable policy. In India, where the digital divide still exists, these challenges must be addressed to ensure that AI-driven economic models are accessible and useful to all sectors of society.

Blockchain for MSMEs: Micro, small, and medium enterprises (MSMEs) are the backbone of India's economy, contributing nearly 30% to GDP, employing a large portion of the workforce, and driving innovation. Despite their importance, MSMEs face several challenges, including limited access to finance, inefficient supply chain management, a lack of transparency, and limited market access. These challenges are exacerbated by the traditional business ecosystem's reliance on paper-based processes, intermediaries, and opaque systems. *With its decentralized and secure nature, blockchain technology* presents an opportunity to address many of these barriers, enabling MSMEs to thrive in a digital economy.

Blockchain's most notable feature is its ability to provide transparency, security, and immutability in transactions. By creating a tamper-proof record of all transactions, blockchain can eliminate fraud, enhance trust, and improve the efficiency of business processes. For MSMEs, which often operate in informal or semi-formal sectors, blockchain can ensure that transactions are recorded in a way that is secure, verifiable, and accessible. This can significantly reduce the risks of fraud and disputes, especially in industries like agriculture, manufacturing, and retail, where intermediaries or unfair trading practices often exploit MSMEs. For example, in the agriculture sector, farmers and agribusinesses can use blockchain to track the entire supply chain of productsfrom farm to table. This traceability ensures that consumers receive fresh, quality products while guaranteeing that farmers receive fair prices. Blockchain can also streamline payment systems by enabling direct, peer-to-peer transactions between farmers and buyers, removing the need for intermediaries. This would increase the transparency of transactions and reduce transaction costs, often a significant burden for small businesses in rural areas.

Access to finance remains among India's most pressing challenges for MSMEs, especially in underserved regions. In combination with smart contracts, blockchain can facilitate easier access to finance by creating decentralized financial platforms that provide MSMEs with alternative funding options. By using blockchain to verify the authenticity and creditworthiness of MSMEs, financial institutions and investors can more confidently lend to these businesses, even those with limited credit history or formal documentation. Additionally, blockchain can enable quicker, more secure cross-border payments, which is particularly beneficial for MSMEs that engage in international trade.

One of the most promising uses of blockchain for MSMEs is in the realm of supply chain management. Blockchain's ability to provide real-time, transparent, and immutable transaction records can dramatically improve supply chains' efficiency and reliability. For MSMEs involved in manufacturing, logistics, or retail, blockchain can enable them to track raw materials, finished goods, and inventory more accurately, reducing inefficiencies, delays, and disputes. This leads to better decision-making, improved product quality, and cost savings.

Blockchain has the potential to foster greater collaboration between MSMEs, large corporations, and governments. By participating in blockchain-based networks, MSMEs can access resources, services, and expertise previously beyond their reach. Blockchain's decentralized nature ensures that MSMEs are not locked into a single proprietary system but can connect to multiple interoperable networks that cater to their needs. This collaborative approach can help MSMEs build resilience, improve their bargaining power, and navigate the challenges of digital transformation. While the potential for blockchain to revolutionize MSMEs is immense, its adoption remains relatively low due to several barriers. First, the lack of awareness and understanding of blockchain technology among MSMEs is a significant challenge. For blockchain to be adopted at scale, there is a need for widespread education and capacity-building initiatives. Second, the regulatory environment around blockchain is still evolving in India, and uncertainties regarding its legal framework can deter businesses from exploring its potential. Lastly, there is a need for significant investment in infrastructure, as many MSMEs may lack the technical resources to implement blockchain solutions.

Integrating AI and Blockchain for Future-Proofing India's Economy: The future-proofing of India's economy lies in the convergence of AI and blockchain. While AI excels in predictive analytics, decision-making, and data processing, blockchain provides the transparency, security, and decentralization needed to create a more efficient and trustworthy digital ecosystem. Together, these technologies can *drive innovation and* provide solutions to the long-standing challenges faced by Indian businesses and policymakers. For example, AI-based economic models can help predict demand, market trends, and consumer preferences, enabling MSMEs to make more informed decisions. Blockchain can then execute these decisions transparently and securely, ensuring that transactions are recorded and verified *to reduce fraud and enhance* trust. In supply chain management, AI can optimize inventory management, predict bottlenecks, and enhance product quality, while blockchain ensures that every transaction along the supply chain is transparent and immutable.

The synergy between AI and blockchain also extends to improving public policy and governance. AI models can analyze vast datasets to predict the outcomes of various policy interventions, while blockchain can ensure that government transactions are transparent, secure, and auditable. This combination can lead to more accountable governance, more efficient allocation of resources, and a stronger, more resilient economy.

Case Study: India's "Make in India" initiative, launched in 2014, represents a strategic effort to transform the country into a global manufacturing hub, fostering innovation, enhancing skill development, and improving the business climate. A core objective of this initiative is to increase the share of manufacturing in India's GDP from 16% to 25% by 2025, generating millions of jobs and attracting foreign direct investment (FDI). To achieve these ambitious goals, integrating advanced technologies such as blockchain in the manufacturing and supply chain sectors can streamline operations, reduce inefficiencies, and enhance transparency and trust, ultimately driving the success of "Make in India." In this case study, we explore the application of blockchain technology in supply chain and trade, focusing on its potential to solve some of the key challenges in India's manufacturing landscape.

Blockchain Technology for "Make in India": Blockchain technology, known for its decentralized, immutable ledger system, has seen significant adoption across various sectors. It promises to transform the supply chain and trade landscape by providing real-time, transparent, and secure tracking of goods, reducing fraud, lowering costs, and increasing operational efficiency. For India, a country with a vast and complex manufacturing ecosystem, blockchain offers a unique opportunity to overcome several hurdles inherent in traditional supply chains, including inefficiency, fraud, and lack of traceability. Blockchain technology can help manufacturers improve the efficiency of their operations, ensuring smoother workflows and timely deliveries. India's manufacturing sector is diverse, from textiles and automobiles to electronics and pharmaceuticals. With such a broad spectrum of goods being produced, managing the supply chain effectively is a significant challenge. From sourcing raw materials to ensuring timely deliveries and tracking finished products, manufacturers face delays, document fraud, counterfeiting, and product quality verification challenges.

Blockchain provides an elegant solution to these challenges by enabling decentralized, tamper-proof records of transactions. This ensures transparency in every stage of the supply chain. For example, raw materials used in producing a car or mobile phone can be traced back to their source through a blockchain-enabled system. Each transaction, whether a shipment of raw materials, the manufacture of a part, or the final delivery to a retailer, is recorded on a blockchain ledger accessible to all relevant parties. This ensures that all stakeholdersfrom suppliers to manufacturers to consumerscan access real-time data on the product's provenance, quality, and movement through the supply chain. This reduces fraud and counterfeiting and boosts consumer confidence in manufactured products.

Blockchain in Supply Chain Transparency and Traceability: One of the key challenges in Indian manufacturing is the lack of transparency across various supply chain stages. This often leads to inefficiencies such as delays, incorrect inventories, and disputes over quality or product specifications. In pharmaceuticals, food processing, and agriculture industries, these inefficiencies can have serious consequences, including substandard products reaching the market and even health risks to consumers. By providing an immutable and decentralized record of each transaction, blockchain technology addresses these issues by allowing all parties involved to view the same information in real time. For instance, in the food processing industry, blockchain can track the journey of food products from farm to table. This traceability helps ensure that products meet safety standards, can be recalled quickly in the event of contamination, and helps prevent fraud in labeling. For pharmaceutical products, where counterfeiting is a major issue, blockchain enables secure tracking and verification of drugs as they move through the supply chain, thereby improving product safety and consumer trust.

The transparency blockchain provides can also streamline customs and clearance processes in trade. In India, where cross-border trade is often encumbered by bureaucratic delays, customs inefficiencies, and fraud, blockchain can significantly reduce paperwork, enable faster clearance times, and ensure more accurate record-keeping. By digitizing trade transactions and allowing for real-time data sharing across borders, blockchain can create a more efficient and transparent system, reducing administrative costs and enhancing global competitiveness for Indian manufacturers. The decentralized nature of blockchain ensures that no single entity controls the entire ledger. This shared responsibility helps build trust between stakeholders who might otherwise be hesitant to share sensitive information, such as pricing strategies or production schedules. Blockchain-based systems ensure that all parties, whether suppliers, manufacturers, or logistics companies, work from the same data, leading to fewer misunderstandings, better coordination, and more efficient production cycles.

Blockchain and Trade Efficiency: The trade landscape for India's manufacturers has traditionally been characterized by complex and slow-moving processes.

Cross-border trade is often hampered by a lack of standardization, poor infrastructure, and inefficiencies in clearing customs. Blockchain has the potential to address many of these issues by digitizing trade documents such as invoices, bills of lading, and letters of credit. Through smart contracts, blockchain allows for automated verification and approval of these documents, reducing the time and cost associated with paperwork and bureaucratic processes.

India's trade and export ecosystem benefits enormously from blockchain. For example, blockchain-enabled platforms like TradeLens (developed by IBM and Maersk) provide real-time visibility into the status of goods in transit. This system not only improves transparency and reduces errors but also minimizes delays by enabling faster processing of documents and better coordination between all stakeholders. In a nation like India, which is a major player in global trade, adopting blockchain for trade facilitation could be a game-changer, making it more competitive in global markets and aligning it with the needs of modern digital economies. Also, blockchain can facilitate access to trade finance for small and medium-sized manufacturers, a critical issue that has long hindered India's MSMEs. MSMEs in India face significant barriers in accessing affordable credit and trade financing, often due to the lack of adequate collateral, credit history, or transparency in their operations. With blockchain, MSMEs can establish their credibility by recording transactions transparently, offering potential lenders a clear and verifiable picture of their business operations. Through this, blockchain can open access to trade finance, enabling smaller businesses to participate more fully in domestic and international trade.

Case Studies: Several companies and initiatives are already leveraging blockchain technology to enhance supply chain operations and trade efficiency within the framework of the "Make in India" initiative. One such example is the application of blockchain in the Indian pharmaceutical sector. India is the world's largest supplier of generic drugs, and counterfeiting is a serious issue that threatens consumer safety. The pharmaceutical company, MediLedger, is working with a consortium of manufacturers and distributors to build a blockchain-based platform for tracking the movement of drugs throughout the supply chain. Using blockchain, stakeholders can track and verify each drug's journey from manufacturer to distributor, ensuring that only authentic drugs reach consumers.

Another example is the pilot project launched by the Indian government in collaboration with the blockchain technology firm, Zebi, to enhance the transparency of government procurement processes. Zebi is using blockchain to streamline the procurement of agricultural goods and services, ensuring fair and transparent pricing and reducing fraud in the system. This initiative aligns with the objectives of the "Make in India" campaign by ensuring the efficiency of India's domestic supply chains, which is critical to both manufacturing and export. Additionally, the logistics sector, a crucial component of supply chain operations, has begun experimenting with blockchain to reduce inefficiencies and improve transparency. India's transportation and logistics companies, such as Mahindra Logistics, are increasingly exploring blockchain for better tracking of goods, monitoring inventory levels, and ensuring timely delivery. Blockchain allows for seamless integration between suppliers, transporters, and manufacturers, reducing bottlenecks and improving the reliability of supply chains.

Future Prospects: While blockchain technology offers immense promise, its integration into the Indian manufacturing landscape is not without challenges. India's manufacturing sector is characterized by a wide range of small-scale producers, especially in rural areas, who often lack the technical expertise or digital infrastructure to adopt such technologies. Blockchain requires robust digital infrastructure, internet connectivity, and skilled workforce, which may pose hurdles for small manufacturers, especially in remote areas. India's regulatory framework around blockchain is still evolving, and there is a need for clear policies and standards to ensure the seamless integration of blockchain in supply chains. Concerns regarding data privacy and the legal recognition of digital transactions must also be addressed to foster confidence in blockchain systems.

Despite these challenges, the future of blockchain in India's manufacturing sector looks promising. With the Indian government's focus on digital transformation, infrastructural improvements, and the increasing adoption of Industry 4.0 technologies, blockchain is poised to play a crucial role in the success of the "Make in India" initiative. By improving transparency, reducing fraud, enhancing traceability, and boosting trade efficiency, blockchain can help Indian manufacturers scale up operations, enter global markets, and foster a more competitive and sustainable manufacturing ecosystem.

Future Scenario

The rapid evolution of digital financial systems over the past decade has prompted a significant shift in the global economic landscape. Central to this transformation are cryptocurrencies and decentralized finance (DeFi), which, together, have the potential to redefine how wealth is distributed and how financial systems operate. Cryptocurrencies, such as Bitcoin, Ethereum, and their evolving counterparts, offer a decentralized approach to currency and asset management, free from the constraints of traditional banking systems. Decentralized Finance (DeFi) is a suite of financial services built on blockchain technology, allowing users to access financial products such as lending, borrowing, and trading without relying on intermediaries like banks or other financial institutions. Combined with the power of artificial intelligence (AI) for market forecasting and prediction, these technologies offer an innovative approach to building more equitable, resilient economies. This section delves into the future scenario of cryptocurrencies, DeFi, and AI-driven market systems, exploring their potential to promote equitable wealth distribution and economic resilience.

Cryptocurrencies and Decentralized Finance: Cryptocurrencies and DeFi have emerged as powerful forces that challenge the traditional financial system's long-standing dominance. At their core, these technologies promise a future where wealth and financial services are more accessible, transparent, and inclusive. Cryptocurrencies, underpinned by blockchain technology, allow individuals to store and transfer value across borders without relying on centralized banks or financial institutions. This opens up opportunities for people in underbanked regionsparticularly in developing economies like India, Sub-Saharan Africa, and Southeast Asiawho have long been excluded from formal financial systems due to infrastructural or institutional barriers.

The decentralized nature of cryptocurrencies means that wealth is not concentrated in the hands of a few intermediaries. Instead, it is distributed across a network of nodes, ensuring that all participants in the system have equal access to resources without gatekeepers controlling the flow of capital. For instance, DeFi platforms, which are built on blockchain and run smart contracts, eliminate the need for traditional banks, enabling anyone with an internet connection to participate in lending, borrowing, and investing activities. This democratization of finance could facilitate access to credit, capital, and investment opportunities for marginalized communities, helping to reduce inequality and foster more equitable wealth distribution.

In particular, cryptocurrencies enable cross-border transactions, which are often slower and more costly in traditional financial systems. Cryptocurrencies can facilitate international trade and economic exchanges by removing the need for currency exchange and minimizing transaction fees, further bridging gaps between developed and developing nations. For example, individuals in low-income regions can access stablecoinscryptocurrencies pegged to fiat currencies such as the US Dollarthus mitigating the risks associated with local currency devaluation and providing a store of value that protects their wealth. This can safeguard against hyperinflation and economic instability, a significant concern in countries like Venezuela or Zimbabwe.

The potential of decentralized finance systems to reduce the reliance on traditional financial intermediaries could significantly lower the costs of accessing financial products. DeFi platforms can offer high-yield savings accounts, loans with lower interest rates, and more secure investment opportunities, all without the overhead costs associated with traditional banking. This could benefit individuals with limited credit access or those trapped in exploitative lending cycles. DeFi platforms can empower individuals otherwise excluded from the formal financial system by providing decentralised credit scoring and identity verification systems, improving their ability to save, invest, and grow wealth.

AI-Based Market Forecasting: While cryptocurrencies and decentralized finance systems offer revolutionary potential for equitable wealth distribution, their success and widespread adoption will also depend on the resilience and adaptability of the global economy. In this regard, artificial intelligence (AI) is crucial in market forecasting, enabling more accurate predictions of economic trends and shifts in financial markets. AI-powered tools, including machine learning algorithms and predictive models, can analyze vast amounts of datafrom historical market trends and economic indicators to real-time news and social media sentimentthus providing policymakers, businesses, and individuals with the insights needed to make informed decisions in an increasingly volatile and complex economic environment.

AI-based market forecasting tools can significantly enhance economic resilience by enabling early detection of potential economic crises, such as recessions, inflationary spikes, or financial bubbles. By analyzing patterns in financial data and identifying emerging risks, AI can help central banks and governments adjust monetary policies, ensuring that economies remain stable and that sudden economic shocks do not unduly impact wealth distribution. This predictive capacity can also be applied to the crypto market, which has been known for its volatility. AI algorithms can forecast price movements based on historical data, macroeconomic indicators, and sentiment analysis, enabling investors to make better decisions and reduce exposure to market risks.

Regarding wealth distribution, AI-based market forecasting can ensure that economic policies and investments align with vulnerable communities' most pressing needs. For instance, AI can identify regions or sectors most at risk of economic downturns, helping governments and institutions target interventions to reduce inequalities. Additionally, AI-driven investment platforms can offer tailored advice to small investors, enabling them to make sound financial decisions and optimize their portfolios, thus empowering individuals to grow their wealth in a data-driven and informed manner. By providing personalized insights into the best investment opportunities based on individual financial goals and risk tolerance, AI can help mitigate the wealth gap by giving marginalized populations greater access to wealth-building tools. One significant area where AI and DeFi intersect is the creation of algorithmic stablecoins, which are designed to maintain a stable value through AI-driven mechanisms that adjust supply and demand automatically. These stablecoins can hedge against inflation, offering a more reliable and equitable store of value for communities prone to economic instability. AI's ability to analyze economic conditions and adjust the supply of stablecoins in real-time ensures that these digital assets can function as resilient tools in the global economy, further supporting equitable wealth distribution.

Blockchain and AI for Sustainable Economic Development: Blockchain and AI technologies are also key to fostering sustainable economic development, particularly in equitable wealth distribution. Blockchain's transparency and immutability offer a powerful mechanism for monitoring and ensuring the responsible use of resources, preventing corruption, and ensuring that financial flows are directed toward social and environmental goals. For example, AI can be used to track and optimize resource allocation for sustainable development projects, ensuring that investments in green energy, infrastructure, and social programs reach the intended beneficiaries.

AI models can also be leveraged to forecast the impact of economic policies on income inequality and social welfare, helping governments create policies that reduce disparities and ensure that the benefits of economic growth are widely shared. By modeling different policy scenarios, AI can provide valuable insights into economic decisions' long-term effects, helping policymakers craft policies that support inclusive growth and poverty reduction. In this way, AI not only improves market forecasting but also enhances the capability of policymakers to design systems that promote social equity.

Through decentralized applications (dApps) and smart contracts, blockchain technology can facilitate peer-to-peer transactions, further enhancing economic participation for individuals excluded from the formal economy. Blockchain can reduce transaction fees by reducing the need for centralized intermediaries, making economic exchanges more efficient and less costly. This, in turn, makes it easier for marginalized communities to participate in the global economy, access credit, and invest in business ventures.

Considerations in the Future Scenario: While the integration of cryptocurrencies, decentralized finance, and AI-powered market forecasting holds immense potential for building more equitable and resilient economies, several challenges need to be addressed. Firstly, the regulatory environment surrounding cryptocurrencies and DeFi is still in its infancy. Governments worldwide grapple with the complexities of regulating decentralized systems operating across borders. In India, for example, the lack of clear regulation on cryptocurrencies has created uncertainty, hindering their potential adoption. Ensuring a balanced regulatory framework that fosters innovation while preventing illicit activities such as money laundering and fraud will be crucial for the long-term success of cryptocurrencies and DeFi. Secondly, adopting AI in market forecasting and wealth distribution raises concerns regarding data privacy and bias. AI models are only as good as the data they are trained on, and biased data can lead to biased predictions and decisions that reinforce existing inequalities. It is essential that AI systems used for wealth distribution are trained on diverse, representative data sets and that safeguards are in place to ensure that AI systems do not perpetuate biases based on race, gender, or socioeconomic status. Finally, blockchain and DeFi promise greater economic inclusion, but the technological divide remains a significant barrier. Many individuals, particularly in rural or underserved areas, lack access to the internet, smartphones, or digital literacy, which limits their ability to participate in the digital economy. Bridging this divide through education, infrastructure development, and access to affordable technology will ensure that these technologies truly benefit marginalized communities.

The future of equitable wealth distribution and resilient economies lies in integrating cryptocurrencies, decentralized finance, and AI-based market forecasting. By offering decentralized financial systems, cryptocurrencies empower individuals in underserved regions, enabling them to access capital, protect their wealth, and engage in global trade. Combined with the power of AI for predictive market analysis, these technologies offer significant potential for more equitable wealth distribution, better economic forecasting, and more resilient economies. However, challenges remain, particularly in regulation, data privacy, and access to technology. Addressing these challenges will be key to unlocking the transformative potential of these technologies and building a more inclusive global economy.

16

Ethical Challenges in IT and Sustainability

Introduction

As information technology (IT) continues to evolve and integrate into every aspect of human life, the intersection of IT and sustainability presents immense opportunities and significant ethical challenges. These challenges arise from the complex relationship between technological advancement, environmental responsibility, social equity, and economic growth. While IT has the potential to drive progress toward sustainability, it also introduces new risks and dilemmas that need to be carefully navigated. The ethical concerns surrounding IT and sustainability can be broadly categorized into data privacy, digital divide, environmental impact, accountability, and the potential unintended consequences of technological innovations.

Environmental Impact of IT Infrastructure: One of the most significant ethical challenges in IT and sustainability is the environmental impact of the technology itself. The production, operation, and disposal of IT hardware such as servers, computers, smartphones, and data centers contribute to resource depletion, energy consumption, and electronic waste (e-waste). Manufacturing processes for IT devices require raw materials like rare earth metals, which are often mined under exploitative conditions with significant environmental degradation. The extraction of these materials can result in habitat destruction, water pollution, and soil contamination. The growing demand for data storage and processing power has led to expanding energy-intensive data centers, which often rely on non-renewable energy sources, contributing to carbon emissions.

The disposal of obsolete or broken devices also creates a significant environmental burden. Electronic waste is often poorly managed, especially in developing countries, where it is informally recycled under unsafe conditions, releasing harmful chemicals into the environment. In sustainability, there is a growing ethical imperative for IT companies to adopt circular economy principles, ensuring that their products are designed for longevity, repairability, and recyclability. The energy consumption of IT infrastructureparticularly in cloud computing and data centersrequires careful consideration. A major ethical challenge is balancing technological innovation with sustainable energy practices, such as using renewable energy sources to power data centers and reducing e-waste through better recycling practices.

Digital Divide and Social Equity: Another ethical dilemma at the intersection of IT and sustainability is the digital dividethe gap between those with access to information technology and those without. This divide is not only an infrastructure issue but also one of access to digital skills and literacy. In developed countries, IT is often taken for granted, but access to the internet and modern computing devices remains limited in low-income and rural areas. This creates a divide between those who can benefit from the advancements in digital technologies, such as e-governance, online education, and remote work, and those who cannot.

The digital divide exacerbates existing inequalities, hindering the ability of marginalized communities to participate in the digital economy. For instance, lack of access to the internet limits people's ability to access healthcare information, job opportunities, and financial services. Additionally, the expansion of IT-based systems often favors urban areas, where infrastructure is more developed, leaving rural areas disadvantaged. The ethical challenge is ensuring that IT solutions that promote sustainability are inclusive and accessible to all, irrespective of socioeconomic status or geographic location. In addressing the digital divide, the ethical challenge lies in ensuring that efforts to promote digital inclusion do not inadvertently further entrench inequalities. IT policies must prioritize equitable access, including affordable internet services, digital literacy programs, and the provision of devices to underserved communities. Design of IT solutions for sustainability must consider the diverse needs of different populations, ensuring that technological solutions do not perpetuate existing social disparities.

Data Privacy and Security: As IT becomes more integral to achieving sustainability goals, the collection and use of data have become crucial. Data is at the heart of many sustainability initiatives, from smart city infrastructure that optimizes energy use to precision agriculture systems that enhance resource efficiency. However, the increasing reliance on big data, IoT devices, and AI-driven solutions raises significant ethical concerns about privacy and data security.

One of the key challenges is how to balance the use of personal and sensitive data for the greater good while safeguarding individuals' privacy rights. For instance, data collected from IoT devices in smart homes or wearable technologies can provide valuable insights into resource usage, environmental impact, and public health. However, this data can be exploited or misused if not handled responsibly. The potential for data breaches, surveillance, and the erosion of privacy is a significant ethical challenge that needs to be addressed through robust data protection laws and practices. The ethics of consent are also crucial for data collection for sustainability purposes. Individuals may not be fully aware of the extent to which their data is being collected, processed, or shared. Ethical IT practices in sustainability require transparency in how data is gathered, how long it is retained, and for what purposes it is used. Ethical responsibility is to ensure that data is anonymized and aggregated to protect individual identities and prevent exploitation.

AI, Automation, and Job Displacement: Artificial intelligence (AI) and automation technologies are central to many sustainability initiatives being developed today, from optimizing supply chains to reducing energy consumption in manufacturing. However, these technologies also present significant ethical challenges, particularly regarding their impact on employment and social equity.

AI and automation can potentially displace many workers, particularly in manufacturing, transportation, and customer service. The shift towards automated systems may lead to job losses, particularly for low-skilled workers, exacerbating income inequality and social unrest. The ethical dilemma here is how to ensure that the benefits of AI and automation are shared broadly across society, rather than concentrating wealth and power in the hands of a few. There is also the issue of ensuring that the transition to an AI-driven economy is fair and just. Governments, businesses, and other stakeholders must consider how displaced workers can be retrained for new roles in a more technologically advanced economy. Sustainable economic practices require a commitment to social equity, and ethical IT solutions for sustainability must involve strategies for mitigating job displacement, fostering reskilling programs, and ensuring that the labor market remains inclusive.

Accountability and Governance: As IT systems become more integrated into sustainable development efforts, the ethical challenge of accountability becomes increasingly important. Who is responsible when a technological solution designed to promote sustainability fails or has unintended negative consequences? For instance, when an AI algorithm used in agriculture or urban planning results in unfair outcomes or exacerbates existing inequalities, it is critical to determine who is accountable for these consequences. The ethical challenge lies in establishing clear governance frameworks that ensure accountability for the actions of IT systems. In many cases, technology developers, governments, and private companies are all stakeholders in the deployment and management of IT systems. However, the decentralized nature of many IT systems, particularly blockchain and decentralized finance, complicates the identification of responsible parties. Ethical governance frameworks must prioritize transparency, traceability, and oversight to ensure that IT systems are used responsibly and in ways that align with sustainability goals.

Unintended Consequences of Technological Advancements: Another ethical challenge in IT and sustainability is the potential for unintended consequences. While IT solutions for sustainability, such as renewable energy technologies, smart grids, and automated resource management systems, hold great promise, their long-term effects may not always be predictable. For example, the widespread adoption of electric vehicles (EVs) may reduce carbon emissions in the transportation sector but could also lead to increased demand for electricity, potentially straining energy grids that rely on non-renewable sources. Similarly, the growth of the Internet of Things (IoT) in smart cities may result in more efficient resource management, but it may also increase e-waste, data privacy risks, and system vulnerabilities. The ethical challenge here is to anticipate and mitigate potential negative consequences through responsible design, testing, and regulation. This requires a proactive approach to risk assessment, stakeholder engagement, and long-term planning to ensure that unforeseen impacts do not undermine the benefits of IT-driven sustainability.

Ethical Dilemmas in Data Use, Privacy, and AI Governance

The rapid advancement of data-driven technologies, particularly artificial intelligence (AI), has revolutionized various sectors, ranging from healthcare and finance to governance and transportation. While the potential benefits of these technologies in improving efficiency, decision-making, and overall societal well-being are vast, they have also introduced a set of profound ethical dilemmas that need careful examination. Among these, the ethical challenges of data use, privacy, and AI governance are particularly pressing. In the current scenario, the pace at which these technologies are evolving far outpaces the development of regulatory frameworks, leading to concerns about misuse, bias, and the erosion of personal privacy. The ethical implications of AI and data use are multifaceted, encompassing issues related to data collection, consent, surveillance, discrimination, and accountability.

Ethical Challenges in Data Collection and Consent: The foundational element of AI is datalarge, high-quality datasets that power machine learning algorithms and enable AI systems to make predictions and decisions. However, the collection of this data raises significant ethical concerns. Data is often gathered through online platforms, social media, and various IoT devices without users' informed consent, leading to questions about the autonomy of individuals over their personal information. While some AI systems rely on openly available data, many are powered by private or proprietary data sources. In these cases, the issue of consent becomes complex: Are users fully aware of the data being collected and how it will be used? Can they meaningfully opt-out of data collection without losing access to services?

In many instances, individuals unknowingly surrender vast amounts of personal data to companies, using this information to develop AI models. For example, digital assistants, social media platforms, and fitness trackers often collect sensitive data such as location, communication, and health information. The lack of transparency around data use raises critical ethical questions about data ownership and individuals' rights. Consent mechanisms often rely on opaque privacy policies that are difficult for the average user to understand, leading to concerns about exploiting personal information for commercial gain. One of the key ethical dilemmas here is the tension between the value of data for technological innovation and the right of individuals to control their personal information. As data-driven technologies become increasingly pervasive, the need for clearer, more accessible consent processes becomes urgent. Ethical AI development calls for ensuring that users are not only asked for consent but are fully informed about the nature and scope of data collection, its potential use, and the risks involved.

Privacy and Surveillance in the Digital Age: As AI and data collection become more sophisticated, the ability to track individuals in real-time and analyze their behavior has expanded dramatically. One of the most significant ethical concerns is the balance between leveraging data for societal benefitssuch as improving public health, transportation, or urban planningand respecting individual privacy rights. Growing concerns over surveillance by governments and private corporations characterize the current scenario. With the proliferation of cameras, sensors, and AI-powered monitoring systems, the line between public safety and personal privacy is increasingly blurred. For instance, AI-driven surveillance systems in smart cities use facial recognition and other biometric technologies to monitor citizens' movements. While proponents argue that such systems can enhance security and streamline urban management, they also raise serious ethical questions about consent, anonymity, and potential misuse. In several cases, facial recognition has been used to monitor protesters, track dissidents, or maintain social control, presenting significant risks to freedom of expression and personal privacy.

The ethical issue of "surveillance capitalism" arises, where private companies gather vast amounts of personal data to sell targeted advertisements or influence consumer behavior. These practices can be seen as an infringement on individuals' autonomy, as they subtly shape decisions without full disclosure or consent. The power imbalance between large corporations that control data and the individual users who provide it further exacerbates concerns about exploiting personal information. Ethical governance demands clear regulations around data collection, transparency, and accountability for public and private sector surveillance.

Bias and Discrimination in AI Systems: Another critical ethical dilemma in the current scenario is the issue of bias in AI systems. Machine learning algorithms, which power many AI systems, are trained on data that often reflect historical inequalities, societal biases, and discriminatory practices. If these biases are not adequately addressed, AI systems can perpetuate and even exacerbate these issues. For instance, AI algorithms used in hiring processes have been shown to favor candidates from certain demographic groups over others, often discriminating against women, people of color, or individuals from marginalized communities. Similarly, AI systems used in the criminal justice system, such as predictive policing tools or risk assessment algorithms, have been found to disproportionately target minority groups, contributing to racial profiling and reinforcing existing societal inequalities.

These biases are often embedded in the data used to train AI models, but they can also stem from biased design choices made by developers or a lack of diversity in development teams. The ethical dilemma here lies in the fact that AI systems, which are often viewed as objective and impartial, can reflect and even amplify human biases, resulting in unfair outcomes. Using biased AI systems raises significant questions about accountability: Who is responsible when an AI system makes a biased decision? Is it the developers, the organizations deploying the technology, or the AI itself? The challenge of addressing bias in AI is twofold: first, it requires more inclusive and diverse data collection practices, ensuring that the datasets used to train AI systems are representative and free from historical biases. Second, it demands ethical AI design principles prioritising fairness, transparency, and accountability. AI systems must be subject to rigorous audits and oversight to detect and mitigate bias at all development and deployment stages.

Accountability in AI Decision-Making: As AI systems are increasingly used to make high-stakes decisions in healthcare, finance, law enforcement, and hiring, accountability becomes critical. AI models, particularly deep learning algorithms, are often described as "black boxes" because their decision-making processes are not easily understood, even by the developers who create them. This lack of transparency creates a significant ethical challenge when AI systems make life-altering decisions, such as determining creditworthiness, diagnosing diseases, or sentencing individuals in criminal trials. In the current scenario, it is unclear who should be held accountable when an AI system makes a harmful or erroneous decision. Is it the developer who designed the algorithm, the company that deployed it, or the AI system itself? Ethical governance in AI requires that responsibility be clearly defined and mechanisms be implemented to challenge AI decisions and seek redress for affected individuals. In addition, the use of AI in decision-making must be transparent, with clear explanations of how decisions are made and the data used to support them.

Ethical AI Governance and Regulation: Given AI systems' increasing power and influence, the need for ethical governance and regulation has become more urgent. While governments and international bodies are beginning to develop frameworks for regulating AI and data use, current regulations often lag behind the rapid development of technology. Without clear and enforceable guidelines, companies may prioritize innovation and profit over ethical considerations, potentially harming individuals and society.

The European Union's General Data Protection Regulation (GDPR) represents a significant step in regulating data use and privacy. However, similar frameworks are needed globally to ensure that AI systems are developed and deployed in a way that respects human rights, fairness, and transparency. The challenge lies in striking the right balance between encouraging technological innovation and ensuring it aligns with ethical principles protecting individuals and communities. AI governance must involve diverse stakeholders, including technologists, ethicists, policymakers, and representatives from affected communities, to ensure that AI systems are designed and used ethically, inclusively, and equitably. The future of AI governance will depend on creating a global consensus on the ethical principles that should guide AI development and ensuring that these principles are embedded into the design, deployment, and regulation of AI systems.

Privacy and Security Concerns in Digital Bharat

The advent of Digital Bharat, a vision to harness the power of technology for national growth and citizen empowerment, has brought about transformative changes in India's economy, governance, and society. As the country progresses towards becoming a digitally connected society, many opportunities and challenges arise, particularly in privacy and security. India's digital landscape, supported by initiatives like the Digital India program, relies heavily on vast amounts of personal data collected, processed, and analyzed to deliver services efficiently. However, as the digitalization of services expands, it introduces significant privacy concerns, data protection issues, and the growing risk of surveillance. This delicate balance between the potential benefits of digital transformation and the risks posed by inadequate data protection and security mechanisms is an urgent area of focus.

Data Protection Challenges in Digital Bharat: Data has become one of the most valuable assets. Data collection, use, and sharing drive many services underpin the digital economy, from government welfare schemes to private sector innovations like fintech, e-commerce, and health-tech. However, India's data protection laws remain underdeveloped compared to global standards. The Personal Data Protection Bill (PDPB), which was introduced in 2019, aims to provide a comprehensive framework for safeguarding citizens' data, but it is still under deliberation. Until then, the absence of a robust and enforceable legal framework has led to concerns about how sensitive personal data is being collected, stored, and used by various private companies and government entities.

One of the key issues is the lack of transparency around data collection practices. Often, data is collected without explicit, informed consent from the individuals involved. Platforms offering digital services tend to collect far more data than is necessary for the service, leading to the risk of misuse or unauthorized access. Data collected in India is frequently stored in servers outside the country, raising concerns about data sovereignty and the risks of foreign access to Indian citizens' data. This situation exposes India to the vulnerabilities of a global data ecosystem where cross-border data flows are poorly regulated.

The challenge is compounded by the potential data misuse by both public and private entities. For instance, data collected for one purposesuch as for the distribution of welfare benefitscan be repurposed or sold to third parties without the consent of the data subjects. This lack of accountability and transparency in data practices contributes to the erosion of trust in digital systems. Additionally, the scale and scope of data breaches, as seen in several global incidents, indicate that India is equally vulnerable to cyber-attacks that may compromise sensitive data, including financial, medical, and biometric information. The demand for robust data protection measures is urgent, and data localization laws could be one avenue to address concerns over cross-border data flow. However, data localization has challenges, such as adequate infrastructure and international trade implications. Consequently, a balanced, nuanced approach to data protection that considers technological, economic, and legal aspects is essential.

Surveillance Concerns in Digital Bharat: Surveillance has become an intrinsic feature of the digital age, with governments and private companies utilizing technology to monitor, analyze, and sometimes control individuals' behaviors. With its expansive push towards digital governance and public welfare schemes, India faces significant surveillance concerns, particularly as initiatives like Aadhaar (the world's largest biometric database) are used for identity verification across various services. While the Aadhaar system has brought about positive changes by improving access to services like subsidies and healthcare, it has also raised concerns about the extent of surveillance that the government is capable of and its potential implications for individual privacy.

The government's ability to track citizens through digital systems presents an ethical dilemma regarding the right to privacy versus the need for governance. India has no constitutional right to privacy (until the Supreme Court's ruling in 2017, which read privacy as a fundamental right under the Right to Life and Personal Liberty), which has led to debates about how much surveillance is justified for security and service delivery. The increasing use of facial recognition technology by law enforcement and the growing number of CCTV cameras in urban spaces heighten the concerns around surveillance. While these technologies are often justified to enhance public safety, their deployment without adequate safeguards creates a potential for misuse, especially in cases where data is not anonymized or where surveillance is used to target specific communities, activists, or political opponents. Additionally, the government's increased use of AI and machine learning to analyze vast amounts of data collected through digital platforms has raised concerns about its role in monitoring and controlling public behavior. AI-powered systems are increasingly used in areas such as public health, transportation, and policing, but the lack of transparency in their functioning creates a serious risk of mass surveillance. If left unchecked, these systems could evolve into tools for social control, limiting individual freedom and creating a dystopian society of constant surveillance.

The intersection of privacy and surveillance in Digital Bharat requires a comprehensive framework focusing on data protection and the ethical deployment of surveillance technologies. There needs to be stringent laws that regulate the use of AI and biometric technologies, ensuring that these tools are not misused for unjustified surveillance and discrimination.

Privacy and AI Bias: While the advent of AI technologies holds significant promise for various sectors, such as healthcare, education, and law enforcement, it also introduces a new set of privacy concerns. One of the critical ethical dilemmas in AI is the potential for biased decision-making. AI systems, which are often trained on historical data, can unintentionally reflect and perpetuate the biases present in society. In India, where issues of caste, gender, and economic disparity are deeply ingrained, AI systems have the potential to amplify these biases, leading to discriminatory outcomes in areas like hiring, loan approvals, or criminal justice.

For example, AI-driven welfare schemes, there is a risk that marginalized communities could be excluded or unfairly treated based on biased algorithms. Similarly, facial recognition technologies, which are deployed in public spaces or for identity verification purposes, have been shown to have higher error rates for women, older people, and ethnic minorities. These biases can lead to privacy and civil liberties violations, particularly for those already vulnerable in society.

The challenge of AI bias in privacy concerns is exacerbated by the lack of transparency in how AI algorithms are designed and deployed. These systems are often "black boxes," making it difficult for individuals to understand how their data is being processed or how decisions are being made. Without clear explanations for the outcomes generated by AI systems, individuals may feel powerless to challenge discriminatory decisions or appeal for rectification. Inclusive design practices and transparent AI models that prioritize fairness and accountability are needed to address AI bias for privacy concerns. Ethical guidelines must be developed for AI systems that operate in sensitive areas, ensuring that the privacy of marginalized or vulnerable populations is protected.

Data Security and Cyber Threats: As Digital Bharat becomes more reliant on technology, the risk of cyber threats and data breaches also grows. Cybersecurity has become a critical concern for India's public and private sectors, as sensitive personal, financial, and health-related data is increasingly stored online. The recent surge in cyber-attacks, ranging from ransomware to large-scale data breaches, has exposed vulnerabilities in India's digital infrastructure. Government services, banking systems, e-commerce platforms, and healthcare systems are all frequent targets for cybercriminals seeking to exploit weaknesses in cybersecurity frameworks.

In the current scenario, the lack of robust data encryption, inadequate security protocols, and a shortage of skilled cybersecurity professionals contribute to the growing threat of cybercrime. For instance, incidents like the 2020 breach of Aadhaar data, where the personal details of over a billion people were allegedly compromised, underscore the potential risks posed by inadequate data security measures. Such breaches can have far-reaching consequences, leading to identity theft, financial fraud, and the compromise of national security. To mitigate these risks, there is an urgent need for stronger data security measures, including the adoption of end-to-end encryption, the implementation of multi-factor authentication, and the development of a comprehensive cybersecurity strategy that spans the entire digital ecosystem. At the same time, organizations must prioritize data protection and invest in building robust security infrastructures that can withstand evolving cyber threats.

Developing Ethical IT Frameworks

As Information Technology (IT) continues to play an increasingly central role in shaping societies, economies, and industries, the development and implementation of ethical IT frameworks have become imperative. Ethical considerations in IT, particularly in Artificial Intelligence (AI) and data use, ensure that technological advancements benefit society while mitigating the risks associated with privacy violations, discrimination, and unintended harm. Developing policies for fair AI and data use, AI governance, and responsible innovation is crucial to fostering a technological landscape that prioritizes societal well-being and fairness. Additionally, there is a need for global standards for responsible IT development to create a cohesive and interoperable framework for ethics across borders. This section explores the importance of ethical frameworks in IT, particularly in AI, and examines the need for policies, governance structures, and global standards to ensure responsible development and deployment of technology.

Policies for Fair AI and Data Use: Artificial Intelligence, powered by vast amounts of data, is becoming ubiquitous in almost every sector, from healthcare and education to finance and law enforcement. However, the immense potential of AI comes with significant ethical concerns. AI systems are highly dependent on data to function, and the quality and nature of this data significantly impact the fairness, transparency, and accuracy of AI-driven decisions. As AI systems increasingly influence critical decisions, such as hiring, loan approval, and criminal justice outcomes, ensuring fair and ethical use of AI is paramount. One of the central ethical concerns in AI development is bias in data and algorithms. AI systems are often trained on historical data, which may contain implicit biases that reflect societal inequalities. For instance, if historical hiring data predominantly favors a particular demographic group in recruitment, AI systems trained on this data may perpetuate gender or racial biases. This can lead to discriminatory outcomes that exacerbate existing social inequities. AI policies must enforce fairness standards to address this, ensuring that AI models are transparent and auditable and regularly tested for biases before deployment. In addition, AI should be designed to promote equitable outcomes and sensitive to diverse populations, avoiding reinforcing stereotypes or social divisions.

Data privacy and consent are crucial elements of any ethical AI framework. Data used to train AI systems often includes personal, sensitive, and confidential information. Clear guidelines and regulations around data collection, storage, and processing are essential to protect individual privacy and uphold the fundamental right to consent. In particular, data anonymization practices and data minimization principles must be central to AI policies, ensuring that only the most relevant data is used for AI applications and handled securely. Governments, industry leaders, and academics must collaboratively define and enforce data protection laws that prioritize user consent and mitigate the risk of unauthorized data usage or breaches.

Regulations such as the General Data Protection Regulation (GDPR) in Europe serve as a benchmark for data protection, and similar policies must be crafted in other regions to safeguard the rights of individuals in the digital age. The Indian Personal Data Protection Bill, once passed, will play a pivotal role in shaping the ethical use of AI in India by providing guidelines on data protection, data rights, and the ethical use of AI. However, a comprehensive AI ethics policy that covers the entire lifecycle of AI systemsfrom design and development to deployment and monitoringis needed to ensure fairness, accountability, and transparency in the system's operations.

AI Governance and Responsible Innovation: AI governance is vital in ensuring responsible development and deployment of AI technologies. Governance refers to the frameworks, laws, regulations, and standards that guide AI systems' design, development, and operation. With AI playing an increasing role in critical sectors like healthcare, security, and finance, establishing robust governance structures ensures that AI is deployed to maximize its benefits while minimizing potential harms. Effective AI governance should be built on transparency, accountability, and inclusivity. Transparency involves making the processes and decision-making frameworks behind AI systems open and understandable to both experts and the general public. One of the challenges with AI is its "black box" naturethe inability of individuals to understand how a particular decision was made. To address this, AI governance frameworks must demand that organizations develop and deploy explainable AI models. This will enable users, stakeholders, and regulators to trace the origins of AI-generated decisions, ensuring they align with ethical standards and public expectations.

Accountability is equally important in AI governance. Developers and organizations deploying AI systems must be held responsible for the decisions these systems make. In particular, liability frameworks must be established to clarify who is responsible for cases of harm caused by AI systemswhether it be the developers, the organizations, or the users of the AI system. This will prevent AI from becoming a tool of evasion, where harmful decisions are attributed to the technology rather than the individuals who created or deployed it. In the event of errors or unintended consequences, AI governance should also include remediation and corrective action provisions to ensure that wrongdoings are addressed and affected parties are compensated or redressed. Inclusivity in AI governance ensures that AI systems are designed and used to benefit all segments of society, including marginalized groups. Inclusivity also extends to the governance processes themselvesensuring that diverse stakeholders, including ethicists, sociologists, economists, and human rights experts, participate in decision-making about AI regulations and policies. This broad and multi-disciplinary approach will ensure that AI systems do not inadvertently favor the interests of specific groups at the expense of others.

Responsible innovation in AI is the counterpart to governance. It involves adopting innovation strategies that focus on advancing technology and consider its long-term societal impact. Innovators and developers must proactively identify and address potential risks associated with AI applications. This means that companies and researchers should engage in anticipatory ethics instead of merely reacting to issues as they arise, carefully evaluating AI deployment's potential social, economic, and political consequences. Responsible innovation also requires adherence to sustainability principlesensuring that AI technologies are developed in environmentally sustainable ways that contribute to broader social goals like the Sustainable Development Goals (SDGs).

Global Standards for Responsible IT Development: As digital transformation accelerates globally, there is a growing recognition that ethical IT development, especially in AI and data-driven systems, should be guided by global standards. These standards are necessary to address the cross-border nature of technology and its far-reaching implications. Without international collaboration and alignment, ethical issues related to AI, data protection, and privacy could become fragmented, with countries and regions adopting inconsistent and conflicting policies. This would create challenges for businesses operating in multiple jurisdictions, consumers whose data crosses borders, and governments tasked with regulating global technologies.

International bodies, such as the United Nations, OECD, World Economic Forum, and ISO, have begun to draft frameworks and guidelines that promote responsible IT development. For instance, the OECD AI Principles emphasize fairness, transparency, and accountability in AI systems. Similarly, the Global Partnership on AI (GPAI) is an initiative launched by several countries which aims to support the development of AI systems that are both human-centric and sustainable. These international frameworks guide countries in developing national AI policies and regulations aligned with global norms yet adaptable to local contexts.

The European Union (EU) has been a leader in creating comprehensive standards for AI governance, with the EU AI Act being one of the first attempts to regulate

AI at the level of a supranational body. The Act categorizes AI systems according to risk and outlines specific obligations for high-risk AI applications. For example, AI systems used in healthcare, law enforcement, or employment must adhere to strict transparency, data protection, and safety standards. This approach offers a valuable model for other countries seeking to develop their regulatory frameworks. However, these international guidelines should be flexible enough to accommodate regional AI development and deployment variations.

The development of global ethical standards for IT is about ensuring fairness and accountability in AI and fostering international cooperation. AI and IT technologies are fundamentally transnationalthey cross borders and affect global markets. Developing a shared set of principles for responsible AI ensures that technology continues to be a force for good, mitigating the risks associated with the digital divide, inequality, and technological monopolies.

Towards a Responsible Future for AI and IT Development: The increasing ubiquity of AI and digital technologies in modern society necessitates the establishment of robust ethical IT frameworks that ensure responsible, fair, and transparent use. Policies for fair AI and data use, strong AI governance mechanisms, and the development of global standards are essential for achieving these goals. As AI systems become more integrated into everyday life, they must operate in ways that respect individual rights, promote fairness, and benefit society. A comprehensive, collaborative approach involving governments, industry, and international organizations is needed to establish ethical standards governing AI and other IT technologies, ensuring their development and use contribute to a just, equitable, and sustainable future.

Future Scenario

Integrating Artificial Intelligence (AI) and Information Technology (IT) into almost every facet of modern life, from healthcare and finance to education and governance, brings immense potential for progress. However, the growing role of these technologies also presents significant ethical challenges that must be addressed to ensure they are developed and deployed in ways that benefit humanity as a whole. As we move into the future, the importance of establishing robust, coherent, and universally accepted frameworks for ethical AI and IT development will become even more pronounced. These frameworks must encompass technical considerations and broader societal, environmental, and economic factors. The transnational nature of AI and IT technologies, the future of ethical development in this domain also hinges on establishing global agreements on digital ethics and sustainability. In this section, we will explore the emerging scenario of ethical AI and IT frameworks, the need for global cooperation on digital ethics, and the role of sustainability in ensuring that AI technologies contribute to societal well-being and environmental protection.

Frameworks for Ethical AI and IT Development: The future of AI and IT development will undoubtedly be shaped by the creation of frameworks that guide how these technologies are designed, deployed, and governed. At the core of these frameworks will be ethical principles that ensure AI and IT innovations are aligned with human rights, fairness, accountability, transparency, and sustainability. Given the pervasive influence of AI across sectors, frameworks must be comprehensive and adaptable to various contexts, ranging from governance and education to health care and the environment. A key component of these frameworks will be the promotion of fairness. In AI systems, fairness concerns typically revolve around avoiding bias in data and algorithmic decision-making. For instance, AI systems used for hiring, credit scoring, or criminal justice must be free from gender, racial, and socioeconomic biases. Future frameworks will likely require more diverse, representative data and the implementation of bias detection and mitigation tools throughout the AI development lifecycle. It will become standard practice for AI systems to undergo regular audits to assess and address potential discriminatory outcomes, ensuring they function equitably for all users, regardless of demographic characteristics.

Another critical element will be transparency. As AI models become more complex, there is an increasing need for transparency in design, functioning, and decision-making processes. While AI systems, particularly deep learning models, are often criticized for their "black-box" nature, future ethical frameworks will demand more explainable AI (XAI) development. This means that stakeholdersincluding users, regulators, and developersmust understand the reasoning behind AI decisions. XAI tools will likely become standard in AI systems, and there will be an emphasis on providing clear, accessible explanations of how AI reaches conclusions. This transparency is crucial for building trust and ensuring that AI systems are held accountable for their actions. Accountability will be another pillar of future ethical AI frameworks. As AI systems take on more decision-making responsibilities, it will become essential to establish clear lines of accountability. In cases where AI causes harm or makes erroneous decisions, systems must be in place to determine who is liablewhether the developers who created the algorithm, the companies deploying the system, or others involved in the AI lifecycle. Future frameworks will also emphasize human-in-the-loop (HITL) mechanisms, where AI systems are designed with oversight and control mechanisms to allow human intervention when necessary. These mechanisms will help ensure that critical decisions are not entirely automated and that ethical considerations are consistently incorporated.

Lastly, sustainability will be essential to future AI and IT frameworks. AI technologies, especially large-scale machine learning models, require significant computational resources and energy, raising concerns about their environmental impact. Future frameworks will incorporate guidelines for the environmental sustainability of AI technologies, advocating for energy-efficient algorithms, green data centers, and reducing the carbon footprint of AI models. Sustainability considerations will extend to the long-term societal impact of AI, ensuring that technologies are developed in ways that promote social equity, job creation, and environmental conservation.

Global Agreements on Digital Ethics and Sustainability: As AI and IT technologies transcend national boundaries, global cooperation in digital ethics becomes increasingly urgent. No single country or organization can fully address the ethical challenges posed by these technologies alone. A coordinated, global approach is necessary to ensure that AI and IT developments are ethically sound and benefit humanity. Global digital ethics and sustainability agreements will be crucial in this effort, providing universal standards and guidelines for countries, organizations, and developers. One of the central themes of future global agreements will be protecting human rights for AI and IT. As these technologies evolve, they will increasingly impact fundamental human rights, such as the right to privacy, freedom of expression, and non-discrimination. Global agreements must establish universal principles for safeguarding these rights in the digital age. For example, ensuring that AI-driven surveillance technologies respect individuals' right to privacy and that data used in AI systems is gathered and processed with explicit, informed consent will be key components of any global agreement.

Data protection and privacy regulations will be central to these global agreements. The General Data Protection Regulation (GDPR) in the European Union has set a high data protection standard, but new privacy challenges will arise as AI and IT technologies evolve. Global agreements will likely call for common data protection standards, such as data minimization, anonymization, and secure data storage practices. They will also emphasize the importance of individuals' right to access and control their data and opt out of unnecessary data collection. These agreements will prevent data misuse and exploitation while encouraging responsible innovation in using data for AI.

In addition to data protection, AI governance will be a central feature of global agreements. These agreements will outline the roles and responsibilities of various stakeholders, including governments, tech companies, and civil society, in developing and deploying AI technologies. Global governance mechanisms will ensure that AI systems are transparent, fair, and accountable and that the benefits of AI are widely distributed. Global agreements will likely establish an international AI ethics board or an analogous body overseeing AI deployment, monitoring adherence to ethical principles, and issuing guidelines or recommendations for responsible AI use.

Environmental sustainability will also be a key aspect of global agreements on digital ethics. As AI and IT technologies grow, they will have an increasing environmental footprint. Data centers, AI training processes, and network infrastructures consume vast amounts of energy, and without mitigation strategies, the growth of digital technologies could exacerbate climate change. Global agreements will address these concerns by setting clear standards for energy-efficient computing, low-carbon technologies, and sustainable supply chains in the IT and AI sectors. The responsible disposal of electronic waste (e-waste) and efforts to reduce the environmental impact of the tech industry will be central to these agreements.

One of the challenges of global digital agreements will be ensuring global inclusivity. Different countries and regions have varying levels of technological development, and global agreements must account for these disparities. Future agreements must address the digital divide and ensure that developing countries are not left behind in the digital transformation. This may include provisions for capacity-building in AI and IT education, technology transfer, and affordable access to technology. Global agreements should prioritize the inclusion of marginalized groups, ensuring that they benefit from digital technologies rather than being excluded from their advantages.

Role of Multi-Stakeholder Cooperation: While governments and international organizations will play a critical role in shaping global agreements on digital ethics and sustainability, multi-stakeholder cooperation will be key to their success. Governments, tech companies, academic institutions, civil society organizations, and international bodies must collaborate to develop, implement, and enforce these agreements. Tech companies, in particular, will have a critical role, as they are responsible for developing and deploying AI systems. They must commit to ethical practices in their design, deployment, and maintenance of AI technologies, and they should engage in transparency and accountability concerning the societal impacts of their products. Collaboration between stakeholders will also be essential in fostering innovation with ethics. Instead of focusing solely on accelerating technological progress, stakeholders must be committed to advancing AI and IT in ways that respect ethical standards, sustainability goals, and social equity. This collaboration will require ongoing dialogues, research, and new tools and frameworks that align technology development with public values and goals.

Path Forward for Ethical AI and IT Development: As AI and IT technologies continue to advance, the need for ethical frameworks and global agreements will only intensify. The future will require comprehensive, adaptive, and globally aligned policies that ensure AI and IT are developed and deployed responsibly, focusing on human rights, sustainability, fairness, and transparency. By establishing strong, universal standards for ethical AI and IT development and fostering multi-stakeholder cooperation, we can ensure that these technologies contribute positively to society, the economy, and the environment. The future of AI and IT will depend not just on technical breakthroughs but on how effectively we can navigate the ethical challenges posed by these technologies and create a more just, equitable, and sustainable digital world.

17

Measuring IT's Impact on Sustainability

Introduction

Information Technology (IT) has emerged as a cornerstone of modern economic, social, and environmental systems, pivotal in driving innovation and sustainability across various sectors. As the world grapples with the urgency of achieving sustainable development, IT offers powerful tools to monitor, analyze, and optimize resource use, reduce environmental footprints, and promote social and economic equity. However, measuring the impact of IT on sustainability requires a nuanced understanding of its diverse contributions, challenges, and long-term consequences. It is essential to assess how IT can simultaneously support environmental conservation, economic growth, and social well-being, often called the three pillars of sustainability.

Role of IT in Environmental Sustainability: IT's role in promoting environmental sustainability is multifaceted. From optimizing energy consumption in industrial processes to enabling remote work that reduces commuting emissions, IT offers solutions to reduce resource use and environmental harm. However, quantifying its impact on sustainability requires evaluating both the **direct** and **indirect** effects of IT.

Direct Impact: The direct impact of IT on sustainability primarily relates to the energy consumption of information systems, including data centers, communication networks, and devices. As IT infrastructures grow globally, the energy demands of computing and digital storage have become significant. Data centers, in particular, consume vast amounts of electricity for both computing processes and cooling mechanisms. Research suggests that the global energy consumption of data centers alone accounts for nearly 1-2% of total electricity use. Measuring the environmental impact requires assessing the **energy efficiency** of these data centers, the **source of energy** (renewable or non-renewable), and advancements in technologies to reduce energy consumption, such as **cloud computing** and **green IT** solutions. For example, the transition to **serverless computing** and **cloud-based services** allows companies to pool their computing resources, optimize energy use and reduce the overall environmental footprint.

Indirect Impact: The indirect impacts of IT are perhaps even more significant when considering how digital technologies enable sustainability across industries.

For instance, the Internet of Things (IoT) can monitor real-time environmental variables such as air quality, water consumption, and soil health. This data can drive more efficient use of resources, reducing waste and supporting informed decision-making. IT-enabled systems in agriculture, such as precision farming, allow farmers to optimize irrigation, use fewer pesticides, and reduce energy consumption, contributing to resource conservation and reduced environmental impact. To measure these indirect impacts, metrics can include the reduction in carbon emissions, decreased resource usage, and waste minimization resulting from the application of IT tools.

Role of IT in Economic Sustainability: IT's contribution to economic sustainability centres on creating efficiencies, improving productivity, and fostering innovation that leads to sustainable growth. Key to this is the ability of IT to optimize processes, increase transparency, and enhance supply chain management. Technologies like blockchain have enabled more transparent and efficient financial transactions, ensured fair trade practices, and fostered sustainability in global supply chains. For instance, blockchain has been employed to track the sourcing of raw materials, ensure that they are responsibly obtained, and reduce waste along the supply chain. To measure IT's economic impact on sustainability, efficiency gains and cost savings from adopting digital tools can be analyzed. Metrics such as reduced operational costs, improved resource utilization, and increased output per input unit indicate economic sustainability. Automation technologies enabled by AI and IT have streamlined manufacturing processes, leading to lower labor costs and more sustainable production practices.

Digitalization also plays a key role in financial inclusion, providing individuals in remote or underserved regions access to banking and credit systems. IT has empowered small-scale businesses and entrepreneurs by enabling mobile banking, digital wallets, and peer-to-peer lending platforms, fostering economic sustainability in low-income and marginalized communities. The financial accessibility offered by these digital solutions can be measured through metrics like increased access to financial services, growth in microfinance loans, and economic upliftment of previously underserved populations.

Role of IT in Social Sustainability: In social sustainability, IT contributes significantly to improving quality of life, education, health, and equity. Digital tools have enhanced access to education through e-learning platforms, expanded healthcare access via telemedicine and health monitoring devices, and fostered social inclusion by providing marginalized communities with digital connectivity.

For instance, telemedicine has revolutionized healthcare delivery, particularly in remote and underserved areas, allowing patients to access medical consultations without travelling long distances. Similarly, online education platforms enable students in rural or low-income areas to receive quality education, promoting social equity. To measure the social impact of IT on sustainability, indicators such as improvement in health outcomes, reduction in educational disparities, and increase in digital literacy can be used. These metrics provide insights into the role IT plays in enhancing social sustainability and promoting social equity.

IT's impact on gender equality and empowerment must be measured regarding the opportunities it provides for women and marginalized groups. Initiatives like digital literacy programs and e-commerce platforms can help bridge the gender gap in employment and entrepreneurship. Metrics for social sustainability in this area may include increased female digital participation, growth in women-led businesses, and reduced gender disparities in employment and education.

Evaluating IT's Impact on Sustainability: To evaluate IT's impact on sustainability effectively, it is crucial to establish comprehensive evaluation frameworks, including qualitative and quantitative indicators. These frameworks should consider the three pillars of sustainability: environmental, economic, and social, ensuring that all dimensions are accounted for in the assessment process.

Environmental Metrics: The environmental impact of IT can be measured using various metrics, including:

- **Energy consumption**: Amount of energy used by IT infrastructures such as data centers, networks, and devices.
- **Carbon footprint**: The total greenhouse gas emissions associated with IT hardware and software production, use, and disposal.
- **Waste generation**: Volume of electronic waste (e-waste) produced by outdated or discarded digital devices and the effectiveness of recycling programs.
- **Resource efficiency**: The extent to which IT optimizes resource use, such as water, energy, and raw materials in industries like agriculture and manufacturing.

Economic Metrics: For economic sustainability, relevant indicators might include:

- **Operational efficiency gains**: Cost savings and productivity increases result from implementing IT solutions.
- **Digital economy growth**: The contribution of IT sectors to the national or global economy, particularly in emerging fields like FinTech, e-commerce, and cloud computing.
- **Job creation**: The extent to which IT leads to new job opportunities, particularly in high-tech industries and the gig economy.

- **Financial inclusion**: The increase in individuals or businesses accessing financial services through digital platforms.

Social Metrics: In assessing social sustainability, key metrics include:

- **Access to education**: The reach and effectiveness of digital education platforms in improving access to learning resources.
- **Health improvements**: The impact of telemedicine, mobile health apps, and digital health monitoring on healthcare access and outcomes.
- **Social equity**: Reducing disparities in technology access, particularly for marginalized and vulnerable communities.
- **Gender empowerment**: The impact of IT on promoting gender equality in employment, entrepreneurship, and education.

Defies in Measuring IT's Impact on Sustainability: While the potential for IT to drive sustainability is immense, measuring its impact is fraught with challenges. One major issue is the complexity of data collection and the need for reliable, standardized data across different sectors and regions. Without consistent data on resource consumption, emissions, and other sustainability indicators, it becomes difficult to assess the effectiveness of IT interventions. Additionally, the long-term impacts of IT on sustainability, for emerging technologies like AI, blockchain, and IoT, are still poorly understood. Research is needed to explore the unintended consequences of IT use, such as privacy concerns, job displacement, or increased energy consumption, and how these may affect sustainability goals. Another challenge is the need for integrated sustainability reporting frameworks that account for the interconnected nature of IT and sustainability. Many organizations and governments currently measure sustainability in isolated silos, focusing on environmental, social, or economic metrics separately. A more holistic approach that integrates these aspects is necessary to understand the full impact of IT on sustainable development.

Current Scenario

The importance of Key Performance Indicators (KPIs) and monitoring systems in assessing sustainability outcomes cannot be overstated. Both environmental and economic dimensions require robust tools for measuring progress, ensuring accountability, and informing strategic decision-making. The current landscape of sustainability assessment is characterized by an increasing emphasis on integrating technology, data analytics, and performance metrics to guide and monitor the progress of sustainability initiatives. This shift reflects the growing recognition that achieving sustainability goals requires policy formulation, continuous monitoring, data-driven insights, and transparent performance tracking.

Understanding KPIs for Sustainability: Key Performance Indicators (KPIs) are measurable values that help organizations track and assess the effectiveness of their sustainability efforts. They provide an objective basis for evaluating the success of sustainability strategies and initiatives across multiple sectors, including business operations, public policies, and environmental conservation programs. KPIs are critical for linking actions to outcomes, clearly showing whether specific sustainability goals are being met, and identifying areas that require improvement. These indicators are often tailored to the specific context of the organization or sector, considering industry-specific requirements and sustainability goals. In sustainability, KPIs typically encompass a broad range of metrics organized under three key pillars: environmental, social, and economic sustainability. The overarching aim is to create a comprehensive and integrated approach to sustainability assessment that is inclusive and adaptable to diverse sectors. Below, we explore the current state of KPIs and monitoring systems used to measure sustainability outcomes, focusing on how they are applied in practice.

Environmental Sustainability KPIs: Environmental sustainability KPIs are designed to measure the effectiveness of an organization's actions in reducing its environmental footprint. These indicators focus on resource consumption, emissions, waste management, and the overall impact on ecosystems. The effective use of KPIs in environmental sustainability enables organizations to track their progress toward reducing carbon footprints, improving energy efficiency, and managing waste. Some common environmental sustainability KPIs include:

- **Carbon Footprint**: This is one of the most widely used indicators for tracking greenhouse gas emissions. It is typically measured in terms of the amount of carbon dioxide ($CO2$) emitted per unit of energy produced or per unit of economic activity. Companies and governments often use this indicator to gauge their progress in meeting climate-related targets, such as those set under international agreements like the Paris Climate Accord.
- **Energy Consumption**: This KPI measures the total energy an organization or sector uses and is often segmented by renewable and non-renewable sources. Energy intensity, or energy consumption per production unit, is also an important metric in tracking energy efficiency.
- **Water Usage**: Given the increasing pressure on global freshwater resources, monitoring water consumption is critical for measuring the sustainability of industrial and agricultural practices. Water usage KPIs help track how much water is consumed and whether efforts to reduce waste, reuse, or recycle water are successful.
- **Waste Generation and Recycling Rate**: Waste management is a key concern in sustainability, especially regarding reducing landfill use and

promoting recycling. KPIs related to waste focus on the amount of waste generated, the proportion of recycled waste, and the effectiveness of waste diversion programs.

- **Biodiversity Impact**: KPIs that assess biodiversity impact consider how business operations affect ecosystems and wildlife. These metrics evaluate the preservation or degradation of natural habitats and wildlife populations due to industrial activities, land development, and resource extraction.

Social Sustainability KPIs: Social sustainability KPIs measure an organization's or government's efforts to improve quality of life, reduce inequality, and enhance access to basic services such as education, healthcare, and housing. These indicators reflect the social dimensions of sustainability, ensuring that development benefits are equitably distributed and that vulnerable populations are not left behind. Some key social sustainability KPIs include:

- **Income Equality**: A common social sustainability indicator is the **Gini coefficient**, which measures income inequality within a population. It helps determine how evenly or unevenly wealth is distributed, with a lower score indicating more equitable distribution.
- **Access to Healthcare**: This KPI assesses the availability and accessibility of healthcare services, including metrics like the **number of healthcare professionals** per capita, **healthcare infrastructure** availability, and **health outcomes**.
- **Educational Attainment**: KPIs related to education focus on literacy rates, enrollment in primary and secondary education, and the achievement of educational outcomes, particularly for marginalized groups. These metrics are vital in understanding social mobility and equity in a society.
- **Labor Rights and Fair Wages**: KPIs measuring labor rights track fair wages, working conditions, child labor, and worker rights, focusing on whether businesses adhere to ethical labor practices.
- **Community Engagement**: This KPI measures community participation in decision-making processes and how businesses or governments involve local populations in initiatives that impact their lives. It can be assessed through metrics such as **community feedback** or **level of civic engagement**.

Economic Sustainability KPIs: Economic sustainability KPIs assess the ability of a system, region, or country to maintain and grow its economic activities while ensuring that future generations can meet their own needs. These indicators are designed to measure economic growth, poverty reduction, financial inclusion, and the long-term viability of economic activities. Key economic sustainability KPIs include:

- **Gross Domestic Product (GDP) Growth**: This is a traditional economic indicator that measures the total value of goods and services produced within a country. While not a direct indicator of sustainability, GDP growth is often tracked alongside other sustainability metrics to assess the balance between economic development and environmental or social outcomes.
- **Poverty Rate**: The percentage of a population living below the poverty line is a crucial indicator of economic sustainability, highlighting the success or failure of inclusive economic policies and programs.
- **Employment Rate and Job Quality**: This KPI measures the proportion of the employed population and the quality of jobs available. Sustainable economic growth depends on creating jobs and ensuring that these jobs offer fair wages, security, and opportunities for advancement.
- **Financial Inclusion**: Measuring the accessibility of financial services, such as banking, credit, and insurance, is critical for assessing how well economic systems provide equal opportunities for wealth creation. Financial inclusion is a key driver of economic resilience and sustainability.
- **Public Debt and Fiscal Sustainability**: This KPI measures the level of public debt relative to national income (GDP). High public debt levels indicate economic instability, while well-managed public finances support sustainable economic growth.

Monitoring Systems and Data Integration: The effectiveness of KPIs in assessing sustainability outcomes depends significantly on robust monitoring systems and data integration. With the rise of digital technologies, organizations and governments increasingly turn to advanced data collection tools, IoT devices, and big data analytics to monitor sustainability metrics in real-time. These technologies enable more accurate and efficient tracking of KPIs, providing the data necessary to adjust strategies and ensure sustainability goals are achieved.

IoT and sensors play a crucial role in environmental monitoring, allowing real-time energy usage, water consumption, and tracking waste generation. This technology also enables the collection of localized data vital for assessing sustainability outcomes at the community or regional level. In agriculture, precision farming technologies use sensors to monitor soil health, moisture levels, and crop yields, optimizing resource use and minimizing waste.

Blockchain technology is another innovation that has the potential to improve transparency and accountability in sustainability efforts. By providing an immutable ledger of transactions, blockchain enables better tracking of the sustainability credentials of supply chains, ensuring that goods are sourced ethically and sustainably. Integrating artificial intelligence (AI) and machine **learning** enhances the ability to predict trends and make informed decisions. These technologies can process vast amounts of data to identify inefficiencies, forecast potential risks, and suggest corrective actions to improve sustainability outcomes.

Defies in Measuring Sustainability Outcomes: Despite the advances in monitoring systems and data analytics, several challenges remain in measuring sustainability outcomes effectively. One significant hurdle is the lack of standardized metrics. Sustainability assessments can vary widely depending on the organization, sector, or region, making comparing results across different contexts difficult. Inconsistent data collection methods, subjective interpretations of data, and a lack of global consensus on what constitutes a "sustainable" outcome further complicate the measurement process. Another challenge is the complexity of interconnected systems. Sustainability issues are inherently complex and interconnected, making it difficult to isolate the impact of individual actions. For example, energy efficiency improvements may indirectly affect water use, emissions, or waste generation, and these interrelationships must be carefully considered in any evaluation framework.

Lastly, the long-term nature of sustainability goals poses a challenge for monitoring systems. Many sustainability initiatives, particularly those that address climate change or restore ecosystems, require long-term monitoring to assess their effectiveness. Short-term performance metrics may not adequately capture the full scope of sustainability outcomes, necessitating a more holistic and integrated approach.

Defining Metrics for Success of KPIs for Sustainability and Digital Transformation

The increasing integration of Information Technology (IT) and digital systems into business operations and societal functions has profoundly impacted economic and environmental outcomes. As sustainability becomes an ever more pressing concern, the need to define clear, actionable Key Performance Indicators (KPIs) for sustainability and digital transformation has gained prominence. These metrics are critical in aligning an organization's strategic goals with its sustainability targets while fostering innovation and operational efficiency. KPIs are the bridge between high-level goalssuch as reducing carbon emissions or achieving zero wasteand day-to-day actions and decisions that shape a company's impact on society and the environment. Therefore, developing appropriate, robust KPIs for sustainability and digital transformation is essential to ensure these goals are set and effectively tracked, measured, and realized.

Role of KPIs in Sustainability and Digital Transformation: Key Performance Indicators are quantitative or qualitative metrics that help organizations measure and track their progress toward achieving specific goals. KPIs offer a structured way to assess environmental, social, and economic outcomes while providing the data necessary for continual improvement. Similarly, in digital transformation, KPIs help measure the effectiveness and impact of IT-driven changes, such as the adoption of digital technologies or the shift to a more automated, data-driven business model. The challenge, however, lies in aligning KPIs across these two domainssustainability and digital transformationsince they may appear distinct yet are intrinsically connected.

Sustainability KPIs: Sustainability encompasses a broad range of outcomes, primarily categorized into three key pillars: environmental, social, and economic. Each of these areas requires targeted KPIs to assess the sustainability of a system, organization, or economy. Environmental sustainability metrics focus on resource consumption, waste management, emissions reduction, and ecosystem health. Social sustainability metrics, on the other hand, focus on societal equity, labor rights, community development, and well-being. Economic sustainability KPIs address long-term financial health, income equality, and the resilience of economic systems. Some essential sustainability KPIs include:

- **Carbon Emissions**: One of the most widely used environmental KPIs, carbon emissions (often measured in CO2 equivalents) indicate the environmental impact of an organization's operations. By tracking the total carbon footprint, businesses can determine whether their emissions align with national or international climate goals, such as the Paris Agreement's target to limit global warming to well below 2°C. This KPI provides data for regulatory compliance and helps organizations develop carbon reduction strategies, adopt renewable energy sources, and improve energy efficiency.
- **Energy Efficiency**: Tracking energy consumption relative to output, energy intensity (energy consumed per unit of product produced) helps measure the sustainability of energy use. Decreasing energy intensity signifies improvements in energy efficiency, making this KPI critical for industries with high energy demand, such as manufacturing, transportation, and data centers. A transition to renewable energy sources is often incorporated in energy efficiency KPIs.
- **Water Usage**: With increasing concerns about freshwater scarcity, water consumption metrics are key to understanding how effectively an organization uses and conserves water. This includes tracking water intensity (the amount consumed per unit of output) and the proportion of recycled or reused water. This KPI is crucial in water-intensive agriculture, textile manufacturing, and mining industries.
- **Waste Diversion and Recycling Rates**: Waste management is another critical area in sustainability. KPIs in this domain track the percentage of

waste recycled, reused, or diverted from landfills. This includes monitoring waste reduction targets, such as zero waste to landfill, and fostering circular economy practices.

- **Social Inclusion and Equity Metrics**: Social KPIs assess the inclusivity and fairness of business practices. They include metrics such as employee diversity (gender, ethnicity, age), wage equality, community engagement, access to education, and healthcare. Social KPIs also examine how businesses address local development needs and contribute to poverty alleviation, equitable income distribution, and the enhancement of social mobility.

Digital Transformation KPIs: Digital transformation refers to integrating digital technologies into all aspects of business and society, fundamentally altering how organizations operate and deliver value to customers. KPIs of digital transformation often track progress related to innovation, operational efficiency, customer experience, and the adoption of technologies like Artificial Intelligence (AI), the Internet of Things (IoT), big data analytics, and automation. These metrics are important for evaluating the extent to which technology drives business success, improves performance, and fosters competitive advantage. Key digital transformation KPIs include:

- **Digital Adoption Rate**: This metric tracks how new technologies are integrated into business processes, operations, and products. For example, businesses may track the number of internal processes digitized, the percentage of products or services offered through digital channels, or the extent of automation in manufacturing or customer service. The higher the adoption rate, the more successful the digital transformation strategy.
- **Revenue Growth from Digital Channels**: In an increasingly digital world, businesses are adopting e-commerce, mobile applications, and digital services to increase revenue. This KPI measures the proportion of revenue generated through digital platforms as a percentage of total revenue. It reflects how well an organization leverages digital tools to expand its market reach, improve customer acquisition, and drive growth.
- **Customer Experience Metrics**: Digital transformation directly impacts customer experience, and KPIs in this domain track metrics such as customer satisfaction (CSAT), Net Promoter Score (NPS), and Customer Lifetime Value (CLV). In addition, customer retention rates and response times to digital customer service channels can also be included. These KPIs are essential for measuring the success of digital tools in enhancing customer relationships and engagement.

- **Operational Efficiency**: This KPI is focused on tracking improvements in productivity and cost efficiency due to adopting digital technologies. Key sub-metrics include reductions in operational downtime, improvements in supply chain efficiency, and reductions in transaction costs, all of which can be attributed to AI, machine learning, and automation.
- **Cybersecurity and Data Privacy**: As organizations adopt digital technologies, ensuring the security and privacy of data becomes a critical KPI. This includes tracking metrics such as the number of data breaches, the response time to security incidents, compliance with data protection regulations, and the robustness of cybersecurity frameworks. This KPI is important in fostering trust with customers and protecting organizational assets.

Integrating Sustainability and Digital Transformation: While sustainability and digital transformation have traditionally been viewed as distinct areas, the increasing convergence of technology and sustainability requires an integrated approach to KPIs. Integrating digital technologies into sustainability efforts, such as using AI to optimize energy use or blockchain to track the carbon footprint of supply chains, has led to cross-domain KPIs that measure both environmental impact and technological innovation.

- **Energy Efficiency through Smart Technologies**: The deployment of IoT-enabled smart meters and AI-powered optimization systems allows organizations to monitor energy consumption in real-time, making it possible to measure energy efficiency in ways that were not previously possible. The KPI for this would measure the reduction in energy consumption achieved through digital tools such as smart grids or energy management systems.
- **Sustainable Supply Chains**: As organizations digitize their supply chains through technologies like blockchain, AI, and IoT, they can track the sustainability of their sourcing and logistics practices. KPIs here may include the percentage of sustainable products sourced, the reduction in carbon emissions throughout the supply chain, and the level of transparency and traceability offered by blockchain technologies.
- **Circular Economy and Digital Solutions**: Digital technologies enhance the shift to a circular economy, which focuses on reducing waste and maximizing resource use. KPIs in this domain may measure how businesses can reuse or recycle products using digital tracking systems or the number of items or materials that have been refurbished, recycled, or resold through digital platforms.

Defies in Defining Effective KPIs for Sustainability and Digital Transformation: Despite the importance of KPIs in guiding sustainability and digital transformation efforts, several challenges exist in defining and implementing them effectively. First, aligning diverse stakeholders with a shared understanding of what constitutes success is complex. For instance, what may be considered a successful sustainability outcome for one organization may not be as relevant for another, depending on the sector, region, and resources available.

Second, organizations may struggle to gather consistent, accurate data to measure KPIs effectively. Data silos, inconsistent data collection practices, and a lack of standardization can impede efforts to track progress accurately. Establishing a direct link between digital transformation initiatives and specific sustainability outcomes can be complex, as it may take years for the full impact of digital technologies to materialize. The challenge of dealing with the dynamic nature of sustainability goals and digital technologies exists. The rapid pace of technological advancements means that KPIs related to digital transformation need to be constantly updated to reflect new tools and innovations, while sustainability goals may shift due to changing regulatory environments or new scientific findings.

Defining effective KPIs for sustainability and digital transformation is critical to guiding organizations toward long-term sustainability goals. KPIs can provide clear, measurable targets for success by focusing on environmental and social impacts and the technological advancements driving change. As organizations continue leveraging digital technologies to address sustainability challenges, developing integrated KPIs that reflect both domains will become increasingly essential. However, defining these metrics is not without challenges, and organizations must be proactive in ensuring data accuracy, stakeholder alignment, and continuous adaptation to evolving technological and sustainability landscapes. KPIs are the cornerstone of responsible, measurable, and actionable sustainability and digital transformation strategies.

AI-Driven Monitoring and Evaluation Tools

Artificial Intelligence (AI) has become a transformative force across various sectors, with significant applications in the monitoring and evaluation (M&E) of projects, policies, and interventions. Environmental management, education, healthcare, and business operations, AI-driven monitoring and evaluation tools provide new methodologies to enhance decision-making processes, improve efficiency, and ensure accountability. These tools leverage the power of data analytics, machine learning, and automation to streamline M&E systems, which traditionally relied on manual data collection, subjective analysis, and reporting. AI enables real-time analysis, predictive insights, and optimized resource allocation, all of which significantly enhance the effectiveness of M&E systems, making them more accurate, timely, and scalable. This comprehensive scientific overview will explore the role of AI in M&E, its applications, challenges, and the future potential of AI-driven tools in reshaping how data is used to assess program outcomes.

AI in Monitoring and Evaluation: Monitoring and Evaluation (M&E) are integral to any project or policy to achieve specific objectives. Monitoring refers to the continuous collection of data to track the progress of activities against set benchmarks, while evaluation focuses on assessing the outcomes and impacts of the intervention. Traditionally, M&E relied on periodic surveys, focus groups, and field observations, which could be time-consuming, resource-intensive, and prone to human error. AI-driven M&E tools, however, allow for automated data collection, real-time processing, and the generation of actionable insights through machine learning algorithms. This shift from traditional M&E methods to AI-powered systems is driving more agile, data-driven decision-making.

AI-driven tools offer several advantages over traditional M&E methods. First, they significantly enhance data collection and analysis, reducing the dependency on manual methods and increasing the accuracy of the data. Machine learning models, for instance, can identify patterns in large datasets that may be difficult for human analysts to detect. Second, AI tools enable predictive modeling, allowing for forecasting outcomes based on historical data and real-time inputs. Third, AI systems can automate routine processes, freeing human resources for more strategic decision-making. Finally, AI can improve accessibility and scalability by processing vast amounts of data from diverse sources, providing previously inaccessible insights to decision-makers at all levels.

Applications of AI-Driven Monitoring and Evaluation Tools: The use of AI in M&E spans multiple sectors, each with distinct requirements and challenges. This section explores the major applications of AI-driven M&E tools across different domains, including social development, environmental management, healthcare, education, and business.

- **AI in Social Development and Policy Monitoring:** In social development and policy interventions, AI-driven tools are used to monitor the effectiveness of programs aimed at poverty alleviation, healthcare access, education, and social equity. For example, AI models can analyze data from satellite imagery, census surveys, and social media to track the geographical spread of development programs, identify underserved populations, and evaluate the impact of social welfare programs on vulnerable groups. Predictive analytics, powered by AI, can forecast the long-term impacts of social policies by analyzing historical trends. This includes examining how policy changes, such as introducing a new subsidy or healthcare initiative, might affect specific regions or demographics. Natural language processing (NLP), a subfield of AI, can be used to analyze vast amounts of text data from policy reports, social media, and public forums to gauge public sentiment and assess the effectiveness of government communication efforts.

- **Environmental Management and Sustainability:** Environmental monitoring is another area where AI-driven M&E tools are making a significant impact. AI can be applied to track and analyze various environmental indicators, such as air and water quality, deforestation, biodiversity loss, and carbon emissions. Remote sensing technologies, coupled with AI algorithms, can process satellite images to monitor deforestation, land degradation, and the impacts of climate change. These tools can detect changes in the environment near-real-time, providing policymakers with up-to-date information to make informed decisions. AI can also support evaluating sustainability programs by analyzing data from smart sensors, drones, and IoT devices. For instance, in urban environments, AI systems can assess energy consumption patterns, waste management systems, and water usage to evaluate the sustainability of city infrastructures. AI can identify inefficiencies in resource allocation, highlight areas where interventions are most needed, and predict future sustainability trends based on historical data.
- **AI in Healthcare Monitoring and Evaluation:** Healthcare systems are increasingly adopting AI-driven tools to monitor and evaluate the performance of healthcare programs, interventions, and services. AI-powered systems track patient outcomes, treatment efficacy, and overall system performance, providing data-driven insights into the strengths and weaknesses of healthcare delivery. For example, machine learning algorithms can analyze electronic health records (EHR) to monitor patient health trends and predict outcomes such as disease progression or treatment success. AI can also be used in telemedicine to monitor patients remotely, assess their responses to treatment, and detect potential complications early. Additionally, AI tools can evaluate the effectiveness of healthcare programs by assessing the reach and coverage of vaccination campaigns or tracking the success of maternal and child health programs.

In resource-limited settings, where healthcare infrastructure may be weak, AI-driven M&E tools can bridge gaps in real-time monitoring. AI can process data from mobile health applications, wearable devices, and sensors to provide continuous feedback on patient health, allowing for more timely interventions and reducing the burden on healthcare workers.

- **Education and Learning Outcomes Evaluation:** AI is also revolutionizing the monitoring and evaluation of educational programs by providing data-driven insights into student performance, curriculum effectiveness, and teaching methods. AI tools can assess learning outcomes by analyzing students' performance across various metrics, such as test scores, attendance, and engagement. Using machine learning models, AI systems can provide personalized learning experiences for students, identifying knowledge gaps and suggesting interventions tailored to individual needs. AI-driven tools can evaluate the effectiveness of educational interventions, such as online learning platforms or new curricula. These systems can analyze the progress of students who are part of specific programs and compare them to control groups, providing evidence of the efficacy of educational initiatives. AI tools can also analyze data from educational assessments and feedback surveys to gauge the overall impact of educational policies.

- **Business and Operational Efficiency:** In the business world, AI-driven M&E tools are being increasingly adopted to evaluate operational efficiency, customer satisfaction, and financial performance. AI models can track key performance indicators (KPIs) such as sales growth, customer churn, inventory levels, and employee productivity. By leveraging big data analytics, AI can identify operational bottlenecks, predict customer behavior, and optimize resource allocation.

For example, AI can monitor supply chain performance by analyzing data from various sources, including sensors on production lines, inventory management systems, and transportation networks. This allows businesses to monitor the efficiency of their supply chains in real-time, enabling quicker decision-making and adjustments. AI-driven tools can assess customer satisfaction through sentiment analysis of online reviews, social media posts, and customer feedback, providing businesses with insights to improve their products and services.

Tasks in AI-Driven Monitoring and Evaluation Tools: Despite the tremendous potential of AI in M&E, there are several challenges to successfully implementing and adopting AI-driven tools. These challenges can be broadly categorized into data-related issues, algorithmic limitations, ethical concerns, and technological infrastructure barriers.

- **Data Availability and Quality:** One of the primary challenges in using AI for M&E is ensuring the availability and quality of data. AI algorithms require large, high-quality datasets to function effectively. However, in many sectors, especially in developing countries, data may be incomplete, outdated, or difficult to access. High-quality data is essential for AI systems to produce accurate and reliable results. The lack of comprehensive data can undermine the effectiveness of AI-driven M&E tools, leading to inaccurate insights or poor decision-making.

- **Algorithmic Bias:** AI systems are only as good as the data on which they are trained. If the data used to train the AI models contains biases, these biases can be perpetuated and even amplified in the AI's outputs. This can lead to discriminatory practices in M&E, especially in sensitive areas like healthcare, education, and social development. For example, AI models trained on biased datasets may underrepresent certain demographics or communities, leading to skewed conclusions that fail to address the needs of marginalized groups.

  Addressing algorithmic bias requires ensuring diverse and representative datasets and regular auditing of AI models for fairness and accuracy. It also requires fostering transparency in AI decision-making processes, allowing stakeholders to understand how decisions are made and providing avenues for recourse in case of errors or bias.

- **Ethical and Privacy Concerns:** Using AI in monitoring and evaluation raises significant ethical and privacy concerns. In many cases, AI-driven M&E tools rely on collecting and analyzing personal or sensitive data, such as health records, education performance data, or financial transactions. This raises the risk of data misuse, breaches of privacy, and violations of individuals' rights.To mitigate these risks, it is essential to implement strong data governance frameworks, including secure data storage, encryption, and robust consent mechanisms. Organizations must also comply with international data protection regulations, such as the General Data Protection Regulation (GDPR), to ensure that individuals' privacy is respected and protected.

- **Technological Infrastructure and Skills Gaps:** Implementing AI-driven M&E tools requires robust technological infrastructure, including high-performance computing, cloud storage, and reliable internet connectivity. In many regions, especially in low-income and rural areas, these infrastructures may be lacking, hindering the effective use of AI. Adopting AI in M&E requires a skilled workforce with data science, machine learning, and AI expertise. There is a significant skills gap in many industries, particularly in developing countries, where access to training and education in these fields may be limited.

AI-driven monitoring and evaluation tools transform how organizations, governments, and businesses assess and optimize their operations. These tools enable real-time data collection, predictive analytics, and automated decision-making, providing stakeholders with more accurate, timely, and actionable insights. While AI has vast potential to enhance M&E processes across sectors, challenges such as data quality, algorithmic bias, ethical concerns, and infrastructure limitations must be addressed to ensure that AI is used responsibly and effectively. As AI technology continues to evolve, its role in monitoring and evaluation will become increasingly central to achieving sustainable development goals, improving public policy outcomes, and fostering economic growth. The future of AI in M&E is poised to be transformative, offering new possibilities for evidence-based decision-making and creating more efficient, inclusive, and effective monitoring and evaluation systems.

Case Studies

India's journey towards achieving the United Nations Sustainable Development Goals (SDGs) has been bolstered by the development and implementation of the SDG Dashboard, a digital tool designed to provide real-time monitoring of progress on SDGs across the country. The SDG Dashboard is a crucial element of India's efforts to integrate technology into governance and policy planning, providing comprehensive data across 17 SDGs and enables decision-makers to track achievements at national, state, and district levels. This case study examines several instances of the SDG Dashboard's application and the lessons learned from its real-time monitoring system. It also explores how these case studies highlight the role of data integration, technological innovations, and collaborative frameworks in addressing India's complex sustainability challenges.

Case Study 1 – Monitoring Progress on SDG 1: One of the primary challenges in India is poverty, with a significant proportion of the population still living below the poverty line. SDG 1 seeks to eradicate poverty in all its forms, everywhere. In India, the SDG Dashboard has become a vital tool for monitoring and tracking progress on poverty alleviation. The dashboard offers real-time insights into poverty metrics, such as income levels, employment opportunities, and social security benefits by integrating data from multiple sources, such as the Ministry of Rural Development, the Ministry of Finance, and various state governments.

The dashboard allows policymakers to track the poverty headcount ratio, multidimensional poverty indices, and the number of people receiving direct financial transfers or benefits under various government schemes. This real-time tracking helps identify regions falling behind in poverty alleviation, allowing for more targeted interventions. For example, the dashboard highlighted areas in rural India struggling to meet the income threshold for poverty reduction under the Pradhan Mantri Awas Yojana (PMAY), the national housing scheme for the poor. As a result, the Ministry of Rural Development prioritised these regions, ensuring that the poorest populations received housing assistance promptly. Challenges remain in the quality and granularity of the poverty data collected. For instance, although real-time data is available, there is a need for more detailed local-level statistics to better understand regional disparities. Poverty indices need to be updated regularly to reflect changes in the socioeconomic landscape. The SDG

Dashboard's real-time monitoring has highlighted the importance of consistent data quality and periodic updates to ensure policymakers make informed decisions.

Case Study 2 – SDG 3: The SDG Dashboard has also been instrumental in monitoring health outcomes, particularly SDG 3, which aims to ensure healthy lives and promote well-being for all at all ages. India faces enormous challenges in terms of healthcare access, particularly in rural and remote regions. Real-time data on health indicatorssuch as maternal mortality rates, infant mortality rates, immunization coverage, and the number of healthcare facilitiesare critical for evaluating the success of India's national health programs. The dashboard has integrated data from the Ministry of Health and Family Welfare, National Health Mission (NHM), and other health agencies to track real-time progress on health-related SDG indicators. One notable example is the dashboard's role in monitoring India's vaccination coverage, especially during the COVID-19 pandemic. As the pandemic unfolded, the SDG Dashboard was used to track the number of people vaccinated in each state, allowing the government to identify vaccination gaps and optimize resource allocation. States with lower vaccination rates could quickly be flagged for intervention, leading to targeted awareness campaigns, increased vaccination drives, and better management of healthcare resources.

Additionally, real-time data on healthcare infrastructuresuch as the availability of hospital beds, medical supplies, and essential servicesallowed the government to allocate resources where they were most needed. For instance, during the second wave of the COVID-19 pandemic, the dashboard provided timely information about hospital bed occupancy and oxygen supply shortages, enabling the central government to respond faster and mobilize emergency relief efforts. Despite the dashboard's utility, challenges remain in integrating data from local healthcare facilities. In India, healthcare services in rural and underdeveloped regions are often inadequately documented, making it difficult to get a full picture of health outcomes. Data privacy and security issues are paramount when dealing with sensitive health information, necessitating strong governance frameworks for data management.

Case Study 3 – SDG 7: SDG 7 emphasizes universal access to affordable, reliable, modern energy services. India's energy access has improved over the years, but disparities remain between urban and rural areas, especially regarding access to electricity and clean cooking solutions. The SDG Dashboard has been crucial in monitoring India's progress in achieving SDG 7, particularly the access to clean energy and the share of renewable energy in the national energy mix. One of the critical innovations in this area is the dashboard's integration of satellite-based data from the Ministry of New and Renewable Energy (MNRE) and the Energy Statistics Division. This integration allows for the real-time monitoring of renewable energy projects, including solar and wind energy installations. It also tracks energy consumption data, which helps policymakers understand energy demand patterns and identify areas where energy infrastructure needs expanded.

An example of the dashboard's impact is tracking solar energy projects under the National Solar Mission. Real-time data on the deployment of solar panels across different states allowed the MNRE to assess which states were lagging in the solar energy rollout and provide targeted assistance. The dashboard also helped track the performance of individual projects, ensuring that the energy output met the planned targets. Challenges related to the management and storage of energy data persist. While real-time data on energy generation is useful, understanding the long-term sustainability of energy infrastructure requires more in-depth data analysis, particularly related to grid reliability, storage capacity, and energy efficiency. Additionally, India's energy infrastructure is still fragmented, with some states lacking the technological infrastructure to report data in real-time.

Case Study 4 – SDG 13: India is one of the countries most vulnerable to the impacts of climate change. Addressing climate change through mitigation and adaptation measures is central to achieving SDG 13. The SDG Dashboard has played a pivotal role in monitoring India's progress on climate action, especially by tracking indicators such as greenhouse gas emissions, renewable energy use, and disaster preparedness.

A notable example of the dashboard's role in climate action was its use while monitoring India's Nationally Determined Contributions (NDCs) under the Paris Agreement. The dashboard provided a platform for tracking India's progress towards its climate targets, including the reduction of carbon emissions and the increase in renewable energy capacity. By integrating data from the Ministry of Environment, Forests, and Climate Change (MoEFCC) and other environmental agencies, the dashboard enabled real-time tracking of the country's progress in emission reductions and sustainable land use practices.

The dashboard also played a critical role in monitoring disaster preparedness and climate resilience. India is prone to natural disasters such as cyclones, floods, and droughts, and the SDG Dashboard helps assess the government's disaster response capacity. By aggregating real-time data on climate events and assessing infrastructure resilience, the dashboard supports evidence-based decision-making for disaster risk reduction and climate adaptation. Despite its success, the SDG Dashboard faces challenges in climate data reporting. The lack of consistent and high-quality data on carbon emissions, the impact of climate change on ecosystems, and the effectiveness of adaptation measures has limited the scope of the dashboard's climate monitoring. Improving the accuracy of climate predictions, integrating disaster impact data, and tracking the long-term effects of climate change will require ongoing advancements in technology and data integration.

Case Study 5 – SDG 16: SDG 16 focuses on promoting peaceful and inclusive societies, providing access to justice for all, and building effective, accountable institutions. In India, the SDG Dashboard has been used to track indicators such as crime rates, access to legal aid, and the quality of governance across states. The real-time data on law enforcement effectiveness, judicial backlog, and the number of people engaged in legal proceedings helps to monitor the functioning of India's justice system and its efforts to ensure transparency and accountability.

For example, the dashboard's ability to track the effectiveness of the Right to Information (RTI) Act has been critical in promoting transparency. By providing data on how effectively RTI applications are processed, the dashboard allows citizens and advocacy groups to assess the responsiveness of government institutions. Real-time data on crime statistics helps assess the safety of urban and rural areas and can guide law enforcement agencies in prioritizing high-crime zones for intervention. Data privacy and security challenges are particularly acute for justice and governance data. Sensitive data about legal proceedings and crime can be vulnerable to misuse, so stringent safeguards must be implemented. Additionally, the lack of real-time data from the grassroots, particularly in areas with weak institutional infrastructure, limits the comprehensiveness of the dashboard's monitoring of justice and governance-related SDGs.

Future Scenario

The future of sustainability assessments is increasingly tied to adopting and integrating advanced AI-based tools that offer real-time insights into various dimensions of sustainable development. These tools can transform how we evaluate, track, and act on sustainability metrics locally, nationally, and globally. By combining AI's ability to process large volumes of data with the precision of real-time analytics, future sustainability assessments will provide a more dynamic and comprehensive understanding of environmental, social, and economic factors that influence sustainable development. This evolution is particularly relevant in global sustainability indices, which measure and track countries' progress toward achieving the United Nations' Sustainable Development Goals (SDGs). Integrating AI and IT systems within these indices will enhance accuracy and enable proactive, data-driven decision-making that aligns with the urgent need for sustainable practices worldwide.

AI-Based Tools for Real – Time Sustainability Assessments: AI technologies have already begun to play a transformative role in sustainability assessments, especially with their ability to analyze complex datasets, recognize patterns, and predict outcomes. As we look toward the future, AI-driven tools will significantly enhance our real-time capacity to monitor and evaluate sustainability efforts. The growing ability of AI to analyze large-scale environmental data, combined with IoT (Internet of Things) sensors, satellite imagery, and geographic information systems (GIS), allows for the continuous tracking of key sustainability indicators such as carbon emissions, energy consumption, deforestation rates, water quality, biodiversity, and air pollution levels.

One key application of AI in sustainability is the automation of environmental monitoring. AI algorithms, particularly those based on machine learning and deep learning, can process vast datasets collected from IoT devices, sensors, and remote sensing technologies to identify environmental trends in real-time. These tools can assess factors such as water usage efficiency, energy consumption patterns, and emissions from industrial operations, providing governments, organizations, and industries with immediate feedback. For instance, AI-based predictive models can forecast potential environmental impacts before they occur, such as the likelihood of extreme weather events, helping governments and businesses take proactive steps to mitigate risks. Additionally, AI systems are increasingly used to optimize resource usage, helping industries and agricultural sectors achieve more sustainable practices. Machine learning algorithms can analyze historical data to predict demand for resources like water, energy, and raw materials, thus enabling more efficient allocation and reducing waste. These capabilities will be significant in future cities and urban sustainability, where real-time data can help manage resource distribution, waste management, and public transportation systems to minimise environmental impact and enhance efficiency.

In agriculture, AI-driven precision farming techniques, which use AI tools to optimize planting, irrigation, fertilization, and harvesting, will become integral to real-time sustainability assessments. These tools use data from drones, soil sensors, and satellite imagery to continuously assess crops and soil health, enabling farmers to make data-driven decisions that minimize water and fertilizer usage, reduce pesticide reliance, and enhance overall crop yields. AI will enable these decisions to be made in real-time, optimizing resource use for better environmental outcomes. Also, AI's ability to simulate various sustainability scenarios and model the impact of different policies and practices will be crucial for future sustainability planning. AI-based decision support systems can integrate data from diverse sources, including social, economic, and environmental factors, to forecast the long-term outcomes of sustainability efforts. These predictive tools can help policymakers choose strategies more likely to result in positive sustainability outcomes, both in meeting SDG targets and fostering long-term environmental health.

Global Sustainability Indices Integrated with IT Systems: The global sustainability indices, such as the Environmental Performance Index (EPI), the Human Development Index (HDI), and the Global Sustainability Index, are designed to provide a benchmark for assessing a country's progress in achieving sustainable development. As the world increasingly moves towards digitalization, these indices are being integrated with advanced IT systems, including big data analytics, AI, and blockchain, to enhance their comprehensiveness, accuracy, and timeliness.

Currently, sustainability indices rely on static, periodic data updates, making it difficult to gauge the ongoing progress of nations in real-time. However, these indices will incorporate continuous data streams from various sources, including satellite monitoring, IoT sensors, social media analytics, and government reporting systems. AI and machine learning will be key in this transformation, enabling real-time processing and analysis of this vast array of data to provide up-to-date assessments of sustainability progress. For instance, AI could monitor air quality across various regions and integrate this data into global sustainability indices, offering a more precise, real-time measurement of a country's environmental performance. Similarly, real-time agricultural output, energy consumption, and water usage patterns will be incorporated into sustainability indices, allowing for a more dynamic and comprehensive view of a nation's sustainability efforts. By doing so, countries can monitor their progress more effectively and benchmark their achievements against global best practices and trends.

The integration of AI will also enhance the predictive capabilities of sustainability indices. Based on current data, these tools can model different sustainability scenarios and forecast the likely outcomes of various interventions or policy changes. For example, AI models can predict the impact of implementing stricter carbon emission regulations or expanding renewable energy infrastructure, helping governments assess which policies will yield the greatest improvements in their sustainability scores.

Another promising development is the integration of blockchain technology into global sustainability indices. Blockchain can be used to ensure the transparency, traceability, and immutability of the data used to calculate sustainability scores. As sustainability efforts often involve multiple stakeholders, including governments, private enterprises, and NGOs, blockchain can serve as a trusted, decentralized ledger for recording and verifying sustainability-related data. For example, blockchain can be used to track the carbon footprint of supply chains, ensuring that companies are held accountable for their environmental impact and contributing to a more accurate and reliable sustainability index.

Blockchain's role in providing transparency will also extend to monitoring financial flows related to sustainability projects. Blockchain can improve accountability and reduce the risk of corruption by ensuring that the funds allocated for environmental conservation or social equity projects are used as intended. This enhanced transparency will increase trust in global sustainability indices and motivate governments and corporations to improve their sustainability efforts. One of the critical challenges in integrating AI and blockchain into sustainability indices is ensuring the interoperability of various data sources. Sustainability data is often fragmented, collected by different agencies and organizations, and sometimes lacks standardization. AI can help address this challenge by automating the data integration process, normalizing and analyzing data from disparate sources, and identifying trends and gaps in real-time. Over time, as countries and organizations agree on standardized data formats and reporting protocols, AI and blockchain technologies can help ensure that sustainability indices become more accurate, reliable, and actionable.

AI for Real - Time Sustainability Analytics and Decision Making: Looking ahead, the role of AI will extend beyond monitoring and data analysis to support real-time decision-making in sustainability management. AI tools will help government, business, or civil society decision-makers identify emerging sustainability challenges and respond rapidly with targeted interventions. These tools will facilitate more agile and adaptive governance structures that promptly react to environmental or social shocks, such as natural disasters, economic crises, or pandemics.

AI's role in facilitating adaptive governance systems is especially critical in climate change, where countries must continually adjust their sustainability policies and strategies in response to evolving environmental conditions. For example, AI can monitor climate-related variables, such as temperature, rainfall, and wind patterns, and help predict the potential impacts of climate change on agriculture, water resources, and coastal ecosystems. Policymakers can use this information to adjust agricultural strategies, water management systems, and disaster response plans in real-time, ensuring that resources are allocated efficiently and effectively. The AI-based real-time sustainability assessments could also serve as an early warning system for potential risks, such as biodiversity loss, pollution hotspots, or economic inequality. These systems can detect emerging issues long before they reach a critical threshold, allowing governments and organizations to intervene before these problems escalate. Based on data-driven insights, early interventions could help minimize the negative impacts of unsustainable practices and prevent the depletion of critical natural resources.

18

Collaboration Across Sectors for Sustainability

Introduction

Sustainability, in its truest sense, requires a multifaceted approach where various sectorsgovernment, private industry, civil society, academia, and individualswork together to address complex and interconnected global challenges. Collaboration across sectors is desirable and essential to meet the diverse and far-reaching objectives outlined in the United Nations' Sustainable Development Goals (SDGs). The global challenges of climate change, resource depletion, biodiversity loss, and social inequalities cannot be addressed by isolated efforts from any single entity; they require concerted action across multiple domains. This collaboration hinges on a collective recognition of sustainability as a shared responsibility that transcends traditional boundaries, promoting the integration of diverse perspectives and expertise to develop more holistic and effective solutions.

Need for Multisectoral Collaboration: Sustainability challenges are inherently interdisciplinary, affecting various sectors of society, the economy, and the environment. For instance, tackling climate change requires the integration of environmental science, energy technology, urban planning, social policy, and economic development. Each of these sectors holds a piece of the puzzle, and their effective collaboration is necessary for designing scalable and impactful solutions. One of the main challenges in achieving sustainability is the siloed approach used by different sectors. For example, while government bodies may focus on regulation and policy development, businesses are driven by profit motives, and civil society organizations may concentrate on advocacy and social impact. Each sector tends to pursue its agenda with limited interaction with the others, which can lead to fragmented strategies that fail to address the problem at the systemic level.

A collaborative approach helps bridge these gaps by aligning different sectors around shared sustainability goals, such as reducing carbon emissions, promoting clean energy, ensuring water security, or advancing social equity. The interconnectedness of these goals calls for joint actions that leverage each sector's expertise, resources, and innovation. For example, urban sustainability initiatives require the cooperation of municipal governments, private construction companies, tech firms, environmental NGOs, and local communities. Government entities set the legal framework and regulatory standards, private companies develop innovative technologies for energy efficiency, and local communities and NGOs help implement grassroots solutions that are sustainable and socially inclusive.

Government – Private Sector Collaboration for Green Innovation: Governments play a pivotal role in establishing the policy environment that enables the private sector to invest in sustainable technologies and practices. This relationship is often mutually beneficial, as businesses require supportive regulations and financial incentives to scale up sustainable innovations, and governments can achieve policy objectives through private-sector engagement. The private sector, in turn, brings forward innovative solutions, operational expertise, and the capital necessary to advance sustainability efforts.

One example of this collaboration is the transition to renewable energy. Governments worldwide have implemented subsidies and tax incentives to encourage solar, wind, and other renewable energy technologies investments. However, private sector participation is crucial for these technologies to scale and reach the level necessary to meet global energy demands sustainably. Companies such as Tesla, Siemens, and Vestas have developed breakthrough technologies in electric vehicles, energy storage, and wind turbines that significantly reduce carbon footprints. These companies often partner with governments to implement renewable energy policies, construct infrastructure, and provide financing mechanisms for large-scale renewable projects. Additionally, the private sector's role in green innovation is amplified by its ability to deploy solutions rapidly globally, which is critical to addressing the urgency of climate change.

The collaboration between government and private entities in sustainability is also evident in sustainable agriculture. Governments have introduced policies encouraging farmers to adopt sustainable practices, such as crop diversification, water conservation, and organic fertilizers. Meanwhile, private companies provide the necessary technology, tools, and training to improve yields while minimizing environmental impact. The development of precision agriculture technologiessuch as IoT sensors, drones, and AI-driven systems for monitoring soil health, water usage, and crop growthdemonstrates the synergy between technological innovation and policy support. These innovations can be scaled and applied through public-private partnerships to increase food security while minimizing environmental degradation.

Academia and Research Institutions: Academic institutions and research organizations are essential partners in the sustainability dialogue, as they generate the knowledge, research, and innovations that drivee sustainability efforts forward. These institutions conduct cutting-edge research on environmental systems, social issues, economic structures, and technology developments, all vital for understanding and solving sustainability challenges. Academia also plays an instrumental role in developing the methodologies and models for assessing sustainability outcomes, evaluating policies, and identifying long-term trends.

Collaboration between academic researchers and industry is becoming a core component of sustainable innovation. Through partnerships, universities provide evidence-based insights that help companies, governments, and NGOs craft policies and business strategies grounded in the latest scientific research. Academic partnerships with governments and businesses facilitate the development of new, more sustainable technologies, as academic institutions often can explore high-risk, high-reward innovations that the private sector may not be willing to fund alone.

One example is the development of sustainable materials in construction and manufacturing. Researchers in universities worldwide are working on developing new, low-carbon alternatives to conventional building materials such as concrete and steel. Through collaboration with private construction companies and local governments, these research initiatives can be commercialized and deployed at scale. This can help reduce the carbon emissions of the building sector, which is one of the largest contributors to global greenhouse gas emissions.

Civil Society and NGOs – Advocacy and Grassroots Impact: Civil society and non-governmental organizations (NGOs) play an essential role in the sustainability landscape, particularly in advocating for vulnerable populations, ensuring transparency, and holding the government and the private sector accountable for their actions. NGOs often focus on the social dimensions of sustainability, ensuring that development is inclusive and benefits all segments of society. They act as intermediaries between the grassroots and higher levels of government and industry, providing the necessary link to understand local needs and challenges.

These organizations also serve as watchdogs, monitoring the progress of sustainability initiatives and highlighting instances where policies or business practices are not aligned with sustainability goals. NGOs can influence government policies, private sector practices, and consumer behavior through campaigns and lobbying, creating a demand for more responsible and sustainable products and services. Their role in mobilizing communities, raising awareness, and providing on-the-ground solutions makes them indispensable partners in the quest for sustainability. In water resource management, for example, NGOs often work with local communities to implement sustainable water practices, such as rainwater harvesting, water conservation, and pollution prevention. Their grassroots efforts are critical to ensuring that sustainability efforts are locally relevant, socially inclusive, and accessible. NGOs can help create integrated water management systems that balance environmental, economic, and social needs by working with local governments and businesses.

Role of Digital Technologies in Fostering Collaboration: Digital technologies, including data analytics, AI, and the Internet of Things (IoT), play a central role in fostering collaboration across sectors by enabling the sharing of information, tracking sustainability metrics, and coordinating of efforts. For example, digital platforms can facilitate real-time data sharing between governments, private companies, and NGOs, helping all stakeholders stay informed about sustainability trends and performance. Collaborative platforms can integrate data on environmental metrics, social indicators, and economic outcomes for a more coordinated response to sustainability challenges. In addition, technologies such as blockchain are being explored to ensure transparency and accountability in supply chains, financial transactions, and resource management. By enabling secure and verifiable data exchanges, blockchain technology can ensure that all actors in a supply chainwhether a business, a government agency, or an NGOcan track the origins and impact of products and services, thereby promoting responsible production and consumption.

Current Scenario

In today's interconnected world, the role of Information Technology (IT) in fostering cross-sector collaboration has become crucial in addressing complex global challenges. The increasing need for multi-disciplinary approaches to climate change, sustainable development, public health, and economic inequality requires coordination and collaboration across various sectors, including government, industry, academia, and civil society. Information technology has emerged as a powerful enabler, providing the tools and platforms to facilitate communication, data-sharing, and decision-making across disparate sectors. The potential of IT to streamline and accelerate collaboration is especially evident in multifaceted global challenges that require collective action at local, national, and international levels.

The current scenario reflects a growing recognition of the importance of cross-sector collaboration, which has been made more efficient and effective by the rapid advancement of IT systems and digital platforms. IT has revolutionized how stakeholders from various sectors communicate, share data, and align their efforts toward common goals. For instance, integrating cloud computing, big data analytics, Artificial Intelligence (AI), and the Internet of Things (IoT) has opened up new avenues for sharing real-time information, tracking progress on sustainability metrics, and monitoring outcomes across various sectors. This transformation has significant implications for addressing pressing global challenges, particularly health, environmental sustainability, agriculture, education, and urban development.

Government, Private Sector, and Civil Society Collaboration Enabled by IT: At the core of cross-sector collaboration is the interaction between governments, private sector companies, and civil society organizations, with each sector bringing its unique strengths, capabilities, and perspectives to the table. IT is essential in aligning these diverse sectors, ensuring that data flows smoothly between them and everyone involved is working from the same information set. Governments have long been central to driving cross-sector collaboration through policy-making, regulation, and resource allocation. However, the traditional top-down governance model has evolved with the advent of IT, enabling more transparent, inclusive, and participatory governance mechanisms. Governments now leverage digital platforms to engage citizens and gather data from various sources, including sensors, mobile applications, and social media, providing a more comprehensive understanding of the issues. For example, government agencies use geographic information systems (GIS) to track environmental data, urban infrastructure, and social trends, enabling more informed decision-making and resource allocation. Cloud computing and AI have allowed governments to analyze vast amounts of data, identify patterns, predict trends, and make real-time data-driven decisions.

The private sector, particularly large corporations and startups in the tech and digital space, plays an increasingly prominent role in cross-sector collaboration. Companies bring innovation, financial capital, and technological expertise to the table. IT has enabled companies to share knowledge, integrate new technologies into sustainable business practices, and establish partnerships with governments and NGOs to implement solutions at scale. For instance, multinational corporations in the energy and transportation sectors have collaborated with governments and NGOs to promote sustainable practices such as renewable energy deployment, electric vehicle adoption, and smart grid development. These initiatives are often supported by IT systems that help monitor energy consumption, predict system demands and track emissions. Companies like Google, Microsoft, and IBM have invested heavily in AI, data analytics, and IoT solutions to enhance operational efficiency and support sustainability initiatives. This collaboration is not limited to large corporationssmall and medium-sized enterprises (SMEs) and startups also leverage digital tools to innovate in areas such as waste management, sustainable agriculture, and healthcare.

Civil society organizations (CSOs), including non-governmental organizations (NGOs), advocacy groups, and community-based organizations, are critical in cross-sector collaboration. These groups often represent marginalized or underrepresented populations and bring an essential focus on equity, inclusion, and social justice. IT has empowered CSOs to reach a wider audience, mobilize communities, and influence local, national, and international policy. Digital tools such as online platforms, social media, and crowdfunding have enabled CSOs to coordinate efforts, raise awareness, and advocate for sustainable solutions. Real-time data sharing and communication tools allow CSOs to engage in joint efforts with governments and businesses, ensuring that sustainability initiatives are socially inclusive and community-driven.

Enhancing Collaboration Through Data and Information Sharing: One of the most significant contributions of IT to cross-sector collaboration is its ability to facilitate data and information sharing. The sheer volume of data generated todayfrom environmental monitoring and economic indicators to social media posts and health recordsoffers immense opportunities for collaboration across sectors. IT systems that support big data analytics, machine learning, and cloud-based storage have made it easier for organizations to collect, analyze, and share large amounts of data in real-time.

For instance, in the health sector, cross-sector collaboration is facilitated by sharing patient data securely across hospitals, research institutions, pharmaceutical companies, and government health agencies. IT systems that enable the seamless sharing of medical records, research findings, and public health statistics are critical to improving patient outcomes, advancing medical research, and responding to health crises such as pandemics. For example, the global response to COVID-19 saw extensive use of digital platforms for contact tracing, vaccine distribution tracking, and real-time data sharing between international health organizations, governments, and private companies. AI and machine learning algorithms have been instrumental in analyzing health data to predict disease outbreaks, optimize resource allocation, and provide insights into disease prevention strategies.

In the environmental sector, data sharing platforms integrating information from governments, businesses, and NGOs enable a more coordinated approach to addressing climate change, deforestation, and biodiversity loss. Using satellite imagery, environmental sensors, and IoT devices has provided a more granular understanding of environmental trends, helping stakeholders monitor natural resources, track emissions, and implement targeted conservation efforts. IT-enabled platforms also allow for real-time monitoring of sustainability goals, such as the progress in reducing carbon emissions or improving water quality, ensuring that all sectors have access to the same data and are working toward shared objectives.

Collaborative Platforms and Ecosystems: IT has created a new generation of collaborative platforms to facilitate interaction and partnership among various sectors. These platforms are being used to connect government agencies, businesses, non-profits, and citizens around common sustainability goals. For example, platforms such as the Global Reporting Initiative (GRI) and the UN Global Compact provide a framework for businesses and governments to report on their sustainability efforts, share best practices, and track progress toward achieving the SDGs. Such platforms encourage transparency and accountability, ensuring all sectors are held to the same standards and can learn from each other's successes and challenges.

Blockchain technology has also emerged as a tool for enhancing cross-sector collaboration, particularly in supply chain management. Blockchain's decentralized and immutable nature allows stakeholders to track the origins and impact of products, ensuring transparency and accountability across entire supply chains. This is especially important in food production, mining, and manufacturing industries, where sustainable practices are often difficult to monitor and enforce. By using blockchain to track the movement of goods from source to consumer, companies can ensure that their products are ethically sourced, environmentally friendly, and compliant with sustainability standards.

IT enables the creation of ecosystems that facilitate innovation and collaboration between diverse actors. For example, innovation hubs and accelerators bring together startups, investors, research institutions, and corporations to co-create solutions to sustainability challenges. These ecosystems foster creativity, promote knowledge exchange, and accelerate the development and scaling of sustainable technologies.

Overcoming Barriers to Effective Collaboration: Despite the clear benefits of cross-sector collaboration, several barriers remain that hinder the effective use of IT in promoting collective action for sustainability. One major challenge is the lack of standardization in data formats and systems, making it difficult for different sectors to share and interpret information. For example, data collected by government agencies may be in formats that are not compatible with those used by private companies or NGOs, leading to inefficiencies and data silos. Standardizing data formats and establishing common data protocols is essential for ensuring all stakeholders can access and use the information they need to make informed decisions.

Another challenge is the need for digital literacy and capacity building. While IT tools are becoming more widely accessible, many stakeholdersparticularly in developing countries or rural areaslack the skills and infrastructure to leverage them effectively. Capacity-building initiatives that provide training on digital tools and technologies are essential to ensure all sectors can collaborate and contribute to sustainability goals. Finally, data privacy, cybersecurity, and digital ethics pose significant challenges to cross-sector collaboration. The sharing of sensitive data between sectors raises concerns about privacy violations and the misuse of information. Governments and organizations must establish robust data protection policies and cybersecurity measures to ensure that data is shared securely and that privacy rights are respected.

Public-Private Partnerships in IT Ecosystems

In the world of increasingly digital and interconnected, role of public-private partnerships (PPPs) in fostering innovation within IT ecosystems is becoming increasingly pivotal. These collaborations between government entities, private industries, and academic institutions drive technological advancements, promote research, and develop sustainable solutions for complex challenges.

The rapidly evolving technological landscape, driven by the need for advanced computing, artificial intelligence (AI), big data analytics, and cybersecurity, has made it essential for various sectors to come together, each contributing unique expertise and resources. Through these partnerships, significant advancements in digital transformation, smart cities, healthcare innovation, and environmental sustainability are being achieved, with each player contributing in a way that accelerates progress.

Role of Industry in Innovation: The private sector, particularly in information technology, is often at the forefront of technological innovation. IT companies, ranging from established multinational corporations like Microsoft, IBM, and Google to innovative startups, contribute through investments in research and development (R&D), application of cutting-edge technologies, and commercialization of solutions. Industry players typically develop the core technological frameworks and solutions, bringing innovative products and services to market. In public-private partnerships, private industries bring several advantages, including access to financial capital, business expertise, market knowledge, and a culture of risk-taking and agility.

For instance, technology companies drive innovation in sectors such as cloud computing, cybersecurity, and AI by developing advanced platforms that enable businesses and governments to leverage these technologies for real-time analytics, automation, and decision-making. Private companies are often the key drivers of developing AI-based diagnostics tools, digital health applications, and telemedicine solutions in the healthcare sector. These innovations are essential for addressing growing challenges in healthcare systems globally, such as reducing costs, improving service delivery, and enhancing patient outcomes.

Private sector involvement in PPPs is critical in fostering a competitive environment that drives efficiency and accelerates technological breakthroughs. Corporations often lead efforts to scale new technologies, ensuring that they can be deployed in a manner that delivers value to society at large. The private sector's experience navigating regulatory environments, ensuring compliance, and managing large-scale projects is critical in implementing technology-driven solutions in smart cities, autonomous transportation systems, and renewable energy. As private industries invest in new technologies, they create an ecosystem of innovation that other sectors can harness.

Role of Academia in Innovation: While the private sector drives technological innovation and commercialization, academia plays an equally important role in fostering foundational research, developing new theories, and advancing scientific understanding. Universities and research institutions are hubs for discovering and disseminating knowledge that underpins technological advancements. Academics contribute by conducting high-impact research in various IT fields, from quantum computing and blockchain technology to data science and machine learning. These research activities provide the scientific basis for new technologies, offering a wellspring of intellectual resources that can be translated into practical applications.

One of the key strengths of academic institutions in PPPs is their ability to generate new knowledge through basic and applied research. Universities are often well-equipped to explore novel ideas that may not yet have immediate commercial applications but could lead to groundbreaking innovations in the future. For example, academic AI and machine learning research have laid the foundation for major natural language processing, computer vision, and robotics advancements. Through collaborations with industry, academic institutions can transition their theoretical research into real-world applications, which can have a transformative impact across various sectors.

Academia plays a significant role in training and developing the next generation of tech talent. As industries continue to evolve rapidly, there is a growing demand for a skilled workforce capable of working in data analytics, software development, cybersecurity, and AI. Universities and technical institutions are key players in producing this workforce through educational programs, certifications, and internships. Additionally, academia often conducts specialized training programs to equip individuals with the technical skills required by industries, thus aligning educational outcomes with market needs. Through internships, research collaborations, and direct industry partnerships, academia ensures that students gain practical, hands-on experience that helps them transition smoothly into the workforce, contributing to the overall innovation ecosystem.

Synergies Between Industry and Academia in PPPs: The fusion of academic research and industry expertise in public-private partnerships has created more dynamic and effective innovation ecosystems. Academia provides the foundational knowledge and innovative ideas, while industry ensures the application of these ideas to real-world challenges. This synergy between academic institutions and private sector companies has been instrumental in solving some of the most pressing global issues.

In sectors such as healthcare, this collaboration has been particularly impactful. For example, academic research on genomics, immunology, and personalized medicine has enabled private companies to develop cutting-edge diagnostic tools, vaccines, and therapies. One notable example is the collaboration between universities and pharmaceutical companies in developing vaccines for emerging diseases like COVID-19. The rapid development and deployment of COVID-19 vaccines were made possible through academic research, industry innovation, and governmental policy frameworks supporting PPPs in the biopharmaceutical field. Similarly, integrating big data analytics and AI in healthcare, supported by industry and academic research, has enhanced the ability to predict and manage patient health outcomes more effectively.

In sustainable energy, industry-academia partnerships have led to innovations in renewable energy technologies such as solar power, wind energy, and energy storage systems. Academic research has significantly contributed to improving solar cells' efficiency, developing new materials for energy storage, and understanding the environmental impact of energy systems. Industry players have taken this research and developed scalable solutions for commercial use. Public-private partnerships in renewable energy support the development of clean technologies and contribute to achieving global sustainability goals, particularly in climate change.

Supporting Digital Transformation and Smart Cities: One of the most significant areas where public-private-academic partnerships drive innovation is in developing smart cities and the digital transformation of urban infrastructure. Cities worldwide increasingly rely on data and technology to improve the quality of life for their citizens, reduce environmental impact, and increase operational efficiency. Smart city initiativessuch as developing IoT-based traffic management systems, smart grids for energy distribution, and AI-powered waste management systemsincreasingly depend on collaboration between governments, private technology companies, and academic institutions.

For example, academic institutions research urban sustainability, transportation systems, and integrating renewable energy into city grids, while private sector companies develop and deploy IoT sensors, cloud platforms, and AI-driven analytics to implement these innovations. Governments provide the necessary regulatory frameworks, funding, and policy support to bring these technologies to scale. Together, these partnerships build smarter, more sustainable cities that can adapt to rapid urbanization, environmental degradation, and resource scarcity.

Public-private-academic partnerships are also driving the development of advanced technologies such as 5G, which will be critical to the next phase of digital transformation. In this case, academia provides the fundamental research in wireless communication while industry develops the infrastructure and technology. Governments support this work through policies that encourage investment in telecommunications infrastructure and the regulatory frameworks that support the deployment of 5G technology across urban and rural areas.

Defies and Opportunities in Industry–Academia Partnerships: While public-private partnerships offer immense innovation potential, several challenges must be addressed to realize their full benefits. One of the main challenges is the difference in objectives between academic institutions and industry. Academia is often focused on long-term research goals and theoretical understanding, while the private sector is typically more concerned with short-term profitability and market viability. These sometimes conflicting priorities require clear communication, transparency, and mutual understanding. Another is the potential for intellectual property (IP) disputes between academia and industry. In collaborations where both sectors contribute to developing new technologies or innovations, it is essential to establish clear IP agreements to prevent disputes and ensure that all stakeholders benefit from the outcomes. Similarly, there may be concerns regarding data privacy, especially when personal data is involved in research projects or commercial applications.

Despite these challenges, there are numerous opportunities for strengthening industry-academia partnerships. Creating innovation hubs, collaborative research centers, and public-private accelerators can allow stakeholders to work together on common goals. These hubs can serve as incubators for new ideas, providing industry, government, and academia a platform to co-create solutions that benefit society. Developing shared funding models, such as government grants and joint research initiatives, can also help facilitate these partnerships, making it easier for smaller companies and research institutions to participate in large-scale innovation efforts.

Global Collaborations for Sustainability

In the face of the escalating environmental crises, global sustainability has emerged as a critical concern that requires immediate and coordinated action. One of the most profound ways technology contributes to sustainable development is by fostering global collaborations that leverage information technology (IT). In a world increasingly interconnected through digital technologies, the potential for IT to facilitate regional and global partnerships for sustainability is unprecedented. These collaborations enable nations, industries, organizations, and individuals to work together across borders, pool their resources, knowledge, and expertise to address pressing environmental issues, reduce carbon footprints, and ensure the preservation of natural ecosystems for future generations.

As the digital revolution unfolds, IT is essential in driving the global sustainability agenda by enabling data exchange, fostering communication, and enhancing transparency in sustainable practices. IT solutions are being integrated into various initiatives to promote sustainable development, address climate change, and support the United Nations Sustainable Development Goals (SDGs). These technologies also facilitate the formation of international coalitions, where countries and organizations collaborate to solve complex sustainability challenges through collective action.

Role of IT in Global Collaborations for Sustainability: One of the central ways that IT facilitates global collaboration is by creating platforms that support data sharing and decision-making and monitoring sustainability outcomes across regions and nations. A key example is big data analytics and cloud computing, which provide real-time platforms for processing large-scale environmental data. These technologies allow governments, corporations, and environmental organizations to analyze data from multiple sources, including satellite imagery, sensors, and climate models, to inform decision-making, track environmental trends, and identify areas for intervention. For instance, platforms like the Global Forest Watch, which is powered by cloud computing, offer real-time data on deforestation, enabling governments, NGOs, and private stakeholders to track and act on the health of forests worldwide.

The ability to share real-time data across borders is particularly critical in addressing the global nature of environmental challenges such as climate change, biodiversity loss, and resource depletion. These challenges do not respect national boundaries, so regional and global cooperation is essential. Information technology enables governments and organizations to collaborate on joint initiatives to reduce carbon emissions, protect biodiversity, and promote renewable energy adoption. Platforms such as the UN Global Compact, which brings together businesses, governments, and civil society to align their strategies with sustainable practices, rely heavily on IT to facilitate communication, resource-sharing, and global initiatives. Additionally, IT facilitates the efficient flow of information between nations, allowing them to align their policies with global sustainability targets and track progress toward achieving SDGs.

IT is pivotal in ensuring that sustainability efforts are communicated effectively and measurable. Technologies like geographic information systems (GIS) and remote sensing tools allow for monitoring environmental changes, such as land use changes, forest cover, and urban sprawl. These tools are critical in supporting international environmental agreements, such as the Paris Agreement, by enabling countries to transparently track and report on their emissions and progress toward their climate goals. These tools ensure accountability and provide evidence that can be used to hold governments and corporations accountable for their environmental impact.

Regional Collaborations for Sustainability: While global collaborations are essential for tackling large-scale environmental issues, regional partnerships are equally important for addressing the unique sustainability challenges faced by specific regions. IT solutions have become a powerful tool in regional collaborations, allowing countries with similar environmental and socioeconomic conditions to work together more effectively. For example, in regions particularly vulnerable to climate change's impacts, such as the Pacific Islands, countries have leveraged IT to create shared platforms for disaster risk management, early warning systems, and climate resilience strategies. The Pacific Islands Forum, a collaboration of 18 countries in the Pacific region, uses digital platforms to share data on climate change, implement disaster preparedness programs, and coordinate disaster response and recovery resources.

Similarly, the African Union has launched several initiatives focused on regional sustainability, such as the African Renewable Energy Initiative, which aims to increase the use of renewable energy across the continent. IT is significant in coordinating these efforts, providing the necessary tools for data collection, tracking performance, and sharing best practices. Satellite technology and data-sharing platforms have been instrumental in monitoring Africa's vast renewable energy potential, such as solar and wind power, and enabling partnerships between governments, private sector companies, and international organizations to bring these resources to market.

In other regions, such as Southeast Asia, which faces challenges related to deforestation, biodiversity loss, and water scarcity, IT is used to establish cross-border agreements for sustainable management of natural resources. Collaborative platforms allow countries in the Mekong River Basin to share data on water resources, plan joint flood control and water conservation efforts, and ensure that water use policies are consistent across the region. Establishing these digital networks of collaboration enables countries to tackle regional challenges in a more coordinated and effective manner. Regional digital platforms also enable the mobilization of resources, including financial capital, for sustainable development projects. In many regions, the private sector and international development organizations collaborate with local governments to fund and implement renewable energy projects, reforestation efforts, and climate adaptation programs. These collaborations are increasingly facilitated by digital platforms that allow for the efficient distribution of resources, transparency in funding allocations, and real-time progress monitoring.

Leveraging IT for Multilateral Sustainability Efforts: IT has proven invaluable for fostering multilateral partnerships that address the sustainability challenges that transcend regional borders. One of the most significant multilateral collaborations is the Paris Agreement, in which nearly 200 countries committed to taking action to limit global warming to below 2°C above pre-industrial levels. The Agreement, which requires extensive data-sharing, transparency, and accountability, relies heavily on IT systems to monitor emissions, track progress, and report on climate goals. The United Nations Framework Convention on Climate Change (UNFCCC) uses digital platforms to facilitate communication among countries, share climate data, and provide resources for nations to implement climate action plans.

Another significant example is the United Nations' SDG Agenda 2030, which calls for international cooperation on sustainability goals. Digital tools, including the SDG Tracker and SDG Dashboard, enable real-time progress monitoring on global sustainability targets. These platforms rely on global data sources, including satellite imagery, government reports, and data from civil society organizations, to track progress toward the SDGs. The data collected is accessible to governments, businesses, NGOs, and individuals, allowing for greater transparency and participation in achieving the SDGs.

IT also plays a crucial role in creating global sustainability indices, such as the Environmental Performance Index (EPI), which ranks countries based on environmental performance. These indices are instrumental in shaping international policy and incentivizing countries to adopt sustainable practices. Through IT systems that aggregate and analyze data on air quality, water management, biodiversity, and waste management, these indices provide a clear benchmark for countries to assess their environmental performance and set goals for improvement.

IT has facilitated the development of global marketplaces for sustainability initiatives, where organizations and governments can collaborate on environmental solutions. Platforms such as the Global Environment Facility (GEF) and the Green Climate Fund allow countries and private sector entities to pool their financial resources to fund global environmental projects. Through digital platforms for project management, monitoring, and reporting, these initiatives can ensure that funds are allocated efficiently and used effectively to achieve sustainability outcomes.

Future of Global IT Collaborations for Sustainability: The future of IT-enabled global collaborations for sustainability appears promising as the world faces pressing environmental challenges. Emerging technologies such as blockchain, artificial intelligence (AI), and the Internet of Things (IoT) are expected to play an even greater role in enabling transparency, accountability, and efficiency in global sustainability efforts. Blockchain, for example, has the potential to revolutionize supply chain management by ensuring traceability and sustainability in sourcing raw materials, reducing fraud, and improving ethical sourcing practices. This technology can also facilitate carbon trading systems, where emissions are tracked in real-time, and businesses can trade carbon credits to meet their climate targets.

AI and machine learning are transforming how we approach sustainability challenges by allowing for more accurate predictions, optimized resource management, and targeted interventions. These technologies enable the creation of smart cities, where data from sensors and IoT devices can be used to optimize energy consumption, manage waste, and improve transportation systems, all contributing to a more sustainable urban environment. AI can also play a crucial role in climate modeling, helping scientists predict future climate scenarios and guide policy decisions to mitigate the impacts of climate change. The rise of digital twin technology, which creates virtual replicas of physical systems, allows for more effective monitoring and management of ecosystems, infrastructure, and urban environments. Digital twins of ecosystems, for example, can help monitor biodiversity, water quality, and forest health, providing real-time insights that can inform conservation efforts and policy decisions.

These emerging technologies will increasingly facilitate global collaborations, enabling more efficient coordination, resource-sharing, and collective action. Digital platforms for international sustainability collaborations will continue evolving, providing more advanced tools for tracking, analyzing, and reporting environmental outcomes. These tools will help ensure that the international community remains on track to meet sustainability targets and can address global environmental challenges with the urgency and collaboration they demand.

Future Scenario

In the coming decades, digital ecosystems are set to transform the way governments, businesses, and communities interact and collaborate towards achieving sustainability goals. These ecosystems will act as interconnected platforms facilitating communication, resource-sharing, and coordination between diverse stakeholders, enabling them to work together seamlessly to address global environmental challenges. With advancements in digital technologies, including artificial intelligence (AI), blockchain, and the Internet of Things (IoT), these integrated digital solutions will increasingly drive the future of sustainability partnerships. As the world faces unprecedented challenges such as climate change, biodiversity loss, and resource depletion, digital ecosystems will enable real-time data sharing, predictive modeling, and transparent decision-making processes, allowing for more effective and coordinated global efforts.

A central feature of these digital ecosystems will be the seamless connection between governments, businesses, and communities. These three sectors play critical roles in addressing sustainability challenges, and their collaboration is essential for achieving large-scale environmental goals. Governments are responsible for creating the regulatory frameworks and policies that promote sustainability, while businesses must integrate sustainable practices into their operations and supply chains. Communities, particularly marginalized or vulnerable populations, must be empowered to participate in sustainable practices, from resource conservation to climate change mitigation. Digital ecosystems will provide the platforms necessary for these stakeholders to collaborate, share information, and implement solutions at a global scale.

Role of Digital Ecosystems in Facilitating Collaboration: Digital ecosystems that connect governments, businesses, and communities will thrive on the foundation of advanced technologies such as blockchain, AI, and cloud computing. Blockchain, in particular, will be a central technology in ensuring transparency, accountability, and efficiency in sustainability initiatives. The decentralized and immutable nature of blockchain makes it an ideal tool for tracking environmental performance, verifying sustainable practices, and enabling secure, transparent transactions across borders. For example, blockchain can create verifiable carbon credit markets, where businesses and governments can trade emissions reductions, ensuring that sustainability claims are authentic and verifiable.

Blockchain's transparency features will also support global supply chain sustainability. By providing a tamper-proof record of every transaction, blockchain will ensure that products are ethically sourced, that resource extraction is sustainable, and that companies adhere to environmental regulations. This technology can be integrated into global supply chains to track the provenance of materials, monitor carbon footprints, and ensure that businesses and consumers make informed, responsible choices. For instance, companies in the fashion or electronics industry could use blockchain to prove that their supply chains are free from exploitation, pollution, or illegal deforestation. By verifying these claims, blockchain will build trust between businesses and consumers and incentivize companies to adopt sustainable practices. Blockchain can facilitate global collaborations on renewable energy and carbon reduction projects. For example, it can manage decentralized energy grids, where individuals, businesses, and countries can trade energy credits or surplus energy through smart contracts. This decentralization allows for more efficient distribution of renewable energy resources, promotes energy access in underserved areas, and incentivizes investment in clean energy projects.

Blockchain–Based Platforms for Global Sustainability Partnerships: Looking ahead, blockchain-based platforms are poised to become central hubs for global sustainability partnerships. These platforms will enable international cooperation and coordination on environmental issues, providing a space for governments, businesses, NGOs, and individuals to collaborate and work towards common sustainability goals. A notable example of such a platform is the Climate Ledger Initiative, which uses blockchain to monitor and verify carbon offset projects, ensuring that the carbon credits being traded are legitimate and lead to tangible environmental benefits. Blockchain platforms like this can also enable transparent reporting on sustainability metrics, making it easier for organizations to track their progress toward the United Nations Sustainable Development Goals (SDGs).

Blockchain technology can be leveraged to create global platforms that link governments and international organizations working towards shared environmental goals. These platforms could include digital marketplaces for exchanging carbon credits, renewable energy certificates, and sustainability-linked financial instruments. Blockchain's ability to provide secure, transparent transactions will promote trust in such systems, enabling governments and businesses to collaborate on achieving carbon neutrality and other environmental targets.

Blockchain-based systems will also support international data sharing on sustainability issues. For instance, countries can share environmental data such as emissions levels, air quality measurements, or biodiversity indicators through blockchain platforms, ensuring that data is authentic, up-to-date, and accessible to all stakeholders. This transparency in data-sharing will help monitor progress towards sustainability targets, hold governments accountable, and drive evidence-based policy decisions. Platforms like the World Bank's "Open Data Initiative" or the European Union's "Green Deal" can benefit from blockchain to enhance data accessibility, transparency, and accountability. The future of sustainability partnerships will involve the integration of these blockchain-based platforms with AI and machine learning algorithms to analyze large-scale environmental data. For example, AI could predict trends in carbon emissions, assess the impact of different environmental policies, and recommend actions to achieve sustainability targets more effectively. Blockchain would provide the infrastructure for storing and sharing the data securely and transparently, while AI would analyze the data and suggest optimized strategies for addressing climate change and other global environmental challenges.

Government, Business, and Community Engagement in Digital Ecosystems: The success of digital ecosystems in sustainability will depend on active participation from all sectors of society. Governments must create the legal and regulatory frameworks that promote the use of IT tools such as blockchain, AI, and IoT to achieve sustainability goals. These frameworks must encourage collaboration between countries and ensure the technologies are used responsibly and ethically. Governments must also invest in the infrastructure necessary for widespread digital adoption, ensuring communities in developed and developing regions can access the technology and tools needed to participate in sustainability efforts.

For businesses, the digital ecosystem will provide the opportunity to drive corporate social responsibility (CSR) and integrate sustainability into core business strategies. The global marketplace will increasingly demand transparency and accountability from businesses, with consumers expecting companies to demonstrate their commitment to sustainability through digital platforms. Blockchain, for example, can provide businesses with the tools they need to prove their sustainability claims, such as ensuring the ethical sourcing of materials or tracking the environmental impact of production processes. As more companies adopt these technologies, they will create a more sustainable global economy prioritising long-term environmental and social value over short-term profits.

Communities, particularly those that are marginalized or vulnerable, will benefit from digital ecosystems through greater access to information, resources, and opportunities for empowerment. Community engagement in sustainability efforts can be enhanced through digital platforms facilitating participation in decision-making processes, knowledge sharing, and resource management. Blockchain can provide a mechanism for communities to engage in carbon credit markets or energy trading systems, empowering local stakeholders to actively participate in global sustainability initiatives. Additionally, AI-driven platforms can provide communities real-time data on air quality, water resources, or deforestation, enabling them to make informed decisions about resource use and conservation efforts.

Digital tools can also bridge the gap between rural and urban populations, empowering communities to contribute to sustainability efforts. For example, IoT-based sensors in rural agricultural communities can help farmers monitor water usage, soil health, and crop yields, improving productivity while minimizing environmental impact. IoT systems can help optimize energy consumption, reduce waste, and improve transportation systems in urban areas, leading to more sustainable and resilient cities.

Defies and Opportunities in Scaling Blockchain–Based Sustainability Platforms: While blockchain-based platforms have enormous potential for global sustainability collaborations, several challenges must be addressed to realize their full potential. One of the main challenges is the need for standardization across blockchain platforms, particularly concerning carbon credit markets, renewable energy trading, and sustainability reporting. Without common standards and protocols, interoperability between platforms and organizations could be hindered, limiting the scalability and impact of these initiatives. The development of universal standards for blockchain in sustainability will be crucial for ensuring that these platforms are effective, reliable, and widely adopted.

Another defy is the environmental impact of blockchain itself. While blockchain technology offers significant benefits in terms of transparency and accountability, it is also energy-intensive, particularly for proof-of-work-based blockchains like Bitcoin. As the adoption of blockchain for sustainability initiatives grows, it will be important to explore energy-efficient consensus mechanisms and greener blockchain technologies that do not exacerbate the environmental challenges they are meant to address. The success of blockchain-based sustainability platforms will depend on the willingness of governments, businesses, and communities to collaborate and adopt these technologies. Widespread digital literacy, trust in blockchain systems, and robust regulatory frameworks will be essential to overcoming barriers to adoption and ensuring that blockchain technologies are used ethically and responsibly.

19

Challenges in Scaling IT for Sustainability

Introduction

Scaling Information Technology (IT) solutions for sustainability presents a complex set of challenges that hinder the ability of organizations and governments to deploy these technologies at scale. As the world faces pressing environmental challenges, such as climate change, resource depletion, and biodiversity loss, the potential of IT solutions to drive sustainable development has garnered significant attention. Technologies such as artificial intelligence (AI), blockchain, the Internet of Things (IoT), and big data analytics can play a crucial role in creating smarter, more efficient systems that optimize resource use, reduce waste, and drive environmental stewardship. However, several barriers must be addressed before IT can be fully harnessed to promote sustainability on a global scale. These barriers include technological, economic, infrastructural, regulatory, and social challenges that impact the scalability of IT solutions for sustainability.

- **Technological Challenges:** One of the foremost challenges in scaling IT for sustainability is the technological limitations that hinder the widespread adoption and deployment of advanced technologies. Many IT solutions that support sustainability, such as IoT sensors for monitoring environmental conditions or AI algorithms for optimizing resource use, require robust infrastructure to function effectively. The lack of infrastructure, especially in developing countries, makes implementing these technologies at scale difficult. For instance, deploying IoT-based environmental monitoring systems in rural or remote areas requires reliable internet connectivity, power supply, and physical infrastructure to install and maintain sensors. Without this basic infrastructure, the potential benefits of these technologies cannot be realized.

Many sustainability-oriented IT solutions rely on large-scale data collection and analysis. While big data analytics has immense potential to optimize resource management (e.g., in water and energy consumption), gathering and processing large volumes of data can be resource-intensive. The challenges of data storage, processing power, and the complexity of real-time data analysis can strain existing IT systems, especially in regions with limited technical expertise and infrastructure. Additionally, the diversity and volume of data sources complicate the integration of various IT systems, making it difficult to aggregate data from different platforms or stakeholders for meaningful analysis.

Economic Challenges: The economic challenges associated with scaling IT for sustainability are significant and multi-faceted. First, the initial capital investment required for deploying IT solutions can be prohibitive, especially for small businesses, startups, and governments with limited budgets. For instance, implementing AI-driven systems for energy efficiency or deploying blockchain for supply chain transparency can involve high upfront costs, including hardware, software, and human resources. In many cases, the long-term financial benefits of these technologies, such as cost savings from energy efficiency or reduced waste, may not be immediately realized, creating a barrier to adoption.

In developing countries, economic constraints can be even more pronounced. Governments and businesses in these regions may struggle to finance the technology infrastructure needed for widespread deployment. There is often a lack of financial incentives to adopt sustainability-oriented IT solutions. Without subsidies, tax incentives, or support from development agencies, organizations in emerging markets may be reluctant to invest in advanced IT solutions, especially when the short-term economic returns are uncertain. This results in slower adoption of sustainable technologies in countries that need them the most. The issue of the digital divide further exacerbates these challenges. While developed countries have the necessary resources to scale IT solutions, lower-income countries often struggle with inadequate financial resources, which limits their ability to invest in cutting-edge technologies. This discrepancy in access to technology undermines global efforts to scale IT for sustainability, as it prevents the widespread adoption of solutions that could address pressing environmental concerns on a global scale.

• **Infrastructural Challenges:** Infrastructure is a critical enabler of IT-based solutions for sustainability, yet many regions face significant infrastructural deficits that hinder the scalability of these technologies. This includes access to reliable electricity, internet connectivity, and physical infrastructure for installing and maintaining hardware. For instance, rural areas, particularly in developing nations, may lack the stable power grid to support the continuous operation of sensors, IoT devices, or other technology solutions. In such regions, power outages or unreliable energy sources can undermine the effectiveness of IT solutions, causing data loss or system failures.

Similarly, the lack of high-speed internet connectivity can be a major barrier to scaling IT solutions, especially in remote areas. Many sustainability technologies, such as AI-driven analytics or real-time monitoring systems, require continuous data exchange between devices, cloud storage, and data processing centers. Slow or unreliable internet connections can disrupt this communication, leading to inefficiencies in resource management and limiting the impact of these technologies. Another infrastructural challenge is the need for specialized physical infrastructure to support deploying advanced technologies. For example, installing smart meters for energy and water management, waste collection sensors, or IoT environmental sensors requires specialized hardware and equipment. The lack of trained personnel and insufficient maintenance capabilities can make sustaining these technologies difficult, especially in less developed regions.

Regulatory and Policy Challenges: The regulatory landscape for scaling IT solutions for sustainability is another significant hurdle. Governments play a critical role in creating the policy environment that enables or restricts the adoption of sustainable technologies. In many cases, existing regulations do not align with the requirements of new IT solutions. For instance, *data privacy and security regulations* may need to be updated to accommodate the growing use of IoT devices and AI systems, especially as these technologies collect vast amounts of personal or environmental data. Ensuring these technologies comply with data protection laws and privacy regulations is crucial for ensuring trust in IT systems.

Policies around sustainability, climate change, and resource management vary significantly from country to country. Inconsistencies in regulations across regions make implementing uniform IT-based sustainability solutions challenging for global companies. For example, different carbon pricing mechanisms, waste management regulations, and renewable energy targets can complicate the development of scalable technologies that rely on cross-border cooperation. Additionally, the lack of standardized frameworks for sustainability metrics makes it difficult to evaluate the performance of IT solutions or compare their effectiveness across different regions. In some cases, the absence of strong governmental support for sustainable development initiatives can limit the adoption of IT-based solutions. Governments may hesitate to prioritize sustainability technologies due to political considerations, lack of awareness, or competing priorities. This can delay the implementation of critical infrastructure or financial incentives needed to drive the adoption of IT solutions that promote sustainability.

- **Social Challenges:** Social challenges also play a significant role in limiting the scaling of IT for sustainability. Public perception and acceptance of new technologies are crucial for their widespread adoption. In many regions, particularly in rural areas or developing countries, communities may be unfamiliar with the benefits of IT solutions for sustainability, leading to resistance or skepticism. Additionally, concerns about job displacement due to automation and AI can create social pushback against these technologies, as workers in sectors such as agriculture or manufacturing may fear that new technologies will replace their jobs.

A lack of digital literacy and technical skills can hinder the ability of communities to utilize IT solutions effectively. Many sustainability technologies require a certain level of technical expertise to operate, maintain, and interpret data. In regions with low levels of digital literacy, there may be resistance to using these technologies, or communities may struggle to understand their potential benefits fully. Capacity-building initiatives and public awareness campaigns will be necessary to address these gaps and ensure that communities are empowered to take full advantage of IT solutions for sustainability. Another task is the uneven distribution of technology access, with wealthier, urban regions often benefiting from advanced IT systems, while rural or impoverished areas are left behind. This digital divide reinforces existing inequalities and limits the scalability of IT solutions to address sustainability challenges at a global scale. Efforts to bridge this divide, such as investing in infrastructure, providing digital literacy programs, and ensuring equitable access to technology, will be essential to scaling IT for sustainability.

Current Scenario

The current scenario surrounding the application of Information Technology (IT) to sustainability initiatives reveals a complex landscape marked by significant financial, infrastructural, and cultural barriers. These barriers hinder the scaling and effective implementation of IT-driven solutions that have the potential to address global sustainability challenges such as climate change, resource depletion, and social equity. While IT can optimize resource management, enhance efficiency, and enable data-driven decision-making, its widespread adoption and impact are constrained by the interplay of these barriers. It is necessary to examine each factor in detailfinancial limitations, inadequate infrastructure, and cultural resistancewhile exploring their compounded effects on sustainability outcomes.

Financial Barriers: Financial constraints represent one of the most significant challenges in deploying IT solutions for sustainability, particularly in low- and middle-income regions. The high upfront costs associated with acquiring, implementing, and maintaining advanced IT technologies such as artificial intelligence (AI), the Internet of Things (IoT), and blockchain systems can deter their adoption. For instance, developing and maintaining AI-driven systems for energy optimization or building blockchain-based solutions for transparent supply chains requires considerable investments in hardware, software, and skilled personnel. While the long-term savings or environmental benefits may justify these investments, the immediate financial burden can be prohibitive, particularly for small and medium-sized enterprises (SMEs) or developing nations with constrained budgets. In addition to the direct costs of technology deployment, organizations may face challenges securing funding for sustainability-oriented IT initiatives. Traditional financing mechanisms, such as loans and grants, may not be readily available or appropriate for such projects. Lack of robust economic models for assessing the financial returns on sustainability-related IT investments complicates decision-making for businesses and governments. For example, an AI-based system that reduces energy consumption may take several years to produce measurable financial benefits, while system installation and operation costs are incurred immediately. This mismatch in financial timelines can make it difficult to justify investments in such technologies without external financial support, such as government subsidies or international development aid.

Infrastructural Barriers: Alongside financial constraints, inadequate infrastructure remains one of the most pressing barriers to scaling IT solutions for sustainability. This issue is particularly pronounced in rural, remote, or economically disadvantaged areas where the physical and digital infrastructure required for deploying advanced technologies is either insufficient or non-existent. The foundation of many IT-driven sustainability solutions relies on robust, reliable, and high-speed communication networks, access to electricity, and the ability to collect and process large volumes of data. In regions lacking basic infrastructure, *deploying* IoT sensors for environmental monitoring, AI-based systems for resource optimization, or smart grids for energy management becomes a significant challenge.

For example, deploying IoT-based monitoring systems for water management or air quality monitoring requires a reliable power source to support continuous data collection. In rural or off-grid regions, the lack of a consistent electricity supply can undermine the functionality of these systems, resulting in incomplete data sets and inaccurate assessments. Similarly, smart agriculture technologies that rely on real-time data to optimize irrigation or fertilization require high-speed internet connectivity to transmit data to central servers for processing. In rural areas where internet penetration is low or unreliable, the full potential of these technologies cannot be realized. Many developing nations face challenges related to the physical infrastructure required for installing and maintaining advanced IT solutions. For instance, installing renewable energy technologies, such as solar panels or wind turbines, often requires significant investments in specialized infrastructure, including energy storage systems and grid connectivity. Even the most advanced and well-designed IT solutions cannot function effectively without the necessary infrastructure, resulting in inefficiencies and limiting scalability.

A major component of this infrastructural gap is the digital divide, which exacerbates inequalities between developed and developing regions. Wealthier countries, particularly those in North America, Europe, and parts of Asia, possess the infrastructure and resources to implement IT solutions for sustainability at scale. In contrast, many countries in Africa, Latin America, and parts of Asia face significant challenges in providing the requisite infrastructure for digital solutions. This divide further compounds global sustainability efforts, as regions most needing IT-driven sustainability solutions are often the least equipped to implement them.

Cultural Barriers: Cultural barriers also play a pivotal role in hindering the adoption and scaling of IT solutions for sustainability. The effective deployment of technology solutions depends on their technical feasibility and the willingness of communities, businesses, and governments to embrace new approaches. Cultural attitudes toward technology, environmental conservation, and sustainability vary significantly across regions, and these differences can create substantial resistance to adopting IT-driven sustainability initiatives.

One major cultural challenge is the resistance to change that often accompanies the introduction of new technologies. In many parts of the world, especially in rural or traditional communities, people may be unfamiliar with the benefits and functionalities of advanced technologies. This lack of awareness can result in skepticism or reluctance to adopt these technologies, even if they offer clear environmental or economic advantages. For example, farmers may be hesitant to adopt precision agriculture technologies, such as AI-driven systems for crop management or drone-based monitoring, due to concerns about the complexity of the technology, cost, and potential disruptions to traditional farming practices.

Another cultural defy arises from the perceived value of technology concerning local customs and practices. In some regions, there is a deep-rooted belief in the efficacy of traditional knowledge and methods, which may clash with high-tech solutions. In such cases, adopting IT solutions requires significant effort to bridge the gap between modern technology and traditional practices. This may involve education and outreach campaigns that highlight the potential benefits of technology while respecting local cultural values. In some cases, cultural barriers also manifest in the reluctance to embrace collective action for sustainability. Many IT-driven sustainability solutions, such as energy efficiency, waste management, or resource conservation, require coordinated efforts among stakeholders. This may include collaboration between governments, businesses, and local communities. However, in some regions, fragmented governance structures, mistrust among stakeholders, or a lack of social cohesion can make fostering collaboration for scaling IT solutions difficult. Cultural factors such as individualism, political polarization, and distrust in government can undermine collective action efforts, reducing the effectiveness of IT solutions for sustainability.

Compounding Barriers and Global Implications: These barriersfinancial, infrastructural, and culturaloften interact and compound one another, creating a complex web of challenges that hinder the scalability of IT solutions for sustainability. For example, financial constraints may limit the ability to build the necessary infrastructure to support IT solutions, while cultural resistance can delay the adoption of technologies that could help address pressing environmental challenges. The cumulative effect of these barriers can result in slower progress toward achieving global sustainability goals, particularly in developing regions where the need for such solutions is most acute.

The global implications of these barriers are profound. As countries and organizations work toward achieving the United Nations Sustainable Development Goals (SDGs), scaling IT solutions for sustainability will be critical. However, the financial, infrastructural, and cultural barriers identified above threaten to impede progress. Bridging these barriers requires coordinated efforts from governments, international organizations, businesses, and local communities to create an enabling environment for IT-driven sustainability. This includes developing financing mechanisms that support sustainability-oriented IT projects, investments in infrastructure to ensure equitable access to technology, and cultural outreach programs that promote awareness and acceptance of these technologies.

Challenges in Implementation

Implementing IT-driven sustainability solutions faces significant challenges, often rooted in financial, infrastructural, and cultural barriers. These obstacles impede the scalability and efficacy of innovative technologies designed to address environmental, economic, and social sustainability goals. While Information Technology (IT) holds vast potential to transform various sectors and enhance sustainability efforts, the real-world implementation of these solutions remains complex due to systemic issues that vary across regions, industries, and communities. This section explores the multifaceted challenges of deploying IT solutions for sustainability, focusing on the financial, infrastructural, and cultural barriers that limit their widespread adoption and impact.

Financial Barriers: Financial challenges stand as one of the primary obstacles to successfully implementing IT solutions for sustainability. While the potential long-term benefits of IT adoption in sustainability are substantial, the initial investment required to develop, deploy, and maintain these technologies can be prohibitively high. For example, implementing advanced systems such as AI for predictive maintenance, IoT for smart agriculture, or blockchain for supply chain transparency demands considerable upfront capital. These technologies require sophisticated infrastructure, ongoing training, and continuous technical support, all of which incur substantial costs.

The financial capacity to fund such projects is limited in many developing countries and emerging economies. Governments and businesses often face budgetary constraints, which are further exacerbated by the absence of clear financial models to evaluate the return on investment (ROI) for sustainability-focused IT solutions. For instance, implementing a smart energy grid might initially require large investments in sensors, data analytics, and cloud computing systems. While these systems can significantly reduce energy consumption and improve grid efficiency over time, the ROI is not always immediate, and many stakeholders may find it difficult to justify such investments in the short term, particularly in regions where financial resources are already stretched.

Additionally, the lack of specialized financing mechanisms and incentive structures for sustainability initiatives further compounds the financial barrier. Traditional financial models, such as loans or equity investments, may not be well-suited for funding sustainability projects due to the extended payback periods and the perceived risks of new technologies. As a result, innovative financing solutions such as green bonds, blended finance, and climate-specific investment funds have been proposed as alternatives to bridge this financial gap. However, access to such mechanisms remains a critical challenge, particularly in countries heavily reliant on international aid or government subsidies for development projects.

The financial burden on small- and medium-sized enterprises (SMEs), crucial to economic growth and job creation, is another significant concern. These businesses, especially in developing countries, often lack the capital to invest in advanced IT solutions that can enhance their sustainability practices. Without financial support mechanisms or government-backed programs, these enterprises may be excluded from the benefits of IT innovations, further deepening the divide between large corporations and SMEs regarding technological capabilities and sustainability performance.

Infrastructural Barriers: Infrastructure is another fundamental barrier to implementing IT solutions for sustainability. Many sustainability technologies rely on robust and reliable digital infrastructure, including data storage and processing capacity, communication networks, and access to electricity. However, significant infrastructural gaps exist, particularly in rural, remote, or economically disadvantaged areas, where the basic infrastructure needed to support IT solutions may be either insufficient or non-existent. These regions face limited or unreliable electricity supply, low internet connectivity, and outdated communication networks, undermining the potential for implementing advanced sustainability technologies.

For example, the application of IoT-based systems for water management or agricultural monitoring. These systems require a sensor network that continuously collects data, which must then be transmitted to central processing systems. In areas with poor internet connectivity or intermittent power supply, the data transmission and processing may be unreliable, leading to incomplete or inaccurate data that can undermine the system's effectiveness. Similarly, energy-efficient technologies like smart grids depend on a stable and high-capacity energy infrastructure to function optimally. Without the requisite infrastructure to support these innovations, the promise of improved energy management remains out of reach for many communities. In many developing regions, the overall digital infrastructure necessary for effective implementation of IT-driven sustainability solutions is still in the early stages of development. The digital divide between developed and developing nations continues to widen, with the latter often struggling to meet the technological demands required for advanced IT solutions. Rural communities, in particular, face disproportionate barriers to accessing essential services like high-speed internet, which limits their ability to adopt technologies that could improve energy efficiency, water usage, or waste management. This infrastructure gap is one of the most pressing challenges to achieving the widespread adoption of IT for sustainability.

The installation and maintenance of the necessary infrastructure are also cost-prohibitive for many governments and businesses. Developing infrastructure for sustainability technologies such as energy-efficient buildings, smart transportation systems, and renewable energy projects requires substantial investments in physical assets like roads, power plants, and data centers. In many regions, especially in low-income countries, governments may lack the resources to build and maintain such infrastructure, while private enterprises may be hesitant to invest without sufficient guarantees of return on investment. This creates a vicious cycle, where the lack of infrastructure hinders the adoption of IT solutions, preventing the development of the infrastructure necessary to support them.

Cultural Barriers: While financial and infrastructural barriers are often more tangible, cultural barriers to IT adoption for sustainability are equally important and can have profound effects on the success of these initiatives. Cultural resistance to new technologies, traditional practices, and socio-political factors play a critical role in shaping the acceptance and implementation of IT solutions. The widespread adoption of new technologies, particularly in developing countries, often faces skepticism and distrust, especially when these technologies are perceived as foreign or incompatible with local customs and practices.

For example, in many rural areas where agriculture is a dominant sector, farmers may be hesitant to adopt digital tools such as AI-based crop monitoring systems or precision irrigation technologies due to a lack of familiarity with such technologies. Farmers may sometimes perceive these technologies as overly complex or unnecessary, preferring to rely on traditional farming methods passed down through generations. Cultural attitudes toward innovation and technology often play a key role in determining whether new solutions can be successfully integrated into daily life. Resistance to change can be exacerbated by concerns about job displacement or disruptions to traditional livelihoods, particularly in regions that rely heavily on agriculture or manufacturing. In addition to resistance to new technologies, a lack of trust in digital systems and governance can further hinder the adoption of IT solutions for sustainability. Many communities are wary of sharing personal or business-related data, especially in regions where data privacy laws and regulations are underdeveloped. Concerns about surveillance, data breaches, and the misuse of information can compound this mistrust. Without clear guidelines and transparent policies addressing privacy and data security, adopting IT solutions for sustainability is likely slow, as individuals and businesses may be unwilling to share the data required for these technologies to function effectively.

Socio-political factors, including political instability, corruption, and weak governance structures, can create significant barriers to implementing IT-driven sustainability solutions. In countries where governance is fragmented or inefficient, the necessary political will and institutional capacity to implement large-scale IT projects may be lacking. Political leaders may also prioritize short-term gains over long-term sustainability goals, resulting in a lack of investment in infrastructure or support for innovative technologies. In such contexts, achieving consensus on sustainability initiatives can be difficult, particularly when it involves balancing the interests of multiple stakeholders, such as government agencies, businesses, and local communities.

Interconnected Barriers: Recognizing that these financial, infrastructural, and cultural barriers do not exist in isolation is essential. They are often interconnected, creating challenges that exacerbate each other and hinder progress. For instance, financial constraints can limit the ability to develop the necessary infrastructure for IT-driven sustainability solutions, while cultural resistance can make it difficult to garner public support for such investments. Additionally, regions with poor infrastructure may struggle to attract investment in new technologies, creating a cycle of underdevelopment that perpetuates the status quo.

The interaction of these barriers means that addressing them requires a multi-dimensional approach that considers the technical aspects of implementing IT solutions and the financial, infrastructural, and cultural contexts in which they are deployed. Solutions tailored to the specific challenges of each region, industry, and community are more likely to succeed. For example, developing financing models that account for the long-term savings and environmental benefits of IT solutions, while also addressing the immediate financial constraints of businesses and governments, is essential for overcoming financial barriers. Similarly, investing in digital literacy programs and fostering cultural acceptance of new technologies can help overcome resistance and facilitate the adoption of sustainability-driven IT solutions.

Scalable IT Solutions for India

With its vast geographical expanse, diverse socioeconomic landscape, and varied regional challenges, India requires scalable IT solutions that can address the needs of its population across different sectors. Whether for enhancing agricultural productivity, improving healthcare access, enabling smart city initiatives, or optimizing resource management, India's IT ecosystem must be flexible enough to cater to urban and rural needs. Modular designs and adaptable frameworks are central to the successful implementation of IT solutions for such a large and diverse nation. These approaches ensure that technology systems can scale efficiently while meeting the unique requirements of varied communities and sectors. This section explores how modular IT solutions and adaptable frameworks are key to driving sustainable development in India, providing the necessary scalability, customization, and resilience to meet the challenges of the country's rapidly evolving needs.

Understanding Scalable IT Solutions for India: Scalable IT solutions refer to technology systems capable of growing and evolving in capacity and functionality to meet increasing demands. In a country like India, scalability is essential to address the diverse needs of its population. With over 1.4 billion people, combining high-density urban areas and sparsely populated rural regions, scaling IT solutions to address sector-specific challenges is crucial. For example, the needs of a densely populated city in states like Maharashtra or Uttar Pradesh will differ significantly from those in remote tribal areas in Madhya Pradesh or Assam. IT solutions must cater to local requirements and accommodate future expansion, ensuring they remain effective as population dynamics change or new challenges arise.

One key feature of scalable IT solutions is their ability to grow vertically (increasing capacity, performance, and resource allocation) and horizontally (expanding to new areas, sectors, or use cases). This means creating systems that can be adopted nationwide, with the flexibility to operate effectively in different regions with varying levels of infrastructure, literacy, and technology access.

Role of Modular Design in Scalable IT Solutions: Modular design is an approach to system development in which a large system is broken down into smaller, independent modules or components that can be developed, tested, and upgraded independently. These modules can be added or removed based on specific requirements, allowing the overall system to grow and evolve without requiring a complete redesign. Modular IT solutions are particularly relevant for India, as they allow for the customization of technological interventions based on local conditions and needs, ensuring that solutions are not one-size-fits-all. In India's urban and rural landscapes, modular systems allow for the incremental roll-out of technologies. For example, in smart grid development, a modular approach enables the deployment of initial pilot projects in select urban or semi-urban areas, before expanding to rural and remote regions. Similarly, in the case of healthcare systems, modular software platforms can initially support basic telemedicine services in underserved areas, with the ability to add more sophisticated features such as AI-driven diagnostics, remote monitoring of chronic diseases, or healthcare data integration as infrastructure improves.

The flexibility inherent in modular design also allows IT solutions to be adapted to sectoral needs. A modular framework enables solutions to be customized to different industries, such as agriculture, healthcare, education, and transportation. For instance, in agriculture, farmers could access an ecosystem of modules like weather forecasting, precision irrigation, and crop health monitoring connected via an integrated platform. These modules can be activated or expanded based on individual farm requirements, making it easier for the farmer to access only the necessary tools without being burdened by complex or irrelevant features. Modular designs allow for the seamless integration of legacy systems into newer IT solutions. In India, many sectors rely on legacy infrastructure or processes that cannot be entirely replaced in the short term. Through modularity, new technologies can be layered onto these existing systems, offering enhanced capabilities without disrupting operations. This approach is particularly beneficial in the public sector, where government departments and agencies might already have established systems for administration, taxation, or public services but need to integrate newer technologies such as AI, data analytics, or blockchain for better efficiency and transparency.

Adaptable Frameworks: While modular design enables scalability and flexibility, adaptable frameworks ensure that IT solutions can be customized to meet local needs, priorities, and contexts. An adaptable framework can be described as a structured yet flexible platform that provides the foundation for developing, deploying, and integrating different technologies in various contexts. These frameworks allow for the reconfiguration of systems in response to regional, cultural, technological, and economic factors, enabling solutions to remain relevant and effective over time.

In India, adaptable frameworks are critical because of the country's stark differences across urban and rural areas and within its various states and union territories. For instance, while urban areas might benefit from smart city technologies such as integrated transportation systems, waste management, and energy management, rural areas may require simpler IT solutions, such as mobile apps for agricultural advice or water resource management. The adaptability of IT systems enables such varied applications, ensuring that technology can be tailored to local resources, infrastructure, and needs. One of the most significant advantages of adaptable frameworks is integrating with diverse data sources and stakeholders, enabling cross-sectoral cooperation. IT solutions must interface with government bodies, private organizations, civil society, and local communities in agriculture, health, and education sectors. Adaptable frameworks ensure these diverse actors can contribute to the system's success while accommodating differences in data formats, access levels, and communication methods. For example, in the healthcare sector, an adaptable framework can link private hospitals, government clinics, telemedicine providers, and pharmacies, each with different data formats and technology platforms, into a cohesive and effective ecosystem for public health management.

Additionally, adaptable frameworks are crucial in managing the dynamic and fast-evolving technology landscape. As new technologies emerge AI, blockchain, IoT, or 5Gadaptable IT systems can evolve to incorporate these innovations without requiring a complete overhaul. This dynamic adaptability is significant for India, where technological change requires systems that can quickly integrate and leverage new solutions. For instance, AI-driven analytics can be integrated into existing platforms for agriculture or healthcare, offering personalized solutions based on large-scale data analysis, while blockchain can enhance transparency and efficiency in areas such as supply chains and governance.

Addressing Regional Diversity through IT: India's regional diversity spanning varied climates, cultures, languages, and economic conditions requires IT solutions that can be tailored to local needs. Scalability is about technical growth and customizing solutions to work effectively in different regional contexts. Designing IT solutions with modularity and adaptability makes it possible to address the unique challenges different regions face. For example, in states like Punjab and Haryana, where agriculture is highly mechanized, precision farming technologies (such as IoT-based sensors for soil health monitoring or AI-based yield prediction) can be adopted at scale. In contrast, a different approach might be needed in states like Bihar and Uttar Pradesh, where agriculture depends on small-scale farmers with limited access to technology. Here, IT solutions could be tailored to provide mobile-based services for crop management, weather alerts, and market price information, making it easier to implement and more effective for local farmers. The same scalable IT platform could be used across regions, but with customized applications depending on the unique needs of each area.

Similarly, in healthcare, urban centers such as Bengaluru and Delhi can implement advanced telemedicine and AI-driven diagnostics systems, while rural areas in the North-East or central India may need simpler solutions, such as mobile health applications for primary care consultations or community-based health workers using tablets to record and track health data. An adaptable IT framework enables these differences by offering scalable solutions that can be configured to meet local realities.

Overcoming Implementation Barriers: For scalable IT solutions to succeed, it is critical that the systems developed are not only efficient and adaptable but also accessible and cost-effective. In India, the barriers to adoption often stem from a lack of awareness, inadequate infrastructure, and financial constraints. IT solutions must be designed to focus on affordability and ease of access. One way to enhance accessibility is through cloud-based solutions, which offer scalable storage and computing power at a fraction of the cost of on-premise infrastructure. Cloud solutions allow for the centralization of resources, making it easier to deploy IT solutions across regions without massive upfront investments in physical infrastructure. Solutions that leverage mobile technology are particularly effective in India, where mobile phone penetration is widespread, even in rural areas. Mobile platforms can deliver a wide range of IT services whether for agriculture, healthcare, education, or governance ensuring that technology reaches users regardless of their geographical location or economic status. For large-scale adoption, these systems must be designed with user-friendliness and inclusivity. IT solutions that require specialized knowledge or infrastructure may not be widely accepted, especially among rural or less-tech-savvy populations. Training programs, user manuals in local languages, and community-based education initiatives can play a pivotal role in ensuring that new technologies are accessible and easy for everyone.

Future Scenario

The future of scalable IT solutions, particularly in addressing sustainability and development challenges, hinges on innovations incorporating modular design principles, robust policy frameworks, and dynamic public-private partnerships. As the world rapidly accelerates towards a digital economy, the role of IT in facilitating growth and sustainability will only intensify, especially in developing nations like India, where scalability is a key challenge due to diverse socioeconomic conditions, infrastructure constraints, and varying regional needs. IT systems' scalability is not just about expanding infrastructure but ensuring that technologies are flexible enough to cater to local needs while being economically viable and resource-efficient.

Innovations Addressing Scalability Challenges through Modular IT Solutions: Modular IT solutions are central to addressing the scalability challenges in the evolving technological landscape. These solutions, which break down complex systems into smaller, interchangeable components or modules, provide a flexible and cost-effective framework that allows for the gradual and adaptive implementation of technologies. One of the primary advantages of modular systems is their ability to scale incrementally. This is especially critical in developing countries, where infrastructure development can be uneven, and resource limitations might constrain the ability to implement new technologies at a national or regional level.

In the future, modular IT solutions are expected to become even more integral to driving scalable development. For example, modular technologies will allow for localized, targeted interventions addressing specific challenges in agriculture, healthcare, and education. For instance, a modular platform for precision farming could start with basic weather forecasting and soil health monitoring modules. As more sophisticated needs arise, additional modules such as AI-based crop yield predictions, automated irrigation systems, and blockchain-enabled supply chains can be integrated seamlessly. This "plug-and-play" approach offers both scalability and flexibility, allowing technology to expand and evolve based on the specific needs of different regions or sectors.

In the healthcare domain, modular solutions could allow for incremental deployment of digital health systems. Basic telemedicine services could initially be deployed in remote rural areas. More advanced modules, such as AI-driven diagnostic tools, remote surgery, and patient management systems, could be added to create a fully integrated digital healthcare ecosystem as infrastructure improves. Such modularity ensures the system can grow with demand without requiring a full overhaul or massive upfront investment. Modular IT solutions also facilitate innovation by making testing, evaluating, and implementing new ideas in specific areas easier before scaling them. For instance, pilot projects in urban areas using smart grids can test energy management solutions, such as demand-response systems or solar energy integration, and refine them before introducing them to rural or peri-urban areas. This incremental scaling reduces the risk of failure, as the system is continuously tested and iterated on a smaller scale before full-scale deployment.

Modular IT solutions provide a strategic advantage in ensuring interoperability across various sectors. A major challenge in scalability is the integration of disparate systems and technologies. Modular design allows standardised modules to be integrated into existing infrastructure without complete system overhauls. For example, smart city applications such as traffic management, waste management, and public safety can each be developed as independent modules. Over time, these can be integrated into a cohesive, city-wide ecosystem that operates seamlessly without requiring the wholesale replacement of older technologies. This modular approach reduces the financial burden of adopting new technologies and ensures that the systems remain adaptable to future innovations.

Policy Support and Regulatory Frameworks: As modular IT solutions evolve and are adopted across different sectors, the role of policy and regulatory frameworks in fostering scalability will be paramount. Governments and policymakers must provide the necessary support to ensure that IT solutions are scalable but also sustainable, equitable, and inclusive. In the future, this will require the development of policies that encourage innovation, address the unique needs of developing economies, and ensure that the benefits of IT advancements are widely distributed.

Policy support can take many forms. One of the most crucial aspects is the creation of frameworks that encourage the adoption of modular IT systems. Governments can provide incentives for private companies and public sector organizations to invest in modular technologies that are designed to scale efficiently. These policies can include tax incentives for companies that design scalable technologies, grants for research and development in modular IT systems, and regulations that ensure interoperability, security, and data privacy standards. Such policies would create a conducive environment for the private sector to innovate while ensuring that technological solutions adhere to public goals such as sustainability, equity, and accessibility. Another important policy area for scaling IT solutions is the development of public-private partnerships (PPPs). In India and other developing economies, PPPs have proven effective in driving large-scale technological interventions. The future scenario will see these partnerships becoming even more central to scaling IT solutions. Governments, businesses, and non-governmental organizations will need to collaborate closely to design and implement solutions that are not only scalable but also equitable and sustainable.

Public-Private Partnerships (PPPs): Public-private partnerships are essential to the success of scalable IT solutions because they combine the strengths of both sectors: the agility, innovation, and expertise of the private sector and the scale, reach, and policy influence of the public sector. PPPs can help bridge the gap between technology development and deployment, ensuring that solutions are both technologically advanced and socially inclusive.

For example, in the energy sector, a PPP model could see the government partnering with private companies to develop modular solar power systems that can be deployed incrementally in urban and rural areas. The private sector can bring in the technological know-how and financial resources, while the government can provide the regulatory support and infrastructure necessary to scale these solutions. Public-private collaborations can help overcome the barriers posed by limited access to capital and lack of technical skills in rural areas. Government support can help incentivize the private sector to invest in underserved regions, thus ensuring that the benefits of scalable IT solutions reach even the most remote and marginalized communities.

The role of academia and research institutions is also critical in scaling IT solutions through public-private partnerships. Universities and research centers can play a pivotal role in the R&D process, providing the foundational knowledge necessary to develop scalable modular technologies. Collaborations between industry and academia can result in developing next-generation technologies designed to meet the specific needs of India's diverse regions, whether through energy-efficient solutions, rural connectivity, or affordable healthcare innovations.

Driving Inclusive Growth through Scalable IT Solutions: One of the most promising aspects of scalable IT solutions is their potential to drive inclusive growth. By designing modular solutions that can be customized to meet local needs, governments can ensure that underserved populations, including rural communities, women, and marginalized groups, are not left behind in the digital transformation. Future policies should prioritize inclusivity by ensuring that deploying scalable IT solutions is aligned with social welfare goals. The future of IT in agriculture could see the implementation of modular platforms that provide farmers with customized access to information on crop management, weather forecasts, and market prices. These solutions can be scaled from small clusters of farmers to entire regions, and local governments can play a key role in providing the necessary support infrastructure. Policies promoting digital literacy and skill-building among marginalized populations will help empower them to benefit from these scalable IT solutions fully, ensuring that technology becomes a tool for social and economic inclusion rather than exclusion.

As modular IT solutions evolve, their potential to address pressing issues such as climate change, resource management, and urbanization will become more apparent. Governments, industries, and international organizations must work together to create policies that support the development of scalable solutions and ensure that they are sustainable, equitable, and accessible. Public-private partnerships will be central to this process, enabling the mobilization of resources, expertise, and technology to create a future where scalable IT solutions can drive growth, equity, and sustainability.

20

Road Ahead: Vision 2047

Introduction

As the world navigates the complexities of the 21st century, the pivotal role of information technology (IT) in bridging change and fostering sustainable development cannot be overstated. Vision 2047, marking a centenary of India's independence, represents a forward-looking commitment to leveraging IT as a transformative tool to achieve global connectivity and sustainability. This chapter delves into the contributions of IT in driving this vision, outlines aspirations for a globally connected future, and reflects on the guiding principles of Panch Parivartan as a framework for transformative change.

Role of IT in Bridging Change and Achieving Sustainable Development: Information technology has become the backbone of modern society, enabling rapid innovation, connectivity, and efficiency across all sectors. IT offers solutions addressing critical economic growth challenges, environmental preservation, and social equity.

- **Facilitating Green Technologies**: IT-driven innovations have enabled the development and deployment of green technologies, such as smart grids, renewable energy systems, and efficient waste management tools. These advancements align with the Sustainable Development Goals (SDGs), particularly those targeting affordable and clean energy (SDG 7) and climate action (SDG 13).

- **Promoting Digital Inclusion**: Bridging the digital divide is central to achieving equitable development. IT provides the infrastructure for digital literacy programs, access to online education, and remote healthcare services, ensuring that marginalized communities are not left behind in the march toward progress.

- **Boosting Economic Growth**: By automating processes and enabling e-commerce, IT has significantly contributed to job creation and economic resilience. Artificial intelligence (AI), blockchain, and big data analytics empower businesses to innovate, optimize operations, and create value chains that are both competitive and sustainable.

- **Enhancing Governance**: IT has revolutionized governance by making it more transparent, efficient, and citizen-centric. E-governance platforms ensure services are accessible to all, reducing corruption and fostering trust between governments and citizens.
- **Supporting Environmental Monitoring**: Advanced IT tools, including satellite imaging and IoT-based sensors, provide real-time data on environmental parameters. These tools enable better resource management, disaster preparedness, and climate adaptation strategies.

Aspirations for a Globally Connected and Sustainable Future: As we approach 2047, the vision of a globally connected and sustainable future hinges on several key aspirations that harness the potential of IT:

- **Universal Digital Access**: A globally connected future demands that digital access becomes a fundamental right. By 2047, the aspiration is to ensure affordable, high-speed internet for everyone, fostering inclusivity and breaking geographical barriers.
- **Smart Cities and Communities**: IT-enabled smart cities will play a pivotal role in sustainable urban development. These cities will leverage IoT, AI, and blockchain technologies to optimize energy usage, manage waste efficiently, and provide seamless mobility solutions, contributing to SDG 11 (Sustainable Cities and Communities).
- **Sustainable Agriculture and Food Security**: Food security is paramount with a growing global population. IT-driven precision agriculturedrones, sensors, and AIwill optimize resource usage, reduce waste, and increase yield, addressing SDG 2 (Zero Hunger).
- **Global Collaboration on Climate Action**: IT facilitates global collaboration by enabling knowledge sharing, fostering innovation, and creating platforms for coordinated action. Digital tools will continue to support efforts to combat climate change, with nations working together toward a carbon-neutral future.
- **Empowered Human Capital**: Technology and innovation will drive The future workforce. Equipping individuals with digital skills and fostering a culture of lifelong learning will empower them to contribute meaningfully to a sustainable economy.
- **Ethical AI and Technology**: Ensuring its ethical use will be critical as technology advances. By 2047, global frameworks must be in place to address issues such as data privacy, algorithmic bias, and the ethical implications of AI and machine learning.

Final Reflections: The Panch Parivartan framework, rooted in transformative change across five key dimensions, is a guiding beacon for achieving Vision 2047. Let us examine its principles for IT and sustainable development:

- **People-Centric Development**: Placing people at the heart of progress is vital. IT can drive initiatives that empower communities through education, healthcare, and livelihood opportunities. The vision ensures that no one is left behind by prioritising inclusivity and equity.
- **Policy Reforms**: Robust and adaptive policies are essential to fostering innovation and sustainability. Governments must collaborate with stakeholders to create policies that support research, innovation, and the ethical deployment of IT for societal benefit.
- **Partnerships**: Achieving Vision 2047 requires collaborative efforts across sectors, including public-private partnerships (PPPs). IT can enabler these collaborations, providing platforms for shared knowledge, resources, and solutions.
- **Processes and Institutions**: Strengthening institutions and streamlining processes is critical for long-term success. IT-enabled governance systems can enhance efficiency, reduce bureaucratic hurdles, and improve service delivery, ensuring that institutions are agile and responsive to changing needs.
- **Planet Sustainability**: Environmental stewardship must remain a core focus. IT can support sustainable practices through green technologies, efficient resource management, and climate-resilient strategies, ensuring a balance between development and ecological preservation.

Key Takeaways from Panch Parivartan

The concept of **Panch Parivartan** serves as a transformative framework for envisioning and realizing a self-reliant and sustainable Bharat. Derived from India's indigenous philosophies and modern developmental needs, this framework outlines five pivotal domains of change that collectively aim to strengthen India's socioeconomic fabric. Panch Parivartan, meaning "Five Transformations," encapsulates a multidimensional approach encompassing economic reform, environmental stewardship, technological innovation, governance improvement, and cultural resurgence. Together, these dimensions provide a structured roadmap for sustainable development while aligning with the ethos of self-reliance, or *Aatmanirbharta.*

Framework as a Guiding Philosophy: At its core, Panch Parivartan acts as a guiding philosophy that integrates India's cultural wisdom with contemporary global standards for sustainability. Unlike fragmented policy approaches, it promotes holistic development by focusing on interconnected reforms across various sectors. The framework's philosophical roots draw from ancient Indian concepts such as *Sarvodaya* (upliftment of all) and *Lokasamgraha* (collective welfare), emphasizing inclusivity and community-driven progress. By weaving traditional knowledge systems into modern governance, Panch Parivartan redefines development as economic growth and a harmonious balance between human well-being, environmental health, and societal equity. This paradigm shift ensures that policies and initiatives are evaluated based on their immediate outcomes and long-term sustainability and contribution to national resilience.

IT as a Cornerstone for India's Sustainable Future: Information Technology (IT) emerges as a critical enabler within the Panch Parivartan framework, underpinning India's aspirations for a self-reliant and sustainable future. The digital revolution has transformed how societies function, and India is uniquely positioned to harness this potential. With a robust IT sector, affordable internet connectivity, and a burgeoning digital economy, India can leverage technology to address pressing challenges such as poverty, education, healthcare, and climate change. For instance, e-governance initiatives like *Digital India* demonstrate how IT can enhance transparency, efficiency, and citizen engagement in public service delivery. Technologies such as artificial intelligence, blockchain, and the Internet of Things (IoT) can catalyze innovation in agriculture, energy, and urban planning, fostering sustainable practices across industries.

The role of IT extends beyond economic growth; it is also instrumental in environmental conservation. Through tools like satellite imagery, predictive analytics, and real-time monitoring systems, India can better manage its natural resources, mitigate disaster risks, and combat climate change. Additionally, the proliferation of digital platforms enables knowledge dissemination and skill development, empowering individuals and communities to participate actively in nation-building. By integrating IT into every facet of the Panch Parivartan framework, India can establish itself as a global leader in sustainable development, showcasing a model that combines technological prowess with ethical governance.

Roadmap to Self-Reliance and Sustainability: Panch Parivartan provides a clear and actionable roadmap to achieve self-reliance and sustainability in a globalized yet increasingly polarized world. The framework recognizes that self-reliance does not imply isolation but strategic autonomy, where India strengthens its internal capacities while engaging constructively with global partners. This roadmap includes investing in indigenous industries, reducing dependence on imports, and promoting local entrepreneurship under initiatives like *Make in India* and *Vocal for Local*. By enhancing domestic manufacturing capabilities and fostering innovation, India can reduce supply chain vulnerabilities, especially in critical sectors like healthcare, defense, and technology. Simultaneously, the roadmap emphasizes environmental sustainability as a non-negotiable component of self-reliance. From transitioning to renewable energy sources to implementing sustainable agricultural practices, the framework advocates for a green economy that prioritizes ecological balance. The Panchamrit goals announced at COP26, which include achieving net-zero emissions by 2070, align seamlessly with the Panch Parivartan vision. These commitments underscore India's proactive stance in addressing global climate challenges while pursuing its developmental objectives.

Synergizing Governance and Grassroots Participation: Effective governance forms the backbone of the Panch Parivartan framework, ensuring that reforms are implemented efficiently and inclusively. The framework stresses the importance of decentralization, where local governments and community organizations play a pivotal role in decision-making and resource allocation. By empowering *gram panchayats*, fostering public-private partnerships, and leveraging technology for data-driven governance, India can create a robust administrative ecosystem that caters to the diverse needs of its population.

Grassroots participation is equally crucial for the success of Panch Parivartan. The framework advocates for a bottom-up approach, where citizens are not merely beneficiaries but active stakeholders in the developmental process. Community-led initiatives in areas such as water conservation, waste management, and renewable energy have already demonstrated the potential of collective action. Scaling such efforts across the country can amplify their impact, fostering a sense of ownership and responsibility among citizens.

Cultural Resurgence as a Pillar of Transformation: Cultural resurgence is an integral pillar of Panch Parivartan, emphasizing the need to reconnect with India's rich heritage while embracing progressive values. This dimension instils pride and identity in India's history, traditions, and ethical principles. By integrating cultural education into mainstream curricula and promoting indigenous art forms, languages, and practices, the framework ensures that development does not come at the cost of cultural erosion. Cultural values such as *Vasudhaiva Kutumbakam* (the world is one family) and *Ahimsa* (non-violence) resonate with global aspirations for peace, sustainability, and harmony. These values can guide India's diplomatic and developmental strategies, positioning the nation as a moral and cultural leader on the world stage.

2030 and 2045 Perspectives for a Global Leader

India stands at a pivotal crossroads in its journey toward becoming a global leader, with sustainability and self-reliance as central tenets of its developmental philosophy. The 2030 and 2045 perspectives present a vision for India's transformation into a benchmark for sustainability, emphasizing systemic changes that align economic growth with environmental stewardship, social inclusivity, and technological innovation. By 2030, India aims to solidify its foundation for sustainable development, and by 2045, it seeks to establish itself as a global leader and exemplar for a green, self-reliant, and equitable future. These timelines reflect India's commitment to achieving long-term goals that resonate with global aspirations, particularly the United Nations Sustainable Development Goals (SDGs), while fostering a uniquely Indian model of progress rooted in resilience and ethical governance.

2030 – Laying the Foundation for a Sustainable Bharat: By 2030, India's roadmap envisions laying the groundwork for comprehensive sustainability across sectors through policy interventions, technological advancements, and societal shifts. The following are key dimensions of the 2030 vision:

- **Environmental Sustainability as a Core Agenda:** A significant focus is placed on achieving ambitious climate goals, such as reducing carbon emissions intensity, increasing the share of renewable energy, and enhancing forest cover. Programs like the National Solar Mission and the Atal Mission for Rejuvenation and Urban Transformation (AMRUT) are expected to expand, driving the adoption of clean energy technologies, sustainable urban infrastructure, and resource-efficient practices. India's leadership at COP26, with its commitment to achieving net-zero emissions by 2070 and installing 500 GW of renewable energy capacity by 2030, underscores this dedication. These targets will position India as a leader in global climate action, inspiring other nations to adopt similar commitments.
- **Economic Transformation with Localized Self-Reliance:** The *Atmanirbhar Bharat* initiative will mature into a dynamic framework for fostering indigenous industries, reducing import dependencies, and enhancing global competitiveness. By prioritizing circular economy principles, India can integrate waste-to-value systems into manufacturing and agriculture, ensuring resource conservation and economic efficiency. Additionally, investments in digital infrastructure, such as expanding 5G connectivity and artificial intelligence applications, will catalyze innovation in key sectors like healthcare, education, and agriculture, empowering rural and urban populations.
- **Inclusive Social Development:** By 2030, India aims to eradicate extreme poverty, achieve universal healthcare coverage, and provide quality education, aligned with SDGs 1, 3, and 4. Programs focusing on women's empowerment, such as Skill India and Beti Bachao Beti Padhao, will be scaled up, fostering gender parity and social equity. These advancements will ensure that sustainability is not limited to environmental metrics but includes human well-being and social justice as integral components.

- **Strengthening Global Partnerships:** India's role in multilateral platforms, such as the G20, BRICS, and the International Solar Alliance (ISA), will grow significantly. India can build coalitions that drive global sustainability by championing climate finance, technology transfer, and equitable trade policies rooted in cultural diplomacy and ancient practices like yoga and Ayurveda, will complement its geopolitical strategies, projecting India as a global leader with a unique developmental narrative.

2045 – Realizing a Global Sustainability Benchmark: By 2045, India envisions evolving into a fully self-reliant and globally recognized sustainability benchmark. This transformation will rest on achieving comprehensive milestones across governance, technology, environment, and cultural integration.

- **Becoming the Epicenter of Green Innovation:** By 2045, India aspires to lead global innovation in green technologies. Investments in renewable energy technologies, energy storage systems, and hydrogen-based fuels will establish India as the hub of next-generation energy solutions. Smart cities, powered by AI, IoT, and blockchain, will operate as prototypes of sustainable urban living, integrating renewable energy grids, zero-emission transportation, and efficient waste management systems. India's agricultural sector will embrace precision farming, agroforestry, and sustainable irrigation methods, ensuring food security while conserving critical ecosystems. This approach will also make India a key exporter of sustainable agricultural technologies and practices, offering solutions to global food and water crises.
- **Achieving Complete Energy Independence:** Energy independence is a cornerstone of India's sustainability benchmark. By 2045, India envisions becoming entirely self-reliant in energy production, with renewable sources accounting for most of its energy mix. Decentralized energy systems, including microgrids and community solar programs, will empower rural areas, reducing the energy divide and fostering inclusive development. Advanced battery storage technologies and a hydrogen economy will ensure energy security while facilitating clean energy export to neighboring countries. This regional leadership in energy sustainability will enhance India's geopolitical influence and economic stability.
- **Redefining Global Governance Structures:** India's role in shaping global governance frameworks will peak by 2045. As a member of the United Nations Security Council and other international bodies, India will champion equitable and sustainable development policies. Its leadership in crafting global climate agreements and advocating for the Global South will

set new benchmarks for collaborative governance, ensuring no nation is left behind in the quest for sustainability.

- **Integrating Cultural Wisdom with Global Progress:** The unique aspect of India's sustainability model lies in its integration of cultural heritage with technological progress. Principles of *Ahimsa* (non-violence), *Vasudhaiva Kutumbakam* (the world is one family), and traditional ecological knowledge will guide India's policies and international partnerships. This approach will preserve India's cultural identity and inspire global audiences to adopt a holistic view of sustainability.
- **Human-Centric AI and Digital Ethics:** By 2045, India will lead the global conversation on ethical AI and digital governance. As technology becomes increasingly integral to governance, healthcare, and education, India's emphasis on inclusivity, privacy, and equitable access will set the gold standard for digital transformation. Innovations in AI-driven healthcare, for example, will ensure affordable, personalized medical care for all, reducing health disparities worldwide.

Pathway to a Globally Sustainable and Self-Reliant Bhara*t:* India's transformation into a sustainability benchmark by 2045 will require meticulous planning and execution, guided by several strategic imperatives:

- **Policy Coherence and Long-Term Vision:** Aligning short-term policies with long-term sustainability goals is essential. Integrating environmental, social, and economic considerations into every policy decision will ensure that immediate gains do not compromise future resilience.
- **Investments in Education and Research:** Transforming India into a global leader demands a skilled workforce and cutting-edge research capabilities. Investments in STEM education, interdisciplinary research, and global academic collaborations will drive innovation and knowledge creation.
- **Public-Private Partnerships:** Collaborative efforts between the government, private sector, and civil society will accelerate progress in critical areas such as renewable energy, infrastructure, and healthcare. Innovative financing mechanisms, such as green bonds and sustainability-linked loans, can mobilize resources for large-scale initiatives.
- **Citizen Engagement and Behavioral Change:** Achieving sustainability is impossible without widespread public participation. Programs promoting environmental awareness, waste segregation, and energy conservation will foster a culture of sustainability, making it a collective responsibility.

- **Global Advocacy and Leadership:** India must proactively engage in global platforms, advocating for equitable resource distribution, climate justice, and sustainable trade practices. Its leadership in the International Solar Alliance and other initiatives will exemplify its commitment to a shared global future.

The 2030 and 2045 perspectives outline a transformative vision for India as a global sustainability benchmark and self-reliant powerhouse. By 2030, the nation will have laid the foundation for systemic sustainability, and by 2045, it will emerge as a leader in green innovation, ethical governance, and inclusive development. India's unique blend of cultural heritage, technological prowess, and policy innovation positions it as an exemplar for nations striving to balance economic growth with ecological and social well-being. As the world grapples with complex challenges, India's journey toward a sustainable and self-reliant Bharat will illuminate a pathway for global progress, underscoring the interconnectedness of humanity's shared future.

Index

Recycling Infrastructure 256, 274, 276, 285